ACC Library
640 Bay Road
Queensbury, NY 12804

DISCARDED

BONES AND OCHRE

BONES AND OCHRE

The Curious Afterlife of the Red Lady of Paviland

MARIANNE SOMMER

HARVARD UNIVERSITY PRESS
Cambridge, Massachusetts, and London, England   2007

Copyright © 2007 by the President and Fellows of Harvard College
All rights reserved
Printed in the United States of America

Library of Congress Cataloging-in-Publication Data

Sommer, Marianne, 1971–
    Bones and ochre : the curious afterlife of the Red Lady of Paviland / Marianne Sommer.
       p. cm.
    Includes bibliographical references and index.
    ISBN-13: 978-0-674-02499-1 (alk. paper)
    ISBN-10: 0-674-02499-0 (alk. paper)
    1. Cro-Magnons—Wales. 2. Antiquities, Prehistoric—Wales.
3. Excavations (Archaeology)—Wales. 4. Paleoanthropology—Wales.
5. Paleoanthropology—Philosophy. 6. Wales—Antiquities. 7. Buckland, William, 1784–1856. I. Title.
GN286.3.S66   2007
936.2'9—dc22   2007021357

To Marc

Acknowledgments

I started work on this book as a two-year Walther Rathenau postdoctoral fellow in the congenial research atmosphere of the Max Planck Institute for the History of Science in Berlin, where I had the opportunity to present and discuss my work in progress. Special thanks are due to Lorraine Daston, who accompanied the progress of the first part of this book. Subsequently, I held a National Science Foundation postdoctoral research fellowship in the Science, Technology and Medicine in Culture program at Pennsylvania State University, where I enjoyed tireless support from its heads, Londa Schiebinger and Robert Proctor. As junior faculty in science studies at the ETH Zurich (Switzerland), where I finished this book, I had the generous backing of the chair, Michael Hagner, whose constructive comments improved the manuscript. With the aid of the Swiss National Science Foundation, I had the opportunity to do concentrated work on the final draft at Stanford University, where in particular Alison Wylie offered critical reflection and inspiration.

Thanks are due to a number of other scholars who have commented on or contributed in other ways to particular chapters, parts, or entire book drafts, and I would like to cite Peter Bowler, James Secord, Robin Dennell, Hans-Konrad Schmutz, Martin Rudwick, Patrick Boylan, Nicolaas Rupke, Claudine Cohen, and Jeffrey Schwartz. In the writing of Part III, I depended significantly on the cooperation of

Stephen Aldhouse-Green and other investigators and artists working on Paviland Cave and/or the Red Lady, particularly Erik Trinkaus, Stephanie Swainston, Gino D'Achille, Anne Leaver, and Trenton Holliday.

Like any book, this one relies on previously published material, and I thank the editors and publishers of *British Journal for the History of Science*, *Journal of the History of Biology*, *Earth Sciences History*, and *Journal for the History and Philosophy of the Life Sciences* for permission to draw on my earlier work. These articles and the present book have built on research in a number of archives mentioned in the list that precedes the Introduction, and I would like to thank the institutions and people behind them. I am particularly indebted to Oxford University Museum's Jim Kennedy and Stella Brecknell and the National Museum of Wales's Tom Sharpe and Elizabeth Walker, the latter of whom patiently and competently answered the steady flow of my questions. Finally, I am grateful for the opportunity to thank my referees, among them Ralph O'Connor, and Harvard University Press, where the editors Ann Downer-Hazell and Vanessa Hayes have contributed their expertise.

Contents

Archives and Abbreviations		xi
Introduction		1
I	*William Buckland, an Ancient British Witch, and the Question of Human Antiquity*	21
	1 William Buckland	33
	2 Man among the Cannibalistic Hyenas?	46
	3 The Romance of the Witch of Paviland	59
	4 Buckland Accused	81
	5 Man as a Crocodile Superior?	91
	6 The Holy Order of Nature and Society	104
	Conclusion: The Red Lady Is No Fossil Man	116
II	*William Sollas, a Cro-Magnon Man, and Issues of Human Evolution*	121
	7 William Sollas	141
	8 Ancient Hunters and Their Modern Representatives	155

	9	The Red Lady Is a Cro-Magnon Man	167
	10	Human Evolution as a Trunkless Tree	175
	11	The "Evolutionary" versus the "Historical" Model	187
	12	The Moral Authority of Nature	197
		Conclusion: Turbulent Times for an "Old Lady"	213
III	*An Interdisciplinary Team, an Early Upper Paleolithic Shaman, and a Definitive Report*		219
	13	The Paviland Project and Its Results	231
	14	There Is Magic at Work	240
	15	Visualizing Paviland Cave	247
	16	The End of the Red Lady's Story: A Definitive Report?	263
		Conclusion: An Unfinished Life	279

Postscript — 282

Appendix A: Archeological and Geological Series — 293

Appendix B: Schematized Views of Human Evolution — 296

Notes — 299

References — 353

Index — 391

Archives and Abbreviations

Archives	William Buckland	William Sollas	Acronym
American Philosophical Society, Library, Manuscripts		William Johnson Sollas Papers, B:So4, Rawlins Autograph Collection (APS B:R199)	Am Phil Soc
Bodleian Library Oxford University, Special Collections and Western Manuscripts	MS Eng lett d 5, HMC Papers of British churchmen 1780–1940, 1987	Correspondence with Sir Aurel Stein, 1924–1927, 107, fols. 175–216, MSS Stein, NRA 26459, Stein	BL
		MS Eng lett d 329, Correspondence with A. Russell Wallace, F. Zirkel (Leipzig); Journey to India (Ceylon)	
British Library Manuscript Collections	Add MS 58995, 1818–1831: 33 letters to Lord Grenville	Add MS 55222, Correspondence with Macmillan and Co., 1909–1934; Add MS 46438; Add MSS 58477, 58480	BrL
Fitzwilliam Museum Cambridge University	Perceval, HMC visit, JRS 1992		FM

(continued)

(*continued*)

Archives	William Buckland	William Sollas	Acronym
London University, Imperial College Library, Archives and Special Collections	B/Playfair, NRA 11556 Playfair, HMC MS papers of British scientists 1600–1940, 1982	Notes of Ramsay's Geology Lectures, 906/Ka 8, Notes on Huxley Lectures, 924 KB8, HMC MS Papers of British Scientists 1600–1940, 1982	IC
National Museum of Wales, Geology Department	HMC MS papers of British scientists 1600–1940, 1982, Buckland, William		NMW
Newspaper Library of British Library			
Oxford University Museum of Natural History	Buckland Papers, NRA 42502 Buckland, HMC MS papers of British scientists 1600–1940, 1982	Geological Collection, NRA 42536, HMC MS papers of British scientists, 1600–1940, 1982	OUMNH
Royal Society of London	MS 251, HMC Papers of British churchmen 1780–1940, 1987		RS
University College London Library, Special Collections		MS Add 152, T. G. Bonney Correspondence, Bridson, Natural history MS resources 1980	UCL
Other			
Bibliothèque Centrale du Muséum National d'Histoire Naturelle	B.34, Boule: Documents relatifs à la découverte de l'Homme de la Chapelle-aux Saints; extraits de presse, correspondance Archives Henri Breuil		MNHN
Bibliothèque Collège de France archives	CDF 16/8, Henri Breuil		CDF

Bones and Ochre

∽ Introduction

THE RED LADY of Paviland is a relatively complete human skeleton from Paviland Cave, southern Wales. The way by which it acquired the curious name that stuck to it through successive interpretations will be discussed in a later chapter, so it suffices to say here that its bones have been reddened through the application of ochre in the course of a ritual interment. It was the illustrious William Buckland (1784–1856), the first reader of geology at Oxford University, who discovered the grave in 1823. Immediately after its resurrection, the mysterious Red Lady provoked jokes, verses, and romances. It was thus embedded in science as culture from the start, and its curious afterlife as a scientific object, which is the subject of this book, is to be understood in this wider sense. How can the same set of bones be read consecutively as postdiluvian tax man, female witch or prostitute of Ancient British times, male Cro-Magnon hunter of the Aurignacian Paleolithic, and male shaman or hero of the Gravettian culture?

The main episodes of the Red Lady's afterlife were predominantly defined by the geologist Buckland in the early nineteenth century, the geologist/anthropologist William Sollas (1849–1936) in the early twentieth century, and an interdisciplinary team under the archeologist Stephen Aldhouse-Green close to the millennium. Thus the reconstruction of the Red Lady's scientific biography is intertwined with chapters in the (pre)history of paleoanthropology. The skeleton was discovered at the dawn of controversy about a human antiquity that by far transcended the few thousand years based on the chronology of the Bible. It was minutely reread at the time when paleoanthropology was coming of age, not least because of hominid finds that indeed in certain respects approached anthropoid morphology. The as yet final major act in the Red Lady's drama serves to illustrate how the immensely complex questions posed by prehistoric sites such as Paviland Cave might today be tackled by interdisciplinary cooperation. At these milestones of the Red Lady's career, the production of scientific knowledge and the personal, institutional, and sociopolitical factors therein will be analyzed. What differentiates the protagonists of the Red Lady's story and their times? What does work on (fossil) human bones mean for them? What are the cosmologies and concepts of humanity that guide their work? Do they work from the armchair, in the field, in the laboratory, or in all three settings? How are these settings organized? Is knowledge produced in communication with "the public"?

The Red Lady's story as told therefore does not strictly belong to one of the genres of scientific biography, disciplinary history, institutional history, or history of ideas in a traditional sense. Rather, this book incorporates aspects of all these genres because the course of the Red Lady's afterlife is determined by the life and work of more than one scientist, cuts across institutions and disciplines, and is bound up with their scientific culture and wider context. Clearly, the emphasis lies on the scientific engagement with fossil human bones and the factors bearing on it. Due to the Red Lady's place of discovery, the geographical emphasis is on Great Britain. Nonetheless, time and again scientists and institutions on the Continent and to a lesser degree in the United States will be of relevance. Also because of the Red Lady, practices and ideas related to questions of modern European origins will take center stage, associated with changing ideas about human racial history and diversity.[1]

Although human fossils seem to be bones of contention by nature, the fact that the Red Lady of Paviland was the first genuinely fossil human skeleton discovered by a scientist renders its postmortem biography particularly varied. Decades before the establishment of human antiquity or evolutionary theory, it suggested questions about human origins to science. In fact, Aldhouse-Green has playfully pointed out that our Paleolithic European forebears should be called *Pavilandians*

Introduction

instead of *Cro-Magnons* because the Red Lady has priority of nearly forty years over the discoveries made in France.[2] This in turn highlights that fossil human bones can be implicated in national rivalries and national founding stories. It further indicates that in the Red Lady's history of interpretation it is also the attributes available to describe it that are in flux. What did *postdiluvian* or *Ancient British* mean? When did the notion of a Paleolithic come into being? What did it mean to belong to the Upper Paleolithic human race of Cro-Magnon? Thus the entangled genealogies of material objects, concepts, and theories that at any one point defined the Red Lady will have to be partially reconstructed. Some of these historically contingent entities still populate current paleoanthropology, while others may no longer be a part of it.

Biography of a Scientific Object

In the history of paleoanthropology, the notion of fossil human bones and other entities that identify the Red Lady first had to come into being and, once in existence, remained unstable. The approach to a science or sciences from their constitutive entities has brought to light that even such basic concepts as scientific objectivity have their histories. The roots of this kind of inquiry coincide with the beginnings of the contemporary understanding of history of science, notably in such works as Michel Foucault's genealogy of the concept of madness. Not only scientific theories and hypotheses but also the core concepts and practices of science ("epistemic virtues"), such as objectivity, and the entities they organize, such as madness, have histories rather than being givens.[3] Furthermore, as in the cases of objectivity and madness, where their being part of binaries is particularly relevant, mirror concepts, such as "the Neanderthals" and "the Cro-Magnons," may sharpen the fuzzy edges of significance.

Material scientific objects, too, have received attention from a historical perspective. Michael Hagner has reconstructed the history of the change from organ of the soul to brain, which acquired its present-day characteristics through attempts at allocating mental functions to specific areas. The process took place in fields as different as physical anthropology, phrenology, and experimental physiology, and the reconceptualization of the brain as defined by purely somatic aspects did not happen at once. Rather, the various characteristics that we associate with the brain today came into being at different times.⁴ Similarly, the main redefinitions of the Red Lady as outlined earlier cannot be the whole story. Between the three major chapters in its biography lie developments in science and other areas of culture that did not leave the reddish bones unaffected, and that set the stage for the incisive (re)interpretations. Its identity did not jump in Sollas's laboratory from postdiluvian within a framework that accepted the Bible as heuristic tool to Cro-Magnon within a secularized and evolutionary model. Rather, it acquired its new characteristics, such as "male," "ontogenetically young," "phylogenetically old," "Paleolithic," "Aurignacian," "a product of evolution," at different times and through the involvement of more than one discipline. Thus what might at first sight appear radical breaks turn out to show some continuities, with a mosaic pattern at work in the acquisition of new attributes. The changes in the Red Lady's identity are neither uniform nor altogether abrupt, once close attention is paid to the story. Although in the horizontal sections of the Red Lady's biography on which I mainly concentrate, different theories, practices, and material objects appear in new formations, the breaks are due rather to my choice of exposed layers—suggested by the Red Lady—than to actual radical discontinuities. Nonetheless, the definitions of the Red Lady vary so greatly that it is at first glance hard to believe that the same object is being discussed. One advantage of material objects might therefore be that they remain traceable across considerable transformations.

Another kind of generic material-organic objects has revealed intertwined economic, political, social, cultural, and scientific histories. New insights into the history of geopolitics have been provided from a focus on such heroic plants as coffee or the cinchona tree (quinine). Londa Schiebinger uncovered the story of the "peacock flower," used as an abortifacient by Caribbean slave women in tacit revolt against their exploiters. But although costly and risky scientific voyages were sponsored by eighteenth-century European imperial powers in search of new medicines from the New World, knowledge of the West Indian plant as abortifacient never reached European societies. This focus on particular plants has therefore allowed reconstructing not only the economic and sociopolitical dynamics of knowledge acquisition and application but also the production of ignorance.⁵ Similarly, in the history of the Red Lady and fossil human bones in general, which in the early nineteenth century were as explosive matter as abortifacients, one has to be alert to strategies of diversion of attention, if not of outright obfuscation, in processes of negotiation about what may pass as reconcilable knowledge.

Obviously, rather than being unambiguously encoded in the skeleton, its age, sex/gender, race, and place in human history/evolution depend on the scientific culture of the time and need to be understood within the meaning system of a society at large. By far transcending the study room, the field, or the laboratory, they involve major ideas about the human past and present. Therefore, I consider vital a dense description of the object at hand and a broad contextualization of its history. The lives and work of the scientists who analyzed the Red Lady's bones, their institutional background, and the methodologies and technologies of their science, as well as its theoretical and social context, will be relevant. The benefits of a detailed engagement with a cultural-scientific object and its context of production, endurance, and displacement have recently been demonstrated by James Secord in his history of the publication and reception of the anonymous *Vestiges of the Natural History of Creation* (1844).[6]

As a comprehensive analysis of the reading practices reconstructed from the traces one particular book has left in archives, newspapers, and memoirs, Secord's work opens up vistas of public and individual engagement with evolution that call into question the standard story of a "Darwinian revolution" as the turning point in the Victorian worldview. Rather than ascribing this kind of power to one person or text, Secord sees books as material objects that are inextricably interwoven with the characteristics of their time and place, in the case of *Vestiges* with new technologies of communication in a society on the verge of major transformations tied to the era of industrialization. Apart from rendering more porous what has come to be seen as a historical break, Secord's microhistory also draws attention to the processes of forgetting and neglecting on yet another level. The question of what is inscribed into cultural memory is not decided by historical actors alone, who, for example, choose not to transfer knowledge on abortifacients, but also in the writing of history, as the case of *Vestiges* vis-à-vis Charles Darwin's (1809–1882) *On the Origin of Species* (1859) illustrates.[7]

Overall, the history of science has increasingly become "histories of traces and of things."[8] As material entities or processes that embody concepts, that is, as epistemic things, to use Hans-Jörg Rheinberger's term, the objects of scientific inquiry expand our focus from a limitation to the immaterial aspects of scientific ideas to the materiality and practices of science. They also draw attention to its visual world. Finally, their hybrid nature renders them full of promise for an endeavor to transcend such nagging binaries as nature/culture.[9] As Bruno Latour, Lorraine Daston, and others have shown, scientific objects from bacteria to dreams are real entities and at the same time historically contingent.[10] Although the bones of the Red Lady possess a physical reality, they too only become alive in the hands, eyes, and minds of the manipulators and contemplators. Their status as scientific objects is dependent on the scientists and the material and immaterial culture of which they are a part, so that the bones' attribution to discovery or invention is not an unequivocal one.

As an ancient human skeleton and part of a burial, the Red Lady wavers between the categories of natural object and artifact, inorganic and organic, or even thing and person. Culturally reproduced in ritual, the bones' reddish color is symbolic of these boundary crossings. The subsequent excavation of the bones and their entering into spaces and networks of scientific representation, investigation, accumulation, distribution, and exhibition were accompanied by a decontextualization and subsequent recontextualizations, by a disappropriation from their time and place and a translation into, or reappropriations to, new times and places. As an anthropological object, the Red Lady is a cue to another time and simultaneously paradoxically timeless. Once removed from the cave, this archive of the annals of history, it remains inertly preserved for subsequent generations of scientists to read.[11]

Reborn as a scientific object, the human skeleton differs from other objects in that we recognize it as the trace of a human life. It vibrates with meaning to the extent of talking. This book shows, however, that it does not tell the same story to everyone. If it speaks, it is in the language of the interpreter, and its story is meaningful in the respective cultural and historical settings. Rather than just yelling at those who handle it, the skeleton is therefore coaxed into whispering, barely audible, and sometimes in a cacophony of its own voice and that of the scientist, as well as background murmur from associated immaterial and material objects and persons. Thus, although the physical anthropologist Jeffrey Schwartz has titled one of his books *What the Bones Tell Us*, he, like many of his colleagues, is in fact as much concerned with what the bones do not tell despite over two hundred years of development of technologies for reading the anatomical details of osseous remains.[12]

Paleoanthropologists and Paleolithic archeologists cherish their rare objects of study, but they do not necessarily trust them. Irresolute about its disciplinary affiliations, archeology in particular has experienced positivist optimism displaced by radical relativism, followed by attempts to strike a balance. Can we ever find the voice of the archeological record? Are we not hopelessly caught up in our presuppositions, forcing our own meanings onto mute objects? Archeologists agree on the danger of projection, but not on the right strategy to establish constraints. The emphasis has shifted from self-restraint to self-reflexivity, demanding the researcher's engagement with presumptions rather than attempting to exclude them from the process of analysis. To adopt an expression by the philosopher of science Alison Wylie, paleoanthropologists and archeologists endeavor to think from things, but their relationship to the objects is an ambiguous and contested one.[13]

In fact, although the bones of the Red Lady show recalcitrance, enabling, enforcing, or resisting particular readings and the larger theories they support, they are at any point so inextricably entrenched in the immaterial and material conditions of their birth and life as a cultural-scientific object that the common denominator in all the stories they have evoked is the identification as human bones. Therefore, Latour's observation that "an entity gains in reality if it is associated with many others [machines, gestures, textbooks, institutions, taxonomies, theories, etc.] that are viewed as collaborating with it. It loses in reality if, on the contrary, it has to shed associates or collaborators (human and nonhuman)," when applied to a specific material object such as the Red Lady, begs the question, reality as what?[14] As human bones, as fossil human bones, as the bones of a Cro-Magnon? Every category added to the definition of the Red Lady suggests that the network of collaborators has considerably changed, that new methods, technologies, instruments, and certainly concepts and theories have entered the equation.

In the shifting networks of mutually determinative entities that are constitutive of prehistoric archeology and paleoanthropology, concepts and theories are powerful components. Thinking from things is difficult in fields that are notoriously theory laden, and where the theories are notoriously underdetermined by data. The anthropologist Adrienne

Zihlman observes that "in a court of law, some evidence is thought to be so clear that 'it speaks for itself' *(res ipsa loquitur)*, but in anthropology an analytical framework is needed for the interpretation of data. That framework is evolution."[15] In the Red Lady's biography, this is complicated by the fact that at the time of its discovery, the majority of scientists did not work with an evolutionary framework. Once in existence, "evolution" has never been a monolithic entity, and the different views of human evolution propagated by the scientists under study in this book were powerful factors in allocating the Red Lady a space in the knowledge system precisely because they were furnished, to speak with Donna Haraway, with sticky psychological, technical, representational, institutional, political, and historical threads.[16]

Chapters in the (Pre)history of Paleoanthropology

The reconstruction of the afterlife of the Red Lady of Paviland with an eye to the concatenations of constitutive entities as discussed so far amounts to what might be called chapters in the (pre)history of paleoanthropology. Obviously, in speaking of a history of paleoanthropology, a term that appeared in the early twentieth century, it is reduced to the lowest common denominator, the scientific study of fossil human bones.[17] Even this is a retrospective description, since although Buckland and his contemporaries began to study fossil human bones scientifically, they might not have considered them as fossils. Nonetheless, although the majority of those who have dealt with the bones of the Red Lady would not have identified or indeed do not identify themselves as paleoanthropologists, their work might be seen as contributing to the history of the scientific endeavors today subsumed under the label *paleoanthropology*. This becomes clear from a current definition of the term that does justice to the complexity of research on hominid evolution:

> Human Paleontology studies human fossils and sub-fossils, both in their morphological aspects, and in those aspects linked to human behaviour. Paleoanthropology is concerned with the origin of the hominid line and with its evolution up to human level. This is revealed not only by the acquisition of distinctive biological attributes, but also by cultural elements such as the lithic,

bone industries, the structure of habitations etc. The approach and the methodology of paleoanthropology are scientific, while other studies of prehistoric conditions may follow historical rather than scientific methods. Since it takes account of the physical and especially the biological environment in human evolution, Paleoanthropology is of necessity interdisciplinary and avails itself of help from various specialties as Quaternary Geology, paleontology and paleoecology.[18]

This definition identifies paleoanthropology as a heterogeneous field in close cooperation with other disciplines from biology, anthropology, and the earth sciences. The Red Lady's life as a scientific object manifests this heterogeneity. Apart from putting into perspective the sharp distinction proclaimed here between historical and scientific methods, however, it is also affected by fields not explicitly mentioned in this definition, for example, technologies from molecular biology.

An inclusive term for those sciences that deal with aspects of hominid evolution is the Anglo-American expression *human origins*. Although the plural in *origins* might point to exactly that plurality of approaches associated with the field, against the background of polygenism one feels a certain unease about joining the adjective *human* to a plurality of roots. Conversely, one might like the emphasis on the fact that there cannot have been a specific moment when the complex set of characteristics we tend to identify with humans came into being. *Physical* or *biological anthropology* also encompasses various facets of the scientific study of humans as biological organisms, such as comparative anatomy and physiology, evolutionary history, and ecology. It may thus incorporate such (sub)disciplines as anthropometry, demography, genetics, growth studies, molecular anthropology, neuroanatomy, evolutionary theory, paleoanthropology, paleoprimatology, and primate field studies, among others. In its nineteenth-century contexts, however, *physical anthropology* mainly refers to the study of human morphological variation and the classification of races (craniometry and anthropometry). Finally, the meaning of terms also differs geographically. For example, in the Anglo-American world *anthropology* subsumes social/cultural anthropology (or ethnology), prehistoric archeology, linguistics, and biological anthropology, but in continental European countries it mainly refers to biological anthropology. *Anthropological sciences* may be used to express the more comprehensive sense. Again, the Red Lady's story evidences these ambiguities and shifts of meaning, and additional designations that pay credit to the importance of nonbiological data in the reconstruction of human evolution, such as *prehistoric archeology*, will appear in its course.[19]

Historical analyses have anchored the beginnings of paleoanthropology "proper" in institutional growth and the discovery and description of particular fossils, have named founding fathers, or have focused on the birth of a new physical anthropology with the theoretical and practical shifts in the aftermath of the evolutionary synthesis. Alternatively, one might consider attempts at the systematization of paleoanthropological and archeological data within a chronological framework derived from geology as relevant. However, none of these important landmarks in the history of paleoanthropology, which are temporally distributed between the second half of the nineteenth century and the decades around the middle of the twentieth century, by itself brought about all the properties of the current field as defined earlier. Indeed, the complexity of the current field renders the mere thought of having to locate its exact origin mind-boggling. Rather, the methods, technologies, theories, concepts, and objects of investigation that make up paleoanthropology, though interrelated, have their own genealogies.[20]

Nonetheless, the chronicling of its own histories is an integral part of paleoanthropology, which might not come as a surprise for a historically oriented field. Indeed, as we will see, the reconstruction of the history of excavation and interpretation of sites is essential for their reinterpretation, including that of human osseous remains. When one opens one of the great books of the paleoanthropological canon, one usually finds at least an introductory chapter on the history of finds and interpretations. Similarly, contemporary scientists tend to provide their frequently large public readership with a history of the field that at the same time situates the author(s) for the more informed reader, who can detect the perspective from which the overview is written. Historical entities such as fossils, anthropologists, and events are enlisted to present the ideas put forward in the book as finally correcting age-old, deeply ingrained, and ideologically motivated errors or as the logical consequence of the progress of science. Retrospectives with this kind of agenda have been termed *preface history*.[21]

This is not to say that there is something like a history from nowhere. Every article and book produced on the history of the anthropological sciences has done so from a covert or overt standpoint, and this one is no exception. Furthermore, not all work by anthropologists on the history of their field can be subsumed under preface history. Rather, some of the most comprehensive histories of paleoanthropology to date, or of episodes and controversies in paleoanthropology, have been written by academic practitioners and museum curators. Another source of histories of human-origins research, and one that further attests the subject's public appeal, is professional writers. A search for "human origins" in any online bibliography provides an impression of the industry of the popular science authors.[22]

The corpus of literature by historians, philosophers, and sociologists on subjects involving paleoanthropology is remarkably moderate. Nonetheless, important work has been done in an attempt to place the anthropological sciences in their cultural and historical context.[23] Concepts and theories of race in physical anthropology and human evolution have attracted particular attention.[24] Anatomical reconstructions of fossil hominids, reconstruction paintings, and museum exhibits have been studied, and so have the narrative structures of human-origins stories.[25] Another important focus of analysis has been the roles ascribed to women in prehistory. Also here, with notable exceptions, archeologists and anthropologists have provided the bulk of the work.[26] Finally, the fields that overlap with paleoanthropology, such as geology, comparative anatomy, molecular biology, and evolutionary theory, have received considerable attention from historians of science.[27]

I share an interest in all these topics that touch on the history of paleoanthropology, and this book is an expression of my fascination with the scientific engagement with human phylogeny and evolution. Therefore, it generally aims at addressing the lack of a more widespread interest of historians in prehistoric archeology and paleoanthropology that has often been bemoaned.[28] With its focus on one particular fossil, my approach resembles publications that deal with one find or forgery, such as the Piltdown fossils or Lucy, and those that have engaged with a certain category of bones, for example, the Neanderthals or the australopithecines.[29] Rather than aiming at the unraveling of a forgery, the resolution of a controversy, or an ontology of current views, this story of the Red Lady is situated in the interest in biographies of scientific objects within the history of science as discussed earlier. It aims at learning how fossil human bones were read, and how the meanings of these objects were codetermined by the human practitioners, theories, methods, and technologies of a science that was part of a culture and a historical setting.

How the Story Unfolds

The Red Lady's drama is made up of three main acts that give this book its three-part structure. Between each of these lie considerable spans of time filled with important changes within the material and immaterial cultures of the anthropological sciences that will be touched on only as far as they seem of relevance to a broad contextualization of the reinterpretations of the Red Lady. In the remainder of this general introduction, I provide a short outline of the main story line as a backdrop for the detailed accounts that follow. It is a streamlined summary that will turn out to be far more convoluted in the ensuing chapters.[30]

The story begins in 1823, when Buckland, then the first reader of geology at Oxford University, discovered the Red Lady in Paviland Cave, southern Wales. Part I deals with the question of how Buckland arrived at his interpretation of the find, and what role his situation at Oxford University played in it, where he had to reconcile the scientific with the ecclesiastical part of his career, a dualism that was institutionalized at Oxford. A closer look at Buckland and his geological activities, publications, correspondence, and notes is embedded in the broader discourse about human antiquity. Thus the opening chapter of the Red Lady's biography as discussed in Part I is simultaneously the beginning of the scientific controversy about human prehistory and evolution.

It is significant that the biographical beginnings of the Red Lady are located within geology, or what I may call the prehistory of paleoanthropology. The discoveries of this discipline were central for the introduction of a new timescale for humanity. A history of the earth that by far surpassed the chronology of the Bible was the first condition for this change in paradigm. The strata in which human bones and artifacts were found in close proximity to the remains of extinct mammals were old; but did the bones and artifacts really originate from these strata? Had they not been brought to these depths in burial rituals or through the action of floods? The response to these questions was neither uniform nor motivated by purely scientific factors. Nothing less was at stake than humankind's place in nature. Besides stratigraphy and paleontology, other sources of knowledge, such as archeology, were central for the acceptance of human antiquity as by far exceeding the reckonings made on the basis of the Scriptures. In the end, human tools rather than bones led to the concept of a human prehistory.

The rebirth of the Red Lady as a scientific object and its persistence as such presupposed local, material, and practical networks in which it could achieve meaning.[31] Because these were not ready for the concept of prehistoric humans, the Red Lady was classified as a historical document. Humankind remained within the frame set by the Scriptures, while only the earth and its nonhuman inhabitants had a pre-Mosaic history. The age and sex of the skeleton were interpreted according to antiquarian standards and folklore. Buckland initially read the bones as having belonged to a male tax man from the time of Roman occupation, the tokens in whose "pockets" might have represented tax money. But already with Buckland the skeleton underwent a change in sex from male to female, now being seen as a witch, if not a prostitute, who had sold her services to the soldiers of the British military camp nearby. The first pedigree established for the Red Lady linked "her" to Eve herself. "She" was enchantress in the sense of both witch and temptress. "She" was lady as well as harlot, and Buckland relied not only on male fantasies of what it may have meant to be a man or a woman in Ancient Britain, but also on what it meant to be a woman or a man in his society.

Although Buckland did not live to witness the acceptance of human antiquity or the breakthrough of evolutionary theory, he was aware of the issues at stake. Objects such as the Red Lady and its grave goods lay and lie at the boundaries between different belief systems and various "disciplines"; they also had and have the potential to contaminate supposedly pure categories of species, sex, and race that were claimed as given by God and by nature. The time revolution in geology, archeology, and anthropology would support, though not condition, an evolutionary interpretation of nature, including humans, and thus provide analogies for a rethinking of the static nature of the social order. Buckland could not silence the Red Lady permanently within his classification system. The skeleton turned out to be a monster in the long run, associating man with a pre- and nonhuman world. The ethical questions arising from the loss of a biblical world order are of a fundamental nature and gave rise to an uncertainty that can still be felt today. The differences between humans and animals, as well as the source of human ethics, had to be defined anew.[32]

After the meeting of the British Association for the Advancement of Science in Aberdeen in 1859, which is for reasons of simplicity taken as the key date for the acceptance of human antiquity by the scientific establishment, and between Darwin's publications *On the Origin of Species* (1859) and *The Descent of Man* (1871), the discourses about the age and the evolution of the human species came to be forever intertwined.[33] Although the mechanism of natural selection was not yet taken on, and although many people continued to believe in the Bible, discoveries such as the Red Lady were viewed in a different light. The pervasive ideology of discourses on the story of humankind in archeology and anthropology, as well as philosophy and history, was one of progress. Evolution was seen as a teleological development in the direction of man of Western civilizations. This belief in morphological and cultural progress seemed verified by the fruits of the Industrial Revolution. It formed the basic principle by which colonialism and imperialism were justified, which brought Europeans in contact with peoples they understood as living illustrations of the cultural stages suggested by the European Paleolithic record. Apart from theoretical changes, anthropology began to be institutionalized, and as more human bones and artifacts were discovered, early attempts were made at a codification of the European Paleolithic.

During these later decades of the nineteenth century, the Red Lady appeared in several publications. Some of these reproduced Buckland's interpretation of the find as of a historical date; others doubted the age and/or the sex as diagnosed by Buckland. A decisive turning point in the Red Lady's biography, which will be the subject of Part II, was achieved through the renewed detailed investigation of Paviland Cave under the supervision of Sollas, a student of Thomas Henry Huxley (1825–1895) and one of Buckland's successors to the chair of geology at Oxford University. In his Huxley Memorial Lecture of 1913 for the Royal Anthropological Institute, Sollas redefined Buckland's female Ancient Briton from the time of Roman occupation as a Cro-Magnon man of the Aurignacian culture. This redefinition took place within a changed scientific and cultural environment. The new scientific practices and technologies, the different theoretical framework, and the concepts of gender and race of the time evoked different meanings

Introduction

from the skeleton. Again, Sollas's scientific biography and work, in particular on the Red Lady, will be situated in the anthropology of his time and culture by drawing on a wide array of archival and published sources.

Sollas, like Buckland, was a central figure for the Red Lady, as well as in geology and anthropology at large. From about 1897, when Sollas arrived at Oxford, his career shifted from an emphasis on geology to anthropology and archeology. Apart from redefining the Red Lady, he engaged in the paleoanthropological controversies around the newly discovered hominid fossils, such as Eugène Dubois's (1858–1940) *Pithecanthropus erectus* (today *Homo erectus*) from Java, and the role of the Neanderthals in human evolution. Illustrative of the changes under way in the young science of paleoanthropology, Sollas moved in the course of his career from an interpretation of both of these fossil forms as direct human ancestors to a dendritic phylogeny, with all known fossil hominids on dead-ending side branches of the human family tree. In this, as Buckland had been strongly influenced in his rejection of any human bones as genuinely fossil by the great Georges Cuvier (1769–1832) at the Muséum d'Histoire Naturelle in Paris, Buckland's successor Sollas looked to a successor of Cuvier at the Muséum, the expert paleontologist Marcellin Boule (1861–1942).

However, the shift from a linear to a diversified model of hominid evolution that began around World War I was linked not only to Boule's reconstruction of the Neanderthals as primitive brutes but also to the scientific forgery of the Piltdown fossils, a fraudulent association of a human skull with an orangutan jaw. The expulsion of the Neanderthals from human ancestry, as well as a Piltdown type with a relatively modern brain, but of considerable antiquity, expanded the possible age of morphologically modern humans. Furthermore, the living human races seemed to be the result of long independent developments, which pushed the date for a common ancestor further back in time. Controversial skeletal remains, as well as eoliths, supposedly human-made tools that were claimed by some as of Tertiary age, were brought forward as evidence of such an early existence of relatively modern humans in Europe. Thus, also at Sollas's time, human antiquity was a subject capable of polarization.

In his work *Ancient Hunters* (1911, 1915, 1924), Sollas embraced a race-succession model in which he compared the human Paleolithic races with what he believed to be their modern representatives.[34] Like others at the time, he thought that higher races had always replaced, extinguished, or marginalized lower forms of humans. This kind of "evolutionary racism" drew on the biological determinism applied to recent human races by means of measurements of skulls and skeletons, and on the theory of polygenism. The European fossil bones, too, were instrumentalized to devalue and distance the living human races that were perceived as lower down the scale of beings through the construction of long independent evolutionary lines. Within this scheme, the Red Lady had belonged to an advanced race, which many anthropologists regarded as a kind of Paleolithic master race. The Cro-Magnons, with their high stature and considerable brain size, as well as elaborate industry and cave art, seemed to have marked the advent of true humanity. They also seemed to have marked the disappearance of the brutish Neanderthals. Thus, in the eyes of Sollas and his contemporaries, warlike and imperialist practices formed part of the human evolutionary past as much as of history and the present.

Sollas compared the bones of the Red Lady's extremities with other Cro-Magnon finds and Native Americans to determine the sex. Although the thighbone was comparatively slender and the muscular attachments not well marked, and although there was only half a pelvis, Sollas found his evidence strong enough to turn the Red Lady into Paviland Man. However, there seemed to have been no tools in the grave but bracelets and necklaces. Furthermore, unfortunately for Sollas, who was a craniometrist with the aim of finding adequate standards for making measurements comparable, and who developed a new machine for obtaining sagittal sections, the skull was missing. The missing skull was part of the Red Lady legend, and everyone who took on the post of professor of geology at Oxford University was asked whether he had found the Red Lady's skull. This may be due to the fact that Buckland represented the skeleton as entire in the section of Paviland Cave he reproduced in his *Reliquiae Diluvianae* (1823), a blemish Sollas, as we will see, removed from Buckland's sketch in *Ancient Hunters* (1911).[35]

Between Sollas's time and the last major reinterpretation of the Red Lady toward the end of the twentieth century, great changes again took place in the anthropological sciences. These concerned theories of human evolution and heredity, the continuing institutionalization and specialization of (paleo)anthropology, and the discovery of unexpected kinds of fossils (with the associated phylogenies and models of human evolution). The stars among these were the australopithecines that began to be reported from South Africa and eventually East Africa from the mid-1920s. In synergy with the impact of the evolutionary synthesis on paleoanthropology, which worked against a non-Darwinian concept of species and evolutionary mechanisms such as use-inheritance and orthogenesis, these small-brained but bipedal hominids changed the focus from Asia to Africa as the scene of early hominization and from the brain to bipedalism as early hominid adaptation. There were also central novelties in the material culture that were applied to the Red Lady. The skeleton was radiocarbon dated in the 1960s (and 1980s) and ascribed an approximate age of 18,500 years. Far from finally settling the question of its origin, this produced the problem of the burial's contemporaneity with the last glacial maximum.

The most recent chapter in the Red Lady's biography, examined in Part III, has been written by a team under the supervision of the archeologist Aldhouse-Green, then at the University of Wales in Newport. I might tentatively call this a happy ending, since the resulting monograph published in 2000 gives evidence of a reexcavation of Paviland Cave and a rigorous reanalysis of its material. The Red Lady, which has gained celebrity in the meantime, is estimated to be the first Pleistocene skeleton discovered by a scientist. Paviland Cave is the richest early Upper Paleolithic excavation site in Great Britain. A dating places the skeleton on the coast of southern Wales some 26,000 radiocarbon years ago and thus removes it safely from the last glacial maximum. A mitochondrial DNA analysis links the Red Lady to the commonest extant lineage in Europe, demonstrating its presence in Paleolithic Britain and thereby contradicting the violent-replacement hypothesis.

The Red Lady has been reproduced in an ethnoarcheological environment, and an exhibit presented the scientific findings to the interested public. In a reconstruction painting at the National Museum of Wales, it seems that Buckland's Red Lady has changed roles in the burial scene. "She" is no longer the object of the burial, but is reincarnated as a woman who takes part in the burial ritual of Paviland Man. Even the Red Lady's powers of witchcraft (which belong to one of Buckland's stories) have been transferred to her red-haired heiress, who holds a wandlike stick in the performance of a magical ceremony (see Figure 15.1 in Part III). Because Stephanie Moser and others have criticized Paleolithic reconstructions for their absence and marginalization of women, the fact that even though we no longer have a Red Lady, we now have a red-haired female who takes center stage in the reproduction (she is at the center of the painting, as well as of the ritual) is a nice way to end the story.[36] The reddish hair of the participants in this perhaps shamanic ritual hints at another thread in the Red Lady's biography: the role of fossils as national icons. Although this topic will not be thoroughly discussed, the Red Lady and "his" race, and their juxtaposition with the Neanderthals, will more than once provide the chance to illustrate how fossil bones become caught up in discourses of national identity and origin. In this process, the identification of one's own nation with a certain set of bones, and with the prehistoric humans they stand for, synergizes with the association of other nations with "a very different" and presumably more primitive set of fossil bones, such as those of the Neanderthals.

The international and interdisciplinary excavation project that began in 1995 culminated in the monograph *Paviland Cave and the "Red Lady": A Definitive Report* (2000).[37] This model scientific project thus raises the question of what we gain by approaching the mysteries of our prehistory from the points of view of many disciplines and of scholars of various national and cultural backgrounds. Is one of the advantages of such projects the arrival at definitive reports? Apart from highlighting our desire for completeness, the claim to have written the final chapter in the biography of the Red Lady opens up further avenues for inquiry. For example, where does narrative fill evidential gaps? Does the definitive report represent a synthesis of all the insights gained into the site's prehistory? How does this most recent rereading of the Red Lady's bones relate to the controversies that have accompanied us throughout the preceding chapters? What kind of magic is at work? These questions are approached through a close reading of the project in the making and of the resulting monograph that strongly draws on the cooperation of those project participants who worked on the human bones and artifacts. The discussion of the individual contributors and contributions to the Paviland monograph and of the visual reconstructions of "what really happened" in Paviland Cave some 26,000 years ago, as well as of the interdisciplinarity and internationality of the Paviland project, allows getting below the surface of the supposed closure. One question will thus be of particular interest: Has the story of the Red Lady been brought to an end?

Introduction

There are several topics of concern to paleoanthropology as defined earlier that were already troubling the successive generations of scientists of relevance to the interpretive history of the Red Lady and Paviland Cave. Among these controversial issues that will permeate the chapters of this book are the antiquity/recency of humans/the human races (and hominids); the correct model of human evolution and geographical distribution (replacement versus local-continuity scenarios); the place of origin of modern humans/the first hominids; the degree of diversity in fossil and living hominids (dendritic versus linear models); and the role of living peoples in the interpretation of the hominid fossil record (ethnological analogy). Throughout the history of its scientific interpretations, the Red Lady was bound up with these questions, although the main emphasis has shifted from one to another. The shift in emphasis is due to the fact that the Red Lady was exemplary for potentially fossil human bones in the controversy around human antiquity (Part I), for the Paleolithic human race of Cro-Magnon in theories of racial origins and diversity (Part II), and for the complex questions of chronology, lineage, and social context posed by fossil finds and archeological sites that are answered by contemporary interdisciplinary approaches (Part III).

Despite the longevity of these issues, the Red Lady's afterlife as discussed in the following chapters points to the fact that these binaries are simplifications, that neither periods nor people in the history of anthropology can satisfactorily be ascribed to a fixed combination of their constituents. Although we have seen how new approaches in the history of science have tended to undermine the notion of radical paradigmatic and epistemic breaks, microhistories such as those concerned with the successive analyses of the Red Lady also warn against encapsulations, as in my case of "the migrationists" versus "the evolutionists." This impression is supported by historical studies such as Hagner's analysis of the history of the brain already mentioned, where, too, the binary of reductionism versus holism turned out to be a violation of the actual complexity once the focus was set on the organ rather than on overall theoretical trends in the medical sciences.

The focus on the Red Lady will also direct attention to strategies in the negotiation of scientific knowledge, such as humor. Buckland, who was a master storyteller and entertainer, used humor as a way of establishing community and excluding nonmembers, but also as a diplomatic strategy for the advancement of unorthodox scientific views. In case of vehement rejection, one could always say, "It was just a joke." Finally, humor could be employed to divert attention from potentially subversive evidence. Like humor, magic will find its way in and out of the Red Lady's narrative and is in the content, as well as in the form, of the science. Artifacts associated with the Red Lady have been interpreted as having been used for magic, while at the same time scientists have claimed magic abilities in the finding and reading of fossils.

The theme of magic highlights that the thread the Red Lady is in my story is indeed red. In agreement with the questions of phylogeny surrounding the Red Lady, the color of ochre is widely held to signify blood. It is worn by Catholic cardinals and is associated with fire, foreshadowing the role of religion and magic ritual in the Red Lady's afterlife. It is also the color of power, decorating kings and judges and pointing to, among other things, the issues of authority and ownership connected to the ochre-stained bones. In fact, it seems as if the successive interpretations of the Red Lady as a person fit well with the stereotypes that have accompanied the color red, and red hair in particular—a fit that transcends the Greek myth about redheads turning into vampires after death.[38] Thus, in the visual reconstruction of the burial scene at Paviland Cave according to Aldhouse-Green mentioned earlier, the red hair of the woman burying the Red Lady denotes not only Welsh affiliation, but also the resurrection of the magic trope. Rather than marking the ritualist as witch and harlot, however, the trait signifies the bestowal of positive spiritual power. I hope that the reader, following this multilayered red thread through time, will find some magic and humor at work in my telling of the Red Lady's story.

I

WILLIAM BUCKLAND, AN ANCIENT BRITISH WITCH, AND THE QUESTION OF HUMAN ANTIQUITY

THE PARTIAL HUMAN skeleton called the Red Lady of Paviland was reborn as a scientific object in the year 1823, when the first Oxford geology reader William Buckland (1784–1856) discovered it in Paviland Cave in southern Wales. This discovery falls within what is often referred to as the golden age of geology, a time of sensational discoveries, identifications, and reconstructions of an antediluvian (pre-Flood) fauna. Besides exotic mammals such as mammoth, rhinoceros, cave bear, and hyena, even much older "terrible lizards" (soon to be known as dinosaurs) were unearthed. Growing insights into geology had led to an expansion of the age of the globe—knowledge that was prerequisite to gain an understanding of human beginnings in Europe and beyond. However, there was as yet no framework of an extended human prehistory within which to interpret the Red Lady. Although the concept of a great human antiquity began to dawn on the scientific horizon, since there were reports of crude stone tools and human bones found in close association with fossil bones of extinct animals, human history was still largely dictated by the chronology of the Bible.

In Britain, antiquaries from the early seventeenth century had sporadically engaged with field monuments and artifacts rather than written documents as clues to the human past, and toward the end of the century this archeological approach became more common. Nonetheless, even though pre-Roman material culture was the object of widespread interest in Buckland's Britain, archeology as a recognizable discipline is a product of the later nineteenth century. Most important in this context, history from the first peopling of Britain to the Roman conquest (what we today would refer to as British prehistory), unraveled through classical texts, the Scriptures, and increasingly antiquities, was conceptualized as a matter of a few thousand years. Thus the terms *Ancient British*, *Celtic*, and sometimes *Druid* referred to antiquities of the pre-Roman period as a whole, and there was as yet no common classification and temporal sequence of cultures.[1]

Where the physical study of humans was concerned, comparative anatomy had been pioneered by the work of Johann Friedrich Blumenbach (1752–1840), professor of medicine at the University of Göttingen.[2] Although attempts at a racial taxonomy have been referred to the age of discovery, which inspired comparative morphology, as well as the comparison of cultures as carried out by early ethnologists, it was around the turn from the eighteenth to the nineteenth century that anthropometry came to refer to a system of measurement of living and dead bodies to determine their proportions at different ages and in different races. Drawing on anatomical, zoological, and medical expertise, the early physical anthropologists, such as Petrus Camper (1722–1789), whose practices in contrast to those of Blumenbach were metrical, established quantitative techniques for the identification and classification of human races. This paved the way for the endeavor to employ skulls for racial taxonomy by means of an amalgam of new instruments, such as calipers, craniometers, goniometers, and craniographs, and the associated measurements, indexes, and angles. In the course of the first half of the nineteenth century, large collections of crania were accumulated for that purpose. The impact of these endeavors and practices, the roots of which reach back to even before Buckland's time, will still concern us in looking at the developments in the second half of the nineteenth and the early twentieth centuries in Part II. It was also the early fruits of these endeavors that the discoverers of human bones in association with those of extinct mammals, as discussed in this part, could draw on in their identification.[3]

This engagement with the differences between living human varieties was associated with speculations about the origin of human diversity. In the eighteenth century and well into the nineteenth, the anthropological debates on racial origins took place between the main outlooks of monogenism and polygenism. Monogenism corresponds to the biblical notion of the unity and common origin of all humankind. Although humans had been created in God's image, original sin and the persistence of sinful ways after the Fall caused them to degenerate and to differentiate in the dispersal across the globe in the aftermath of the Flood and the building of the Tower of Babel. Polygenism, on the other hand, was the belief in a pre-Adamic creation and refers to the notion that the existing human varieties can be traced back to more than one origin. Polygenist thought, too, reaches back at least to the age of European expansion, when many scholars came to consider the newly discovered aboriginal peoples too distinct from Europeans to represent the same species. Nonetheless, the alternative beliefs in either a recent common origin of all existing human varieties or separate origins associated with independent histories of the human races will continue to concern us throughout this book.[4]

Although the controversies about the origin of human varieties were framed by the limited time assumed for the history of humanity, the notion of a prehuman history of the earth was not new to the nineteenth century. It had been theorized in such well-read opera as Thomas Burnet's *Telluris theoria sacra* (1681), Comte de Buffon's *Histoire naturelle* (in particular, *Époques de la nature*, 1778), and Johann Gottfried Herder's *Ideen zur Philosophie der Geschichte der Menschheit* (1784–1791). In his *Theory of the Earth* (1795), the Scottish natural philosopher James Hutton (1726–1797) envisioned an endless cycle of rock erosion and formation in which the magma erupting from the interior of the earth transformed the sea sediments, which had accumulated on the bed of the oceans out of land detritus by the action of rivers, and threw them up to build continents. Because of the notion of a molten interior earth and the importance of volcanic action, this theory was called Plutonism. This image of the earth as a *perpetuum mobile*, in which the same geological forces are forever at work, without beginning or end, laid the basis for the ideas of uniformitarianism.

Increasingly, reconstructions of the history of the earth relied on insights from stratigraphy and comparative anatomy. The mineralogical classification of rock formations was institutionalized at the Mining Academy of Freiberg, and schools of mining followed in other European towns. Toward the end of the eighteenth century, Freiberg became *the* institution for geological teaching. Abraham Gottlob Werner

(1749–1817) was its professor of mineralogy from 1775 and attracted students from different countries, some of whom would become important figures in geology, such as Alexander von Humboldt (1769–1859). Buckland gave Werner credit for having been the first to point to the importance of petrifactions as a basis for the arrangement of geological formations.[5] Werner was also the main exponent of a geognosy that came to be known as Neptunism, according to which the older crystalline rocks had initially been deposited from a primeval ocean through chemical precipitation. The receding of the once-universal ocean to its present locations had exposed the landmasses with their universal rock formations, and detrital sedimentation had set in. This stood in contrast to the Plutonists, who believed in the igneous origin of basalt and granite. Both of these positions—Huttonian Plutonism with its steady-state view of earth history and Wernerian Neptunism with its universal stratigraphic system based on mineral criteria—were advocated by professors at the University of Edinburgh, which was therefore another important center for the new science.

The more historical interpretation of the earth with time conceptualized as an arrow rather than as a cycle became popular in the form of catastrophism. The Swiss-born naturalist Jean-André de Luc (1727–1817), who had chosen England as his home and was reader to Queen Charlotte, explained the abundance of marine fossils in today's terrestrial strata as the result of a water cataclysm, during which land and water mass had been exchanged. He identified this catastrophe, which had engulfed everything with the exception of the highest mountains, with the biblical Flood. The French savant Georges Cuvier also worked with geological revolutions and with the concept of extinction. The bones in caves and open sites of fossil animals such as hippopotamus, elephant, hyena, and lion that no longer inhabited Europe were thus the remains of species that had been extinguished in European regions through the action of a violent geological upheaval. Cuvier became a great expert on vertebrate paleontology and comparative anatomy, which he institutionalized at the Muséum d'Histoire Naturelle in Paris. His knowledge of the functional whole of the organism became so thorough that he bragged that he was able to reconstruct an entire animal from a single fossil bone. Thus the Paris Muséum, which also had Jean-Baptiste Lamarck's (1744–1829) invertebrate zoology, became, after Freiberg, an important institution for earth history.[6]

Cuvier and Alexandre Brongniart (1770–1847) worked on the succession of strata in the Paris basin to arrive at the geological history of the Paris region. They reconstructed the advances and retreats of the sea and its shell inhabitants and used fossils to discriminate between similar strata. This method of the relative dating of strata by means of their fossil content was pioneered by the English civil engineer and mineral surveyor William Smith (1769–1839). Smith was the son of a blacksmith and obtained only rudimentary education. His *Strata Identified by Organized Fossils* (1815), a complete geological map of England and Wales, was initially overlooked by the scientific community. Eventually, however, his work was recognized to the extent that Adam Sedgwick (1785–1873), Woodwardian Professor of Geology at Cambridge since 1818, conferred on him the title "Father of English Geology" when presenting him the Wollaston Medal of the Geological Society of London in 1831.[7]

Smith's achievements indicate how closely the progress of geology was linked to the processes of industrialization. The drilling of holes into the bowels of the earth in search of coal and other riches not only brought to light stratification and fossil content but also allowed insights into geological mechanisms such as erosion, marine sedimentation, and the folding of strata. Generally speaking, the age of improvement provided many opportunities to those interested in geology. The rapid growth of domestic agrarian and industrial production and of the associated national and international markets necessitated and allowed the building of roads, bridges, and canals that facilitated the travel of persons and the transport of goods as the basis of international networks of exchange of information and specimens. The new economic powers were based on the development of a coal and iron technology that augmented production from farming to the cotton industry. However, it was only with the age of the railways—in 1830 a railway connection was established between Manchester and Liverpool—that modern industry became the dominant sector of the entire national economy, and a period of prosperity set in that lasted without any major setbacks until the 1870s. Toward the end of Buckland's life at the middle of the century and about the time of the Great Exhibition of 1851, Britain was the leading economic and financial power.[8]

The fact that the political and social power of the nobility was nearly as strong at midcentury as it had been one hundred years earlier should not obscure the prosperity of the growing middle classes of commerce, finance, industry, and professions. Their political impotence led to the establishment of clubs, unions, and societies, including those of a literary and philosophical nature. Although science began to provide career opportunities, of which Buckland himself was an early beneficiary, gentlemen who financed their scientific enterprise from other resources constituted the bulk of the geological societies. Like the British economy in general, science could profit from the empire, and naturalists tried their fortunes in Canada, New Zealand, the West Indies, South Africa, Australia, and, above all, India. Geologists were needed to locate and exploit natural resources and as welcome government advisors in related issues, with the Geological Survey (founded in 1835) coordinating efforts and storing information. British expansion seemed to open up unknown and unlimited sources for new geological and paleontological research.[9]

Furthermore, after the Napoleonic Wars the second Peace of Paris (1815) facilitated travel within Europe, so that historical geology from Germany and France was increasingly introduced into Britain. However, British geology did not integrate the secular German philosophy of history. Moreover, although Cuvier's geology and paleontology became very influential, they were adjusted to a peculiarly British taste. Although Cuvier referred to his last geologically significant revolution as *deluge*, he did not identify it with the biblical Flood. It was only with the importation of his work into Britain that this identification came about. To Cuvier, the Scriptures were one among several historical texts that could be used in the reconstruction of the history of humankind. Furthermore, Cuvier restricted the chronological authority of the Bible to the human realm and conceptualized the geological processes that led to the current shape of the earth's crust as of much longer duration.[10]

Regency England provided a different environment than French society did to its savants after the assaults on the power of the church in the French Revolution. State, church, and society were still closely intertwined, and a powerful church was considered the backbone of a stable and well-ordered society. Throughout the seventeenth and eighteenth centuries and well into the nineteenth century, sacred chronology, which reconstructed earth and ancient human history by means of the Scriptures and other historical sources and increasingly by means of geology, remained an important tradition with varying degrees of literalism. The adherence to an age of the earth of approximately 6,000 years stayed in place until well into the Enlightenment. Toward the end of the eighteenth century, the days of Creation came to be interpreted allegorically, so that the possibility for a very long history of the earth arose. Growing knowledge of primary and also secondary rock formations, the latter of which often contained marine shells, demanded a timescale that far transcended a literal interpretation of Genesis.

The argument from design in natural theology, which interpreted the order of nature as indicative of the existence of a supreme designer or creator and went back to classical and medieval philosophy, had been taken up in early modern times. A new wave of physicotheology began with William Paley's (1743–1805) *Natural Theology* (1802), which was a centerpiece of the Oxford curriculum in Buckland's times, and culminated in the 1830s in the Bridgewater Treatises on the "Power, Wisdom, and Goodness of God, as manifested in the Creation." These tracts, instigated by the Earl of Bridgewater, were widely circulated and represented the scientific elite's reconciliation between science and religion as dominant in the liberal Anglican stance of the members of the British Association for the Advancement of Science (founded in 1831). As we will see, it was Buckland who wrote the volume on geology as a kind of climax of the belief in the divine design of each organism for its environment. He also demonstrated the overall increase of organic complexity throughout the geological layers, a concept that guided paleontological reconstruction and theory in the 1820s and 1830s.[11]

Historical geology as it became institutionalized at the universities of Oxford and Cambridge in the early decades of the nineteenth century maintained the importance of historical documents (including the Scriptures) in the reconstruction of the last period of the history of the earth, thereby reconciling geology with the classical traditions. Buckland and Sedgwick, who occupied the newly established positions of geology reader and professor, achieved this by means of a catastrophist framework, the argument from design, and the notion of geological progress. Buckland was instrumental in the establishment of a British stratigraphic column through the integration of existing knowledge and his own researches. He became a central figure in Quaternary geology that dealt with the rather superficial deposits of mud and gravel and their mammalian fossil content. He was among those who searched caves on the Continent and in the British Isles for fossils in the attempt to resurrect their antediluvian animal worlds. His hyena-den theory allowed him to reconstruct some of these as prehistoric shelters of packs of wild beasts. These vistas into the deep past through the entrance of caves gained great popularity catalyzed by the spirit of the age of romanticism.

As a cultural movement, romanticism captures the transition from the Enlightenment to "modernity," during which such processes as the instrumentalization of time, the division of labor, and the mechanization of science and society were as yet incomplete. For natural philosophy, romanticism is perceived as an era in which disciplinary specialization has just begun, with most scientists still being polymaths, and in which a nonutilitarian and nonmaterialist approach to nature was mediated through the personality of the inquirer. Indeed, the geologist might have styled himself as romantic hero and knight, as a man particularly qualified for the discovery of fossils and the reading of bones and stones. The age's passion for historicity was favorable to a historical understanding of the cosmos, the earth, and its inhabitants. Even human society and human beings came to be understood as products of historical processes. Not surprisingly, then, biographies and correspondences were published, histories of small and large scope were written, a general grammar gave way to historical philology, and the Bible was increasingly read as a historical document. There was also an interest in the history of science, as masterfully demonstrated by William Whewell's (1794–1866) *History of the Inductive Sciences, from the Earliest to the Present Time* (1837).[12]

In the course of his popular cave researches, Buckland also made discoveries of human bones and tools and was informed of such discoveries by other explorers. These, like the mammalian bones with which they were found, demanded interpretation. The first chapter in the Red Lady's biography therefore touches on the debates on geology as a science in the particular setting Buckland found himself in, and on the history of the earth and its inhabitants, especially that of humans. Not surprisingly, from its very beginning, the Red Lady was surrounded by myth, magic, controversy, and humor:

> Have ye heard of the Woman so long under Ground
> Have ye heard of the Woman that Buckland has found

> With her Bones of empyreal Hues?*
> O fair ones of modern days, hang down your head
> The Antediluvians rouged when dead
> Only granted in life-time to you.[13]

Although the author of these lines jokingly called the Red Lady an antediluvian, Buckland could not agree. *Antediluvian*, as suggested by the term, referred to a being that had lived before the Deluge had drowned humans and animals alike, whereas *postdiluvian* referred to those beings that had either survived the event or had been created anew to inhabit a purified earth. For a geologist such as Buckland, however, *antediluvian* also referred to a time when there had existed strange, huge mammals that were no longer to be found in the British Isles. To call the Red Lady antediluvian would have rendered "her" contemporary with the gigantic mammoth, cave bear, and hyena; it would have placed "her" into a geologically not-yet-modern world. As we will see, this was not acceptable to Buckland because, among other things, he worked within the framework of a progressive history of organisms on earth alluded to earlier. Humankind was in this scheme a latecomer and clearly belonged to the present order of things.

However, Buckland did not lack the necessary humor to find pleasure in the bold poem, speculating that "Cambrians wd readily contend that Adam was a Welch Man,"[14] referring to the cave's location in southern Wales. The Red Lady might thus kindle Welsh pride by providing them with a long line of ancestry, if not by linking them to the very first couple. Buckland's joke gives testimony to a "center-periphery" tension that we will encounter again in Parts II and III with regard to the questions of who owns and who defines the Red Lady, and that indicates the role of prehistoric human finds as national icons. The Red Lady's close relation to the first people in paradise was further supported, Buckland continued the joke, by the skeleton's reddish color: "This problem is at once solved by supposing her a near Connexion of Adam. Perhaps Eve herself, for is it extraordinary when Adam was made of red Earth, that his Rib should have a tinge of ruddle? That the redness lasted some time in the family, we learn from Esau's being a red Man."[15] However, as we will see, the color red at times rather appeared to mark the bones as those of a scarlet woman than as those of a biblical lady.

The poem provides an appropriate starting point for the first episode in the Red Lady's story. It raises the central points of discussion associated with the discovery, which are the skeleton's age, its sex and its having belonged to a lady, and the applicability of the Scriptures as a frame for interpretation. In addition, the poem warns of Buckland's use of humor when confronted with potentially threatening ideas and controversial theories. In order to understand how the Red Lady arrived at "her" first definition, it will thus be necessary to take a closer look at the circumstances of the discovery and at the context of interpretation of the bones. Who is the Red Lady's first lord, William Buckland, who had the power to define "her"? And how did "she" come to be defined as a lady, a red lady, and one of postdiluvian or Roman-British times? To answer these questions, this part will begin with an introduction to Buckland as scientific persona (Chapter 1). For his education and later for his advancement as a professional geologist, Buckland relied on networks of influential family friends and scientists. His position as the first reader of geology at Oxford University, an institution for the education of the Anglican clergy, where he also met with suspicions about the propriety of geology, might have worked to heighten his sensibilities toward the opinions of his superiors and peers.

As we will see, Buckland's stance toward the possibility of fossil human bones and the position of humankind in his theoretical framework changed in the course of his career in response to insights from his fieldwork and developments in the larger scientific community. It will therefore be of interest to look at his work at Kirkdale Cave that was fertile for his vision of antediluvian England and that had a direct impact on his early responses to claims of antediluvian humans (Chapter 2) before moving to his excavation at and interpretation of Paviland Cave (Chapter 3). As the poem quoted earlier suggests, Buckland and his peers were not unreceptive toward the poetics of the cave as a place that might provide glimpses of a former world, and Buckland built on its mystic aura, even if in his scientific publications he described his researches at Kirkdale and Paviland caves in a matter-of-fact style. Besides forming an important part of his scientific achievement in geology, his caving therefore also enhanced the cult around his scientific persona. As in Buckland himself, where the dry scientific writing and the stern rhetoric of the churchman met with a humor verging on the frivolous and an at times very lively imagination, caving was at the same time a dangerous and arduous undertaking and a means to sublime experience and revelation of scenes long gone. Direct experiences of caving and the meaning of caves in poetry and popular culture will therefore help one understand the discrepancy between the way the Red Lady was reconstructed in Buckland's major publications on Paviland Cave and the "romance" he spun around it.

In the aftermath of Buckland's Paviland excavation and his definition of the Red Lady as postdiluvian, the evidence for human coexistence with the extinct antediluvian animals increased, and Buckland was directly involved in the excavation of one of the caves that turned out to be among the most important in the establishment of human antiquity (Chapter 4). His role in the history of the early excavation and interpretation of the evidence from Kent's Cavern has led historians to see Buckland as obscurantist. The fact that Buckland had the power to influence the interpretation of the provincial clergyman who conducted the bulk of the research at Kent's Cavern also points to questions about the social structure of caving. Although Buckland was strongly guided by the opinions of the big names in geology and paleontology in Britain and on the Continent, he might have been dominating toward his social inferiors.

Buckland's continued rejection of the increasing evidence for antediluvian humans can only be understood by taking not only his scientific but also his social background into account (Chapters 5 and 6). More was at stake than the settling of a scientific question. The notion of an age for humankind that transcended the biblical framework seemed to be associated with transmutationist views of the organic world that again were linked to political tendencies that presented a threat to Buckland's conception of the just order of society. In the first episode of its afterlife, the Red Lady was therefore caught up in the controversies around human antiquity and transmutationism. Buckland's stance toward these issues would at any time codetermine his interpretation of the skeleton. He had to rely on ever more evasive strategies to maintain his opposition to the Red Lady as evidence for human antiquity. All to no avail, since as we will see at the end of this part, the scientific communities reached consensus on the contemporaneity of extinct animals with human-made tools within a few years of Buckland's death. The history leading to the acceptance of human antiquity has been covered in full by Donald Grayson and A. Bowdoin

Van Riper and will here be fragmentarily reconstructed from Buckland's involvement. From this particular perspective, the Red Lady stands in for human bones uncovered in the process of geological excavation up to the 1850s that raised scientific, religious, and social questions in a way that makes clear that the three realms were not separate entities.[16]

1

William Buckland

From William Buckland's biography, written by his daughter Elizabeth Gordon and published in 1894, one gets to know him as a bizarre but sympathetic gentleman scientist of wide interests, working endlessly with never-faltering energy, and as a pious practitioner of active charity. He seems to fit the cliché of the eccentric but brilliant mind in accordance with which important scientific figures have often been portrayed. Although some of his practical jokes, such as serving his dinner guests mice or blowing his nose on a portrait of Queen Adelaide, were beyond etiquette, the typical Victorian genre of life and letters idealizes Buckland, and we never have to give up our full-hearted sympathy over serious flaws in his character. It was not least because of Buckland's humor, which could verge on the coarse, that he was perceived as a colorful personality whose idiosyncrasies inspired many anecdotes and even verses, so much so that the author of Buckland's scientific biography, Nicolaas Rupke, speaks of a personality cult.[1]

Buckland was born in Axminster (Devonshire, Great Britain), the son of Elizabeth Oke, daughter of a wealthy landowner, on 12 March 1784. The birth of his interest in matters of natural history is said to lie as far back as his childhood, when he went on hunts for shells of ammonites with his father, the Reverend Charles Buckland, who was rector of Templeton and Trusham in the county of Devon and of West

Chelborough, Dorset. Buckland partly owed his career in science and in the church to the support of persons of influence, as well as "invisible" helpers. Although the early network that procured a good education for him had not been of his own making, he later proved himself a most apt networker, establishing connections to leading scientists at home and on the Continent, to people of influence and wealth, and to collectors and "amateur" naturalists indispensable to a geologist.[2] J. M. Edmonds has recovered the family connections set in motion to secure a good education for William, which included such big names as the member of Parliament Reginald Pole Carew. This made it possible for Buckland to enter St. Mary's College in Winchester in 1798, the public school that had served as the model for Eton and Harrow, after he attended Blundell's School in Tiverton.[3]

At Winchester College, Buckland developed a fledgling interest in rock formations and started a collection of fossils. In 1801, he won the contest for one of the two Exeter Diocese (Devonshire) scholarships at Corpus Christi College, Oxford. After graduation, he continued to the MA by supporting himself on his scholarship and by taking pupils. As was the tradition at Oxford, Buckland was ordained before his election as probationary fellow in 1808. John Kidd (1775–1851) integrated mineralogy into his chemistry classes and thereby inspired many Oxford scholars to become involved with the new geology. They were spellbound by the subject that began to be cultivated by the founders of the Geological Society in London (founded in 1807). Among them, the brothers John Josias Conybeare (1779–1824) and William Daniel Conybeare (1787–1857) were particularly dear to Buckland and would influence his geology, accompany him on journeys, and remain faithful friends. Like Buckland, they were early members of the Geological Society (John Conybeare, like Kidd, was even a founding member). Another geological association was Henry Thomas De la Beche (1796–1855), who would become the first director of the Geological Survey (1835).[4]

Thus Buckland already reached out beyond Oxford, and one of the reasons for his extensive geological travels in Britain and on the Continent was to establish and maintain connections. Between 1808 and 1815, he toured England, Scotland, Ireland, and Wales, collecting sections of strata and their fossil remains, and subsequently, when the end of the Napoleonic Wars made travel easier, he undertook a geological tour of five months to Belgium, Germany, Poland, Hungary, Austria, Italy, Switzerland, and France, accompanied by John Conybeare and George Bellas Greenough (1778–1855), who had been the first president of the Geological Society, and with whom Buckland kept up a close relationship. They visited every noteworthy collection and professor, among them Werner at Freiberg and Johann Wolfgang von Goethe (1749–1832) in Weimar, and purchased maps, books, and prints.[5]

With the prospect of a readership in anatomy (he eventually became regius professor of medicine in 1822), Kidd was looking for a successor and first settled on William Conybeare. However, Conybeare opted for marriage and left Oxford for what turned into a career in the church, successively as rector of Sully, vicar of Axminster, and dean of Llandaff. Fortunately, his independent means prevented a break in his pursuit of geology. Thus it was Buckland who succeeded Kidd as reader of mineralogy in 1813, and the family network was once again set in motion to procure for him a new readership in geology in 1818, which was founded at the prince regent's instigation. The same year, he was elected a fellow of the Royal Society. In the second decade of the nineteenth century, historical geology also began to develop at Cambridge University, with a circle of interested natural philosophers such as John Henslow (1796–1861) as professor of mineralogy (1822–1828), his successor Whewell, and of course Sedgwick, Woodwardian Professor of Geology since 1818.[6]

The universities of Oxford and Cambridge were institutions for the education of the Anglican clergy, and they therefore valued geology according to its synergies with their classical traditions and its ecclesiastical relevance. Nicolaas Rupke has shown how this context was formative for Buckland's work as part of what he referred to as "the Oxford School of Geology."[7] Even when the repeal of the Test Acts of 1828 and the Catholic Emancipation Act of 1829 removed the discrimination of exclusion from public office against Protestant sects and Catholics, the universities of Oxford and Cambridge remained closed to non-Anglicans. Criticism came from the Whigs, who did not approve of the enforced conformity with regard to an unquestioning acceptance of Anglican doctrines or the fact that mainly the classics were taught, with the sciences being of secondary importance.

The colleges controlled the headships, livings, canonries, deaneries, and bishoprics, and the university, where the responsibility for the sciences was located, played a minor role. Most college fellows were clergymen and depended on income from ecclesiastically controlled positions. In letters to the university chancellor Lord William Wyndham Grenville (1759–1834), dated 24 and 27 November 1818, Buckland complained about the amount of the stipend for the readership in geology. As is well illustrated by Buckland's career, besides undertaking trips to sites, a geologist had to establish and maintain an international network among colleagues. This involved reciprocal visits and examination of each other's collections, correspondence, and exchange of specimens, as well as the purchase of books and maps. For the last eight years, Buckland had spent £200 per year on travels alone, so that the ordinary sum of £100 pounds a year seemed hardly enough.[8]

His application to the prince regent for a readership in geology furthermore makes the difficult political position of the new science clear. He cunningly tried to use the potential threat to received religion that some authorities perceived in geology to his advantage when he emphasized

> that Geology is a Branch of Knowledge at this time so much cultivated, of so much National importance, and so liable to be perverted to Purposes of a tendency dangerous to the Interests of Revealed Religion, that it is considered by persons of the highest Authority in the University to be a proper and desirable Subject for a Course of Public Lectures in Oxford.[9]

With its great utilitarian potential, geology was a branch of knowledge a nation should not disregard. Buckland argued that the danger to the status quo would best be met by making geology the subject of the establishment, thereby securing its services for the traditional institutions, including the church. Even though the Scottish University of Edinburgh criticized the universities of Oxford and Cambridge for their mingling of science and belief, the new geology of the 1820s was an important factor in at least forcing biblical literalism out of the English academe.[10]

Those who attended Buckland's geology lectures at the Ashmolean Museum (Figure 1.1), during which he handed around his primeval reptiles and other fossils and used additional visual aids such as charts and maps, were not likely to resent his success. In a letter dated 18 June 1821, Buckland informed Lord Grenville that this term, as well as the last, he had been lecturing on the comparative geology of England

William Buckland

Figure 1.1. Nathaniel Whittock's (1791–1860) ca. 1830 engraving of "Buckland Lecturing in the Ashmolean Museum on 15 February 1823"; courtesy of The Oxford University Museum.

and on the Alps to "an overflowing Class larger than my Room will hold."[11]

Overflowing, too, were his boxes of fossils that further crowded the lecture room. In 1822, some amends were made to his den, as he called it, when his collection was arranged, cataloged, and put into £300 worth of new cabinets.[12] The natural history collection of the Ashmolean Museum, after the mineralogy and geology collections had been moved to the adjacent Clarendon building in the early 1830s, were eventually transferred to the Natural Science Museum on Parks Road, which opened in 1860 under the geologist John Phillips (1800–1874) as the first keeper. It was here that the Red Lady's bones would be stored.

Buckland further emphasized his practical approach to teaching and research by taking his audiences and students on excursions to ossiferous rocks and exposed strata. He also continued his own fieldwork on an international scale, collecting insights into geological phenomena, as well as valuable pieces for the collections at Oxford. He visited the great Cuvier at the Muséum d'Histoire Naturelle in Paris and was invited to his home. Following this acquaintance, Buckland stayed in correspondence with Cuvier and his assistant of Irish origin, who could facilitate any communication problems. Letters, fossils, casts, books, and plates were sent back and forth between Paris and Oxford. In time, Buckland developed an extensive network of informants from the gentleman geologists of scientific societies and institutions, as well as among the lower social circles of provincial excavators and fossil collectors, to stay abreast of the newest cave finds and to receive important fossils. He also addressed some of the central figures throughout the British Empire, among them Lord Henry Bathurst (1762–1834), the secretary of the British colonies. Buckland's instructions on how to collect fossils were published in the *American Journal of Science* and basically advised the finders of precious specimens to send them to him at Oxford.[13]

Buckland's field studies in England tied in with Cuvier's and Brongniart's on the Paris basin when he showed that the London plastic clay, as well as the chalk below and the loam above, corresponded to the French stratigraphy and that their formation must have taken an enormous stretch of time. But it was the work on the quartz pebbles of Lickey Hill (Worcestershire) that became the starting point for his line of evidence for the biblical Flood, since the distribution and orientation of the pebbles indicated to Buckland the direction from which the Flood must have arrived and which valleys had been excavated by it.[14] In his inaugural lecture, Buckland clearly honored the implicit promise made to the prince regent. He attempted a balancing act between old and new that became visible already in its published title, *Vindiciae Geologicae; or, The Connexion of Geology with Religion* (1820). Buckland's "Vindication of [his] favourite Science of Geology in which [he] endeavoured to show it innocent of the charges that have been brought against it, as leading to opinions inconsistent with the Mosaic Chronology and Account of Creation," was influenced to a large degree by the natural theology of William Conybeare, with whom he shared a moderate Anglicanism.[15] This was followed by his wonderful *Reliquiae Diluvianae; or, Observations on the Organic Remains Contained in Caves, Fissures, and Diluvial Gravel, and on Other Geological Phenomena, Attesting the Action of an Universal Deluge* (1823), the title of which once again made an immediate statement in form and meaning. Here he supported the ideas brought forward in *Vindiciae Geologicae* by the results of extensive fieldwork in caves, fissures, and open gravel, as well as by the analysis of their fossil content.

Buckland's works contributed significantly to the ideas of diluvialism. He attempted to dispel the potential dangers to established religion perceived in geology and to unite his two professional worlds of science and church by allowing the Scriptures the status of a heuristic tool. He promoted a universal Deluge at a period not very remote in time and the low antiquity of the human race. This catastrophe, which he equated with the biblical Flood, was responsible for a variety of geological surface phenomena, such as valley formation, the deportation of boulders, and the deposition of the near-universal diluvial mud and gravel, as well as of animal remains:[16]

> The period at which all these caves [in England, Germany, and Austria] were inhabited by the animals in question [e.g., extinct species of hyena, cave bear, rhinoceros, hippopotamus, and elephant], was antecedent to the formation of that deposit of gravel, which it seems to me impossible to ascribe to any other origin than a transient deluge, affecting universally, simultaneously, and at no very distant period, the entire surface of our planet.[17]

Buckland referred to the deposits that had accumulated during the great Flood as *diluvium* in opposition to those sediments that had been laid down after the Flood, the *alluvium*. Although the Deluge was a recent event, and so, by inference, was humanity, Buckland realized that the age of the globe must be considerable and that the earth underwent several revolutions that gave testimony of the continued intervention of an omnipotent architect (even if through natural causes). To reconcile the notion of a high antiquity of the planet with the biblical record, he proposed that the word *beginning* in the first verse of the book of Genesis stood for an indefinitely long time before creation and the Deluge: "We know only that 'in the Beginning God created the Heavens & the Earth.' As to the actual condition of the Elements at that primieval period Science may fairly enquire & is justified in reasoning within the limits prescribed by our moral condition & intellectual Powers."[18] Buckland also interpreted the Scriptures freely enough to allow for some species having been extinguished by the Flood, while others might have been created subsequently; Noah would not have been able to secure a pair of each.[19]

Apart from the fossil bones from Quaternary deposits, Buckland worked on the fossils from older strata that would become known as dinosaurs in the 1840s. Already in 1818, he examined parts of a jaw and teeth of an animal from the Stonesfield Jurassic (Oxfordshire), which he recognized to have belonged to a gigantic reptile. Many stunning finds were made by the legendary Mary Anning (1799–1847) of Lyme Regis, Dorset, who from the age of ten supported herself and her mother by collecting, reconstructing, and drawing fossils. Some of the monsters she found, such as the pterodactyl, the iguanodon, and the cetiosaur, were analyzed and described by Buckland. Apparently, Anning felt ill used by the men of science who took her work for publication without paying due credit to her. Only recently have her specimens displayed in museums acquired her name. However, Buckland was Anning's friend and at times helped raise money for her. A year after her death, De la Beche remembered her in the presidential address to the Geological Society, which had excluded her during her lifetime, primarily because of her sex.[20]

It is impossible to ascertain how much Buckland owed to Anning for the identification of the coprolites as fossil feces, but in his paper on the subject he referred to the information provided by her that they were frequently found in the abdominal regions of ichthyosaurs and plesiosaurs. For Buckland, who, as we will see, was very interested in the ways of life of the curious extinct creatures that were being discovered during his time, the "dung stones" were traces of the past that could tell, for example, what the monster's death meal had consisted of. Buckland was warned about entering too indecent a subject for gentlemanly sensitivity with his enthusiasm for fossilized feces as vicarious evidence for the existence and habits of ancient animals, but for Buckland they were the stuff of "coprolitic visions" (Figure 1.2).[21]

Buckland was president of the Geological Society of London in 1824–1825 and again in 1840–1841 and a member of the Council of the Royal Society from 1827 to 1849. His college gave him the living of Stoke Charity in Hampshire, and Prime Minister Lord Liverpool appointed him to the canonry of Christ Church Cathedral, Oxford, in 1825. This security made a marriage with Mary Morland (1797–1857) possible. As a wedding tour that initiated the inseparability of work and private life during their marriage, Mary and William Buckland visited

Figure 1.2. The caricature by Henry De la Beche titled "A Coprolitic Vision" shows William Buckland, in his professorial robe and hammer in hand, looking into a cave that opens up the entire history of the earth and makes visible creatures of the past, not all of which were actually found in caves; courtesy of The National Museum of Wales.

France, Germany, Austria, Switzerland, Italy, and Sicily in 1826 and 1827. Like other women in Buckland's life, Morland, who had her own fossil collection, proved to be a great help in his scientific endeavors, writing down his papers from dictation before correcting and improving them. She was also a scientific engraver and contributed plates to her husband's publications. All in all, it appears that Mary Buckland's enthusiasm for science matched that of her husband, and that she devoted as much time to this interest as carrying and giving birth to nine children and the upbringing of the five surviving ones would allow. How much Mary Buckland craved scientific knowledge is evidenced in letters written during an illness of her husband, in which, temporarily lacking the source of information her husband was to her, she asked for scientific news.[22]

Buckland's career took another turn in the 1830s. On a positive note, he was chosen president and host of the second meeting of the British Association for the Advancement of Science in 1832, of which he had been one of the founders. The association, which was intended to further truth without infringement from politics or religion, allowed the members of the many local philosophical and literary societies to meet under one roof and to tighten the network and system of exchange of information and specimens between, for example, professional geologists, gentleman geologists of independent means, and local collectors. The BAAS symbolized the new and reformed way of doing science that sought independence from the rigid privileges and limiting orthodoxy of the Royal Society.[23]

On a less positive note for Buckland, toward the close of the 1820s, opinions started to change within the Geological Society from a predominantly diluvial to a fluvial attitude. Charles Lyell (1797–1875), who had been one of Buckland's early students, argued in his *Principles of Geology* (1830–1833) that all geological change had come about by forces still in operation, and that these had not only been the same in kind but also in degree. The controversy was also about the place of religion in the realm of science. Buckland had been acutely aware of the continued criticism voiced by Scottish geologists of a Huttonian as well as Wernerian conviction against the idea that geology could afford evidence of the Mosaic Flood. They tended to disapprove of the combination of science with biblical exegesis in general. As was typical of Buckland's harmony-loving personality, he remarked that arguments would be best avoided by comparing the general points of agreement with those of disagreement. After all, "there are 1000 points on which all agree / there is 1 on which we differ."[24] However, by the early 1830s, diluvialism (if not catastrophism) was running out of favor within the geological community. To add to his troubles, the 1830s also saw the growth of the Oxford Movement that opposed science and natural theology, which it associated with too latitudinarian an attitude. The number of attendees of Buckland's once-fashionable geology classes plummeted.[25]

Nonetheless, Buckland remained adamant in his endeavor to approach geological phenomena from natural theology and in the Bridgewater Treatise titled *Geology and Mineralogy Considered with Reference to Natural Theology* (1836) undertook another effort to illustrate the power, wisdom, and goodness of God in the economy, order, and design of the global system. He outlined the geological, paleontological, and mineralogical facts that gave testimony to a progressive earth history in accordance with a divine plan. The identity in the fundamental principles of the structures of fossil plants and animals evidenced a unity of design indicative of a single omnipotent intelligence, just as God's benevolence was obvious in the fact that each and every one of these organisms had been perfectly adapted for its respective circumstances and network of organic interdependences.

However, Buckland now regarded the last geological cataclysm as much older than the biblical Deluge, which he thought of as having left no visible traces on the surface of the earth. Nevertheless, unlike the uniformitarians, Buckland maintained that certain phenomena, such as valley formation, could be explained only by extraordinary events rather than by the action of rivers alone. Moreover, in contrast to Lyell, he continued to emphasize that the geological past of the globe evidenced a historical, namely, a progressive trajectory; there had been a beginning and a time when no organic beings existed. This picture fitted the theory of the earth's beginning as an incandescent mass that gradually cooled down, so that cold-blooded reptiles were better adapted to a warmer surrounding, while mammals appeared only when the cooling had further advanced and seasonal changes occurred (central-heat theory).[26]

The new version of catastrophism, in which geological revolutions still played an important role but were not connected to passages in the Bible, was a vantage point for the acceptance of the glacial theory that was developed by the Swiss naturalist and expert on fossil fishes Louis Agassiz (1807–1873), whom Buckland visited in the autumn of 1838. In Switzerland, Buckland soon became convinced of the evidence of a former extension of the alpine glaciers. Moreover, he quickly grasped that the action of glaciers could account for such phenomena as the diluvial mud and gravel, boulders, and valley formation, which he had attributed to floods. These revolutionary ideas were discussed at successive meetings of the Geological Society of London. Famously, at the session of 18 November 1840, instead of feeling humbled by the overwhelming rejection of the glacial theory, Buckland "pronounced upon his opponents who dared to question the orthodoxy of the scratches, and grooves, and polished surfaces of the glacial mountains (when they should come to be d—d) the pains of *eternal* itch, without the privilege of scratching!"[27] Although Buckland found the theory plausible enough to plan a second volume of *Reliquiae Diluvianae* with the telling title *Reliquiae Diluviales et Glaciales* to account for the new insights, the project was given up in the face of too much resistance. As Andrew Crombie Ramsay (1814–1891) entered in his diary after the Geological Society meeting of 16 April 1845: "Jolly night at the Geological. Buckland's glaciers smashed."[28] Thus the events and Buckland's role therein illustrate his readiness to accept new theories if they were able to better explain the facts. But they also illustrate his dependence on approval from his colleagues, despite the occasional slip of the tongue. We will be able to witness repeatedly that he was not someone to swim against the tide for long.[29]

Notwithstanding these disappointments, Buckland was approaching the apex of his career in the church. The Tory prime minister Robert Peel (1788–1850), who was connected to Buckland through a long friendship, recommended him to the queen as dean of Westminster in 1845. This was a controversial act, since some orthodox churchmen were in opposition to Buckland's natural theology. Nonetheless, Buckland left an Oxford University hostile to the sciences for the deanery of Westminster, holding the rectory of Islip as a country home. As dean, and in harmony with the increasingly utilitarian times, he became more strongly engaged in questions of agriculture, farming, and manufacturing. In 1848, the year before his last lecture course at Oxford, De la Beche handed him the highest honor known in geology, the Wollaston Medal. Unfortunately, his mental state was deteriorating, and at appearances in public, such as at the Geological Society sessions he continued to attend, even though in moments of great clarity he was still capable of the "most witty speech[es]," he increasingly "made an ass of himself," showed symptoms of "having forgotten his geology," grew "wilder in his orations every day," and indulged in "buffoonery, . . . completely marr[ing] the discussion." Even "in attempting to quote Scripture [he] made a great mull of it, & broke down greatly to the amusement of all."[30] These painful scenes were tragically followed by seven years of apathy and depression spent in Islip. Buckland died in a mental asylum in Clapham in 1856. Presumably to preserve his memory and scientific glamour, this fact does not appear in obituaries and biographies. It is a sad irony indeed that Buckland, whose eccentric personality was perceived as verging on the insane by those of more orthodox sensitivity, eventually fell permanent victim to his own jokes.[31]

~ That Buckland's greatest achievements owed more to his field research and creative experiments than to any grand theorizing should not obscure the fact that his natural theological framework was tightly interwoven with his evaluation of data gathered in the field.[32] One should therefore keep in mind his theoretical background and the situation in which it arose, as well as the considerable changes his geological theory went through. These are essential for an understanding of Buckland's interpretation of the Red Lady and his attitude toward the question of human antiquity in general, a question that gained in momentum during his lifetime, and on which a consensus was reached shortly after his death. The insights gained into Buckland's personality and career in this chapter will also help one understand, on the one hand, the discrepancies between his accounts of the Red Lady and Paviland Cave in scientific publications and at less formal occasions and, on the other hand, his sensitivity to opinions of influential scientists with regard to human antiquity. Apart from a sound reasoning power and a sparkling imagination prerequisite in the reconstruction of antediluvian worlds, his skill at using humor proved helpful in creating bonds even in the face of possible disagreement, but as the roles of Mary Buckland and Anning suggest, this was a man-only affair. Buckland's humor will also be encountered as a strategy to divert attention from potentially uncomfortable topics and to advance ideas that were best tested humorously before they were expressed in all seriousness.

Before moving on to Paviland Cave, it is of interest to learn more about Buckland's cave geology, which was essentially shaped by his work in Kirkdale Cave. The case of Kirkdale Cave also provides a good example of the opposition, as well as approval, Buckland's diluvialism met. Beyond that, it will serve to illustrate the cave as a place of encounter between men of different social strata and their direct experiences in these potential windows into the past, which were mediated by the mysticism surrounding caves in the culture they shared. In front of this background, the "romance" Buckland created around the Red Lady might appear less strange to the contemporary reader. Finally, in the paper Buckland wrote on Kirkdale Cave, he also dealt with supposedly antediluvian human bones that had been brought to his attention. A look at this engagement will shed light on his stand toward the issue when he entered Paviland Cave.

∾ 2

Man among the Cannibalistic Hyenas?

THE DEVELOPMENT OF Buckland's diluvialism makes clear that he approached his object of research with a certain interpretational framework in mind. The visions of prehistory he arrived at from his groundbreaking cave explorations were therefore of a somewhat apocalyptic nature, and they also fueled romantic imagination. Buckland and his colleagues celebrated the fact that they gained valuable insights into the ecology of Quaternary Europe by elevating the cave to a kind of icon for the young science. In verses and sketches they made use of its rich symbolism, with ties to mythology, folklore, religion, and romantic perceptions of nature. This was not only the time of the cave as place of poetic inspiration and prophecy. The cult of the cave in geology also communicated with the public interest in caves. In the last decades of the eighteenth century, the Lake District had become an important tourist attraction to the English gentry. Yorkshire also was visited for its caverns. Both in the quest into the cave in poetry and in tourist guides, the cave was a very effective stimulus to the imagination. This is certainly also true for the caricature by De la Beche reproduced in Chapter 1 and for the cartoon of Buckland's friend William Conybeare (Figure 2.1).

According to Martin Rudwick, this image of Buckland entering Kirkdale Cave was widespread among members of the Geological Society and maybe even on the Continent; it was at least known to Cuvier. In both this and De la Beche's image, the scene from deep time

Figure 2.1. Lithographed cartoon by William Conybeare, showing William Buckland in professorial attire entering Kirkdale Cave; courtesy of The National Museum of Wales.

simply appears when Buckland peeps into the past, as if by magic. The antediluvian vision is conjured up by Buckland's candlelight, or by "Science' wondrous wand," to use an expression coined by John Keble (1792–1866). As we will see in the context of Buckland's work at Paviland Cave, he not only liked to style himself as a conjuror but also used folklore and contemporary practices of magic as interpretation aids in reconstructing the past.[1]

However, the romantic image is partly euphemistic. It stands in contrast to the social hierarchy and the practicality of geological excavation. In a covert way, however, the cartoon alludes to these aspects by showing Buckland as crawling on all fours and by the hammer hidden in his pocket. In other words, the cartoon embraces the discrepancy between the practice of caving and the emotional responses to and iconic use of caves that were often of a romantic nature. Furthermore, the dirty and harsh practice of caving may be viewed as an integral part of the romantic quest as the hardship the geologist needed to overcome. Similarly, by showing the Oxford reader on all fours, the image hints at a subversive quality of the cave. Finally, the light Buckland brings into the cave is not only science's enlightened look through the keyhole at times long past; it not only makes visible one of Buckland's fantastic visions. The candle is also the flag of the conqueror, of the pioneer, who claims new territory for his personal career, for his science, and his nation. Scientific verses are replete with the geologist (especially Buckland) as sage, employing reason and imagination in combination with his geological utensils to gain access to the truths hidden in the earth.[2]

Also in "undergroundology," a wide and dense network of informants and collectors was prerequisite, since caves were often hit upon by chance, as in the process of building and quarrying. The rare occasion of the discovery of a new cave would attract many visitors who could disturb the floor and poach its fossil content. Buckland was well aware of these dangers and spread awareness thereof to forestall the event.[3] In the case of Kirkdale Cave, he was informed of its location near Kirkby Moorside, northeast of York, by the bishop of Oxford and the warden of All Souls, who attended his geology lessons. Although he was also pressed by Cuvier to go and explore the new discovery, Buckland was not the first to arrive at this keyhole into the past.

In July 1821, a native of Kirkby Moorside who was a partner in an Essex chemical firm realized that blocks of limestone used to repair a road contained bones. He traced the bones to a nearby quarry that had just hit upon a cave. He made a considerable collection, which he distributed throughout the London institutions. William Salmond, inspector general of taxes and later parliamentary commissioner and member of the Yorkshire Philosophical Society (founded in 1822), drew a map of the cave. He and his coexplorers, George Young, the Scottish geologist and minister of the United Presbyterian Chapel in Whitby and a central figure of the Whitby Literary and Philosophical Society, which was also founded in 1822, and William Eastmead, Congregationalist minister of Kirkby Moorside and future honorary member of the Yorkshire, Hull, and Whitby literary and philosophical societies, further emptied it of its bones.

Buckland commenced his own excavation in December 1821. Having just returned from Yorkshire, he announced a paper intended for the Royal Society in a letter to Lord Grenville. The new facts, Buckland wrote, not only provided insight into the state of the country immediately before the Deluge, but also proved the biblical Deluge itself. The cave furthermore constituted a kind of natural chronometer, indicating "that the latter [the postdiluvian period] must have begun at or about the point of time assigned to it by our common chronologies."[4] The proportion of the stalagmite layer below the diluvial mud to that above seemed to function as a measure for the proportion of the ante- to the postdiluvial period. In this way, the thin upper layer of stalagmite confirmed the recent date of the Deluge as assigned to it on the basis of the Scriptures. Already in the letter to Grenville, Buckland made it clear that the extinct species of elephant, rhinoceros, hippopotamus, and hyena had inhabited Yorkshire and other regions before the Flood. At this point, Buckland also realized that the climate had been no warmer than it is today. Rather, the species differed from those still found in the lower latitudes. The elephant and rhinoceros discovered in perfect condition under a mass of ice in Siberia showed that these animals wore a coat of wool.

Buckland's pursuit of undergroundology in Britain and on the Continent led him to speculate that the antediluvian bones found in caves were those of animals that had either fallen through fissures in the ceiling before the Flood or had been washed in by the waters of the Deluge.[5] Kirkdale Cave he interpreted as belonging to yet another type of prehistoric cave, the den, which had been occupied in antediluvian times by large carnivores such as hyenas or bears that left behind their own bones and dragged in those of their prehistoric prey. Buckland reconstructed the ecological situation at Kirkdale Cave as that of a den of hyenas, since it was their bones that were by far the most numerous. These animals had dragged the cadavers of the larger animals piecemeal into the cave, a conjecture supported by the fact that the bones were not rolled, which they would have been if they had been carried in by the waters of the Deluge. The bones that made up the stalagmitic breccia had been gnawed by the bestial hunters, a behavior of which even their own bones gave testimony.

As these observations render obvious, and as befitting a man who liked to style himself as equipped with magical faculties, Buckland's work at Kirkdale Cave was a masterpiece of what Carlo Ginzburg has called, with Thomas Henry Huxley, *retrospective prophecy*. The paper on the subject he delivered to the Royal Society was full of samples of the "evidential paradigm" at work, showing Buckland as reconstructing the past through the clues and traces left in the cave.[6] Buckland made ample use of all his senses to learn about the animal bones' color, form, density, incrustation, hardness, flexibility, and weight as clues to their state of decomposition. He asked whether they were ready to fall to pieces at the slightest touch, if, when struck, they rang like metallic bodies falling to the ground, and if they exfoliated and cracked on exposure to air. In the end, the fact that within the same cave sediment the decomposition of the animal bones varied greatly did not jeopardize the theory that they were the remains of animals that had roamed Britain in antediluvian times. Buckland reasoned that the variance in their state was due to the manner and degree of coverage in diluvial mud or stalagmite and to the duration of exposure before the covering had taken place.[7]

Beyond these and many other signs of the state of decay, Buckland, in accordance with common practice, also drew on experiments and chemical analysis. He immersed fragments of bone in acid till the phosphate and carbonate of lime were removed to determine the amount of gelatin they still contained (also called *gluten* or *animal matter*). The chemist and physicist William Wollaston (1766–1828) confirmed Buckland's theory that the *album graecum* found in the cave was the fossilized dung of the hyenas, and the description of the keeper of the Exeter Zoo of the habits of living specimens gave additional support to Buckland's interpretation. His reconstruction of Kirkdale Cave was indeed convincing, if at first sight fantastic. Cuvier accepted it despite the fact that Buckland's scenario did not entirely correspond to his or to de Luc's. In addition to the fact that Cuvier did not identify the flood that had extirpated the large mammals with the one described in the Scriptures, both Cuvier's and de Luc's versions of catastrophism included an exchange of dry land and sea. Since Buckland realized that the now-fossil animals had lived on the spot before having been drowned in water, Yorkshire had to have been continent also before the watershed. On the other hand, Buckland's observations supported the Cuvierian notion of several cataclysms.[8]

As promised to Grenville, Buckland's hyena-den theory appeared in the *Philosophical Transactions of the Royal Society* in 1822, where he compared Kirkdale Cave with other British and Continental caverns. Although Buckland's work in English bone caves and his hyena-den theory turned him into a central figure in the geological establishment of the 1820s, the views brought forward in the Kirkdale paper and *Reliquiae*

Diluvianae (1823), where he integrated Kirkdale into his geological system at large, provoked criticism by biblical literalists. Contrary to Buckland, they argued that the Deluge had also deposited the limestone of the caves and that the animals had been washed—according to some from the tropical regions—into the still-soft limestone sediments or were postdiluvian. Many of the critics believed in only one flood, that of the Mosaic record. Additionally, the Bible did not speak of the extinction of animal species but of the rescue of a pair of each by Noah. Last but not least, the Deluge lasted forty days and was thus not the rapid event pictured by Buckland, responsible only for the rather superficial diluvial gravel.[9]

Thus the provincial explorers of Kirkdale, who stood in the tradition of such scriptural geology, did not agree with Buckland's reconstruction. Salmond and Eastmead believed that the limestone itself had been deposited in the Deluge and that the hyenas and other animals had inhabited the cave in postdiluvian times. Salmond thus reasoned that the diluvial mud had been introduced into the cave at an earlier stage than the bones of hyenas, rhinoceroses, and elephants, which had then been trodden into it. Young, too, interpreted the limestone and its caves as having been formed by the diluvial waters. In his scenario, however, the hyenas and their contemporaries had been washed into the cave at a later point during the Flood. Although the three presented their views in lectures and publications, it was Buckland's rigorous interpretation that was taken up. Only at the museum of the Whitby Literary and Philosophical Society would one have seen Kirkdale bones exhibited according to Young's theory.[10]

Caving had its hierarchy, and the authority of a vision depended also on the status of its receiver. At the same time, as evident in the cartoon by Conybeare, geological field research, and cave exploration in particular, had a temporarily subversive element. As befits the apocalyptic visions, Eastmead experienced his descent into the interior of the earth as dreamlike wriggling through the intestines of a huge subterranean monster. Having undergone this mininglike experience, Eastmead was also aware that the work of caving was better suited to a workman than to a gentleman:

> The reader would have been amused . . . to have seen men of science exchanging the splendid apartments of mansions for a den of

> Hyaenas, creeping on their hands and knees through the slender passes, where once carnivorous animals growled, visiting those abodes to which they brought their prey, and in which they devoured one another . . . you might have beheld a rustic's frock investing a man of letters . . . , equiped with knee caps and trowsers, his head bound about with an handkerchief, his hands and face patched with mud, and nearly assimilated to the colour of the cave in which he had been immured.[11]

This is a wonderful account of the strange social structure of caving. Although the hierarchy of established geologist, provincial "amateur," and workman was strict, it might temporarily have been inverted, as when the workman or collector provided important information or specimens, or when the gentleman geologist performed the part of a workman, a part that, according to the excerpt, turned him into a vermin, a denomination that was often used for miners.[12]

Even though some visions of the former world of the cave counted for more than others, they had one thing in common. According to both the more and the less literalist expressions of the hyena-den theory, the debris the Deluge had left behind should have contained the remains of antediluvian humans. Indeed, some biblical literalists were eagerly waiting for proof of them. The first fact recorded that could have given testimony to the contemporaneity of humans with the antediluvian fauna might have been a Paleolithic implement associated with mammoth bones discovered by an apothecary and antiquarian in a gravel pit in London around 1690. In 1700, a human skullcap that only many years later was recognized as prehistoric was found in the loess at Canstadt, near Stuttgart (Württemberg) in Germany. In 1784, the physician François-Xavier de Burtin of Maestrich (1743–1818) found a flint implement in a deep layer in the surroundings of Brussels. Even though by this time such finds were regarded as of human origin, neither bone nor implements were generally conceptualized as prehistoric. An exception was the English antiquarian John Frere (1740–1807), who brought to light stone implements in association with fossil bones of extinct animal species in the gravel near Hoxne in Suffolk in 1797. He realized that they belonged to a very remote time, beyond the present world, when men had fabricated "weapons of war" out of stone at such manufacturing places.[13]

Buckland addressed the issue in the Kirkdale paper and *Reliquiae Diluvianae*. In fact, the physician and geologist James Parkinson (1755–1824) had reproved him for not having sufficiently discussed the absence of human remains from the debris deposited by the Deluge in *Vindiciae Geologicae*, which represented a lack of crucial evidence in corroboration of the Mosaic account. Parkinson felt that such important issues were better not left to those less sympathetic to natural theology and sacred chronology. If these issues were not treated with the same caution as that with which Buckland had discussed other issues concerning the convergence of the sacred and the geological accounts, the conclusion that Moses had wrongly ascribed humans to antediluvian times might be harmful to both geology and religion.[14] The new publications brought no real alteration, however. Although Buckland now discussed the candidates for a *homo diluvii testis* he had heard of and examined, he regarded none of these as truly fossil and did not dwell on the consequences of this estimate.

He had visited the Gailenreuth cave bear den in 1816 and 1822 and in the Kirkdale paper referred to the literature on the cavern. In 1771, the naturalist and pastor Johann Friedrich Esper (1732–1781), examining the cave at Gailenreuth in the German Jura (near Bamberg and Nuremberg), unearthed a human jaw and shoulder blade associated with the remains of hyenas, lions, and other species now absent from these regions. For Esper, their location seemed to indicate contemporaneity. In 1804, the German physician Johann Christian Rosenmüller (1771–1820) supported Esper's conclusion. So did the German naturalist and Blumenbach student Ernst von Schlotheim (1764–1832), whose careful excavation of the Gailenreuth cave, where he encountered human bones mixed with those of extinct animal species, led him to believe that humans had lived before the formation of the alluvium and thus before the last geological revolution. However, according to Buckland, Rosenmüller had found human bones associated with a more recent fauna of wolf, fox, horse, mule, ox, sheep, and stag and better preserved than those of the bear and hyena. This, in combination with the fact that Esper had also discovered fragments of urns, argued for a postdiluvian origin of the human bones.[15]

In *Reliquiae Diluvianae*, Buckland further referred to Thomas Weaver's translation of Schlotheim's memoir in the *Annals of Philosophy* of January 1823 about the discovery of human bones mixed with those of extinct mammals in the diluvium of the Elster Valley. Although "in one quarry (called Winters), the bones were found eight feet below those of rhinoceros, and 26 feet below the surface," Buckland conjectured that they had been introduced later into the diluvial loam, leaving open the question of when and how: "And thus far the case of Köstritz affords no exception to the general fact, that human bones have not been discovered in any of those diluvial deposits which have hitherto been examined; . . . it is highly improbable that they ever will be found."[16]

Buckland also mentioned the human bones encrusted with stalactite discovered in a limestone cave at Burrington in the Mendip Hills in Somersetshire. Among these was a skull of which even the inside was filled entirely with stalactite. Also, the mouth of the cave was nearly completely sealed with the substance. Buckland explained Burrington as a sepulture or resort for refugees during some military operation. He concluded: "The state of these bones affords indications of very high antiquity; but there is no reason for not considering them postdiluvian."[17] The large number of human bones from a limestone quarry in the Mumbles near Swansea (southern Wales), where the wealthy landowner and naturalist Lewis Weston Dillwyn (1778–1855), later the first president of the Royal Institution of South Wales, had worked in 1805, were interpreted by Buckland as bodies thrown away after battle.[18] In 1810, at Llandebie in Caermarthenshire, about a dozen human skeletons were discovered in a cave nearly entirely closed up. Buckland's verdict was the same for all these finds; he summed up: "It is obvious, that in neither of these cases, are the bones referable to so high an era as those of the wild beasts that occur in the caves at Kirkdale, and elsewhere."[19] Do we hear something like regret in this formulation? Or was Buckland only too ready to dismiss the coexistence of humans with the carnivorous, even cannibalistic hyena and its antediluvian kin?

There were several practical problems associated with cave finds. Caves were often washed out by floods of water, carrying away or mixing up the deposits; human visitors had always found caves attractive and had disturbed the floor deposits; but for Buckland, the main point was that human skeletons could have been introduced into lower strata in burials. For all of these reasons, the association of human remains and artifacts with fossil bones of antediluvian fauna remained nondecisive. Apart from these considerations, it seems that placing humankind within Buckland's reconstruction of pre-Flood Britain as markedly different from the post-Flood state, as populated by gigantic mammals, including cannibalistic carnivores, would have added one more heresy to the concepts of a long history of the earth, several floods, the extinction of species, and a hostile primeval world.[20] That Buckland's hyenas were of a monster pedigree becomes clear from Eastmead's *History of Kirkby Moorside:*

> But no words can give an adequate idea of this animal's figure, deformity, and fierceness; more savage and untameable than any other quadruped, it seems to be for ever in a state of rage and rapacity, for ever growling, except when receiving its food, its eyes then glisten, the bristles of its back all stand upright, its head hangs low, and yet its teeth appear, all which give it a most frightful aspect, which a dreadful howl tends to heighten . . . it might perhaps have been sometimes mistaken for that of a human voice in distress, and have given rise to the account of the ancients, who tell us that the Hyaena makes its moan to attract incautious travellers, and then to destroy them.[21]

Has there ever been a more gruesome monster than "this ferocious and dreadful animal, once the native inhabitant of Britain, but now happily extirpated from our dales and our country"?[22] However, Eastmead was describing recent species of hyena. What Buckland found at Kirkdale were monsters of yet another kind, who were found nowhere on earth at his time. Thus, although Eastmead's postdiluvian hyenas might have been reconcilable with humans, in fact depended on their existence (somewhere) in order to render the picture compatible with a literal reading of the Mosaic account, where death was a consequence of original sin, the case was different within Buckland's framework. According to the argument from design as understood by Buckland, humans were God's last creation, only added to a world fully prepared for their needs. They belonged to the present system of organisms. In the Kirkdale Cave paper, Buckland stated that "the bones which are in most perfect preservation, and belong to existing species, have been introduced during the *post-diluvian* period; whilst the extinct bears and hyaena are referable to the antediluvian state of the earth."[23] Taking into account the difficulty of the state of bones as an indication of their age, as discussed earlier, the main argument against contemporaneity appears to have been the fact that humans were an existing species and did not seem to fit into this charnel house of a world dominated by different, now-absent animals. Thus the paleontologist Hugh Falconer (1808–1865), who, as we will see, was directly involved in the processes that led to the general acceptance of human antiquity in the early second half of the century, observed in retrospect that

> viewed in the light of a strictly physical inquiry, the chief rational argument in support of the opinion that the advent of man upon the earth dates from a very modern epoch, was . . . that taking him in conjunction with the Mammals, with whom he is now associated, they appeared, as a group, to belong to a new order of things strikingly different from that of the immediately preceding period.[24]

Indeed, even without humans, Buckland's reconstruction of prehistoric England was spectacular enough to induce him to exercise caution and to use the mixture of styles typical of him when treading upon new or insecure ground. After his presentation of his vision of the antediluvian Kirkdale Cave as hyena den at the annual dinner of the Geological Society on 8 February 1822, Lyell commented: "Buckland in his usual style, enlarged on the marvel with such a strange mixture of the humorous and the serious, that we could none of us discern how far he believed himself what he said."[25] The function of this use of humor can be to introduce an unorthodox idea while maintaining the possibility of retraction. It is up to the audience, as the quote from Lyell suggests, whether to transpose the content from the realm of the jest to the serious.[26]

Nonetheless, regardless of challenges from biblical literalists and initial disbelief from his geological peers, Buckland's work at Kirkdale earned him acclaim in scientific circles and even the prestigious Copley Medal.[27] Relieved, he could announce this sanction by the Royal Society to his confidante Lady Mary Cole (1776–1855), the sister of one of his pupils, and add that "I am now not much afraid of any further opposition to my Hyaena Story which my friends at first predicted no body would believe a word of it."[28] Beyond these hard-earned credentials, he gained the image of a cave explorer equipped with that extra gift of a strong imagination, an image circulated among those whose work Buckland respected most. The appeal of the hyena-den theory was considerable, and Buckland was well aware that it even entered the poetry of Lord Byron (1788–1824) and Percy Bysshe Shelley (1792–1822). Byron took up the image of the wild beasts drowning in the Deluge in his *Heaven and Earth*, where "The creatures proud of their poor clay, / Shall perish, and their bleached bones shall lurk / In caves, in dens, in clefts of mountains, where / The deep shall follow to their latest lair" (1823:I.iii.173–176).[29]

❦ BUCKLAND AGAIN USED both scientific reasoning and vision when making sense of Paviland Cave; only this time, the vision was strongly colored by his propensity for the improper mediated by a disconcerting sense of humor. For an understanding of Buckland's stance toward the issue of human antiquity and, by inference, the Red Lady, all the threads picked up in this chapter, from his early diluvialism that theoretically demanded antediluvian humans to his authority over researchers of lower scientific and social status, will be of central importance. The question of whose vision of antediluvian Britain weighed most, which in this chapter has been brought into focus through the social hierarchy of caving, will return with greater urgency in connection with Buckland's involvement at Kent's Cavern. As we have seen, when Buckland entered Paviland Cave, he had already been engaging with human bones and human-made stone tools that were claimed by some to be antediluvian. But he was also aware of the problems facing geologists when dating such remains from caves, and of the tensions between a progressive earth history and the possibility of antediluvian humans.

The following longish chapter that deals in detail with the rebirth and first interpretations of the Red Lady will elaborate on the magic trope used by Buckland in the construction of his scientific persona, as well as in the interpretation of Paviland Cave. His excavation there and the help he had from others will be reconstructed before moving on to the diverging half-informal and published accounts he provided of its history. In this context, I will make an excursion into the ideas about British history, and Welsh history in particular, on which Buckland's diverse jokes and stories, as well as his ascription of sex, drew. Furthermore, this antiquarian setting, in which an analysis of the bones of the

Red Lady themselves seems to have played a minor part, will suggest a look at other instances where physical anthropology was consulted to gain information about sex and age. Finally, I will have to take a step away from the cave and to return to possible contextual reasons against using the Red Lady as the *homo diluvii testis* implicated by his diluvial theory.

～ 3

The Romance of the Witch of Paviland

> A lady-witch there lived on Atlas' mountain
> Within a cavern, by a secret fountain.[1]

WHEN GEOLOGISTS VISUALIZED the cave as a place of imagination, inspiration, and vision, they could draw on a long tradition of cave symbolism. Caves were places where monsters and fairies had lived from Greek mythology to British folklore and romantic art. As the quote from Shelley's poem indicates, they were magical places where witches might hide. In keeping with this culture, Buckland created a "romance" around Paviland Cave and its former inhabitant, the Red Lady, even if it was mainly intended for fireside occasions. A romance fit well not only with the romantic revival of the genre but also with Buckland's scientific persona as created by himself and others. He was humorously called "Ammon Knight" (homophone of *ammonite*) by his students, and stories of romantic quest were built around him, such as the quixotic one that his mare was so well trained to carry him to fossil-rich places that it would stop without command at every interesting stone. The glamour of the quest was also claimed by some of Buckland's coadventurers, such as his friend William Conybeare, who wrote of himself in 1811, "I partake more largely of the spirit of the knight of La Mancha than of his craven squire and prefer the enterprise and adventures of geological errantry to rich castles and luxurious entertainments."[2]

Buckland used magic lore to claim a special gift when he boasted in a letter to the Reverend Vernon Harcourt about the excavation of two hyena dens in Devonshire: "I passed for a conjurer by telling them where the bones would lie before the crust was touched, and the more so as in three subsequent experiments hardly any bones were found."[3] In connection with his comprehensive vision of Kirkdale Cave as the den of antediluvian hyenas that roamed the neighborhood for prey, one of Buckland's many critics, who belonged to those who stuck to the word of the Bible literally, had attacked him by ridiculing precisely this gift. He wrote in a letter titled "The Wonders of the Antediluvian Cave" to the *Gentleman's Magazine:* "Perhaps Mr. Buckland, like many of our brethren of the isles of North Britain, may have possessed the gift of *second sight* . . . ; and possibly, ere long, the world may be favoured with some more of his speculations; or as we may say, 'visions, having his eyes open.'"[4] Indeed, Buckland soon had another vision, although also in the case of Paviland Cave, his magic depended on assistants, and his excavation policy and practices were far from romantic. As it turns out, even his romance of the lonely witch in a dreary cave lapsed into the vulgar.

Paviland Cave, or Goat's Hole, is located on the coast of Glamorganshire in the district of Gower, fifteen miles west of Swansea (between Oxwich Bay and Worm's Head), on the former property of the late Christopher Mansall Rice Talbot of Penrice Castle. Lady Mary Cole, who had married Christopher Cole after Talbot's death, and her daughters were accomplices in Buckland's cave science. Although Cole was mainly interested in botany, her daughters ("the Talbots") were engaged "amateurs" in the new science of geology. They had provided information about Glamorgan geology for Greenough's map of England and Wales, published by the Geological Society in 1819. They also brought newly discovered bone caves to Buckland's attention, followed his instructions on how to proceed with them, answered his questions received by letters, and sent him specimens for inspection and for his collection. In return, Buckland sent them dried seeds and fossil bones and shells for their "nascent Museum of Penrice," as well as books from the British Isles and his travels abroad.[5]

When the existence of a cave on their own property came to their notice, they set out to investigate it. On 27 December 1822, Mary Theresa Talbot (1795–1861), the antiquarian and geologist John Montgomery Traherne (1788–1860), and Dillwyn, the wealthy landowner and naturalist whom we have met earlier in connection with the human bones from the Mumbles, visited Goat's Hole. Dillwyn, too, belonged to Buckland's acquaintances and had attended his mineralogy lectures. At Dillwyn's house, Buckland and other famous figures met, such as Wollaston, William Conybeare, and De la Beche, whose daughter later married Dillwyn's second son. Already at that time, Paviland Cave was hard to reach, and the explorers had to climb a steep cliff to get access to the interior. They excavated the elephant tusk that had formerly been discovered, but left in place, as well as part of the skull to which it belonged. On the following day, Talbot and Dillwyn brought away many bones to the collection at Penrice Castle.[6]

By the end of December 1822, Buckland pressed the Penrice women for a detailed account of Paviland Cave: Was the bottom flat or inclined? What was its size, and were there fissures that ascended to the surface? What were the position and depth of the mud? Were there any pebbles? Were there a crust of stalagmite above the mud and a second crust below it dividing it from the rock or any stalagmite investing the bones? Were the bones broken from decay over time, or were they fragments, the edges of which were old as if broken before the new bones got near them by beasts inhabiting the den? Was it an elephant skull they had found? Clearly, Buckland was anxious to figure out the history of Paviland Cave on the basis of the knowledge he had gained from other sites. He was impatient to get a peek into it and jealously requested that Mary Talbot barricade the entrance until his arrival, which was delayed by his work on *Reliquiae Diluvianae*. In this mood, he made use of the metaphor of the cavern as the mouth of the earth that was also a frequent poetic image. In Buckland's case, the mouth needed to be shut so that the cave might not speak to someone unauthorized about the secrets of the past: "Meantime pray have the Mouth *closed* up again to prevent destruction."[7] It was quite common practice to judiciously take "every precaution to secure [a cave] from injury, till some qualified person shall be present to observe, and record the undisturbed appearance presented by its interior."[8]

When Buckland finally arrived on 18 January 1823, he visited Paviland Cave accompanied by Dillwyn, Mary Talbot, and Traherne. On 21 January, Buckland left earlier for the cave, and later in the day, when he was joined by the rest, they explored Hounds Hole and Deborah

Cave. However, he cannot have spent more than one entire day at Paviland Cave altogether. From his observations, Buckland reached the conclusion that Paviland did not belong to the den category of bone caves. Rather, he conjectured that either the remains of the now-extinct animals at Paviland had been washed in at the same time as the diluvial loam by the biblical Deluge (although the bones were not rolled), or the animals had fallen in through the aperture in the cave's ceiling before the catastrophe (although he considered it too small for the elephant and rhino). In combination, these two mechanisms could explain the situation at Paviland Cave. The following year, Buckland was presented with a spectacle that could serve as an analogy for the first mechanism:

> Mr B. recapitulated the effects of the storm of last November upon the town of Sidmouth the barrier of gravel silt which formed the barrier on the beach had been washed into the lower storey of the houses, & carried with it fish pans & all apparatus into the *cellars below*. when the storm subsided the inhabitants descended into their cellars which now resembled his Caves, containing masses of mud & gravel with all these extras mixed with them, out of which the pots & pans were fished. Thus was it he supposes in his dens or caves. hence he calls the lower part of them the *scullery*, into which all the apparatus & bones of the beasts were washed & deposited by the deluge.[9]

Under six inches of earth, Buckland discovered a human skeleton, consisting of the humerus, radius, and ulna of the left arm, the entire left leg and foot, part of the right foot, the pelvis, and many ribs. The bones appeared undisturbed in extended burial position and were covered by ochre. At the place where a pocket would be, he encountered small shells of the *Nerita littoralis* (*Littorina littorea*, or common periwinkle). In contact with the ribs were forty or fifty fragments of small, nearly cylindrical ivory rods. At the same place, he found fragments of ivory rings the size of a teacup handle. All these sepulchral goods were also red with ochre. At some distance from the burial, three further fragments of worked ivory were unearthed, one of them tongue shaped. At the same spot, there was a rude instrument produced from a wolf's metacarpal bone.[10]

It seems that the excavating party discussed the identity of the skeleton after they returned to the warmth of Penrice Castle. From a letter by Buckland to Cole, dated 15 February 1823, that is, before the publication of his *Reliquiae Diluvianae* on 5 June 1823, we know that Buckland and his coworkers at first took the human bones to have belonged to a man, joking that it was probably an early example of an excise man, possibly thinking that he had been murdered and buried by the smugglers for which the coast was notorious. The shells in his pocket thus might have represented tax money. However, as one also learns from this letter, once back at Oxford, Buckland changed his mind as to the sex and occupation of the Ancient Briton whose bones they had exhumed. "The Blade Bone of Mutton" from the cave turned the excise man into a witch:

> The Man whom we voted an Exciseman turns out to have been a Woman, whose History wd afford ample Matter for *a Romance to be entitled the Red Woman or the Witch of Paviland* ... The Blade Bone of Mutton gives grounds for a Conjecture, wh favors the Theory that she was a Dealer in Witchcraft. In Meyricks Hist. of Cardigan, p. 188 of the introduction, are some curious Stories extracted from Giraldus Cambrensis.[11]

In his *The Itinerary of Archbishop Baldwin through Wales*, of which the translation from Latin into English by Richard Colt Hoare (1758–1838) had appeared in 1806, Giraldus Cambrensis indeed told of the ancient custom of divination through reading the secret marks on the right shoulder blade of a ram or goat. Once stripped of their flesh and boiled, the bones told the Flemings (the inhabitants of the province of Rhos who originated from Flanders) about events removed in time or space. Giraldus himself considered "it ... wonderful, therefore, that these bones, like all unlawful conjurations, should represent, by a counterfeit similitude to the eyes and ears, things which are passed, as well as those which are now going on."[12] Thus Buckland imagined his witch to have used the sheep shoulder blade from the cave as a conjuring tool in accordance with Welsh custom, together with the ivory rods and rings, the "tooth pick made of wolf toe," the yellow shells, and the tongue-shaped bit of ivory.[13]

Giraldus was not unsympathetic toward the customs of the Welsh, who made up one-third of his noble pedigree. However, the archdeacon must have been aware of the image of superstitiousness and backwardness that the Welsh shared with the Irish, since he justified his treatment of "so lowly a subject" in the first preface to *The Description of Wales*:

> Some, indeed, object to this my undertaking, and . . . compare me to a painter, who . . . is endeavouring with all his skill and industry to give celebrity to a cottage, or to some other contemptible object, whilst the world is anxiously expecting from his hand a temple or palace. Thus they wonder that I, admidst the many great subjects which the world presents, should choose to describe and to adorn, with all the graces of composition, such remote corners of the earth as Ireland and Wales.[14]

Buckland clearly shared these deep-rooted prejudices and unconsciously reproduced them in his imaginary travel through the marginal regions of Britain (including the tip of his home county) with which he opened his immensely successful magnum opus, the Bridgewater Treatise: "If a stranger, landing at the extremity of England, . . . should make the tour of all North Wales . . . he would conclude that Britain was a thinly peopled sterile region, whose principal inhabitants were miners and mountaineers."[15] He also alluded to the supposed love of high birth and ancient as well as noble genealogy of the Welsh when in the aforementioned satirical remark he anticipated that the Red Lady would be claimed as a Welsh ancestor. He jocularly expressed his fear that through her reddish bones, the Welsh would trace their origins back to the very first couple ("this problem [of the red color] is at once solved by supposing her a near connection of Adam, perhaps Eve herself"), not in the sense that all humanity descended from Adam and Eve, but in the sense that "Adam was a Welch Man."[16]

The letter to Cole about the witch of Paviland was apparently shown to Traherne, who wrote to Buckland on 3 March 1823:

> The subject of the caves having been reviewed in my mind in consequence of your letter to Lady Mary which I read; & *which entertained me much*; I might remind you, (if indeed you were aware of the fact) that there is a small British encampment immediately above the cave—so that it is not unreasonable to suppose that your

enchantress may have stimulated the former inhabitants of Gower to warlike deeds—you say nothing of the bones of the other skeleton—probably they were too imperfect to ascertain the sex.[17]

Traherne was obviously inspired by Buckland's romance and seems to have continued it in his mind, creating a story of a witch in a cave that gave the soldiers of the nearby camp advice on which moves would prove fortunate, thus driving "the former inhabitants of Gower to warlike deeds." In fact, Giraldus had described how the women and men in southern Wales who were of "a prophetic kind of spirit" used the shoulder blade of a sheep, among other things, for prophecy in matters of peace and war.[18]

Since in his letter to Cole Buckland had still been in the dark about a possible "motive for living in such a Place," it may well be that he learned of the British camp through Traherne. Besides providing an explanation for the cave's occupancy, the connection between the cave inhabitant and the military encampment strongly supported Buckland's estimate that the skeleton was postdiluvian: "The circumstance of the remains of a British camp . . . seems to throw much light on the character and date of the woman under consideration; and whatever may have been her occupation, the vicinity of a camp would afford a motive for residence, as well as the means of subsistence."[19] As is obvious from the quote, in *Reliquiae Diluvianae*, the witch of Buckland's letter to Cole had disappeared. All that remained was the ominous hint at an occupation not entirely corresponding to Georgian standards of decency, whether that might be a prostitute, a witch, or both.

Buckland also took up the idea introduced in Traherne's letter that the sepulchral items might have been used in a game:

> The small nerite shells either have been kept in the pocket for the beauty of their yellow colour, or have been used, as I am informed by the Rev. Henry Knight, of Newton Nottage, they now are in that part of Glamorganshire, in some simple species of game. The ivory rods also may have either been applicable to some game, as we use chess men or pins on a cribbage board; or they may be fragments of pins, such as Sir Richard Hoare has found in the barrows of Wilts and Dorset.[20]

Again, in *Reliquiae Diluvianae*, Buckland presented the artifacts in a different light than in his preceding letter, where they had been the tools of witchcraft. In *Reliquiae Diluvianae*, the ivory rings were read as armlets. Buckland supported this interpretation with Hoare's pioneer work on barrows and a passage from the Greek historian, geographer, and philosopher Strabo (63/64 BC–23 AD?), who had reported an Ancient British trade in ivory bracelets and necklaces with the Continent.[21] The rude instrument made from the metacarpal of a wolf, described as "tooth pick made of wolf toe" in his letter to Penrice Castle, was now suggested to have been used as a skewer.

In both the matter-of-fact account and the romance, Buckland took all the artifacts to have been used by the woman, although the wolf bone, the only flint tool appearing in *Reliquiae Diluvianae*, the tongue-shaped ivory fragment, and the sheep scapula had not been in direct association with the burial. In the same way, the charcoal and bones of recent oxen, pigs, and sheep, as well as the modern shells, were suggested to be the remains of her feasts in the cave. Indeed, the presence of marrow bones suggested another "theory" to Buckland, "that they are the very Bones that were stolen by that wicked Ancient Briton whose Peccadillo's gave origin to the old & everlasting Scandal that 'Taffy was a Thief & stole a Marrow Bone.'"[22] On a more serious note, Buckland's estimate that the Red Lady's living in the cave had been causal for the arrangement of artifacts and bones of recent animals in its floor might have been inspired by Hoare's description of the remains of the huts and hearths of Ancient Britons.[23] Be that as it may, it was another strong motive in the decision against the assumption that the human skeleton and the bones of extinct animals were contemporary, because Buckland believed that sheep had been a postdiluvian arrival in Europe (though not a postdiluvian creation, since Abel had been a shepherd). In contrast to other kinds of Pecora, the sheep was so feeble that it had not made it into Europe before the Deluge.[24]

These aspects thus contributed to Buckland's definition of the Red Lady as postdiluvian, albeit as of "very high antiquity." However, there was also rather strong evidence for contemporaneity. Buckland considered the ivory ornaments and utensils to have been produced out of the antediluvian tusks of the same cave and acknowledged that this must have been done while the ivory was still hard (contrary to the crumbling state he found it in). Furthermore, the skeleton's close association with the mammoth skull strikes one as convincing, especially as represented in Buckland's sketch of the cave (see Figure 3.1). The flint tools might also have offered valuable clues as to the Red Lady's age. In *Reliquiae Diluvianae*, Buckland stated that he himself had discovered only one worked flint at Paviland, which he described as "a small flint, the edges of which had been chipped off, as if by striking a light."[25] However, he was informed of Knight's unearthing of several such "slices of flint with sharp Edges." Buckland interpreted those as having been used as knives to cut the ivory rods.[26] Furthermore, according to Dillwyn, numerous worked flints had already been encountered on 23 December 1822, which he took to indicate that the cave had once functioned as a place of concealment for a manufacturer of Celtic implements.

The time had not yet come when such flint tools were generally read as the traces of "antediluvian" humans rather than of comparatively recent Ancient British or Celtic tribes; Frere was clearly the exception. In 1825, Buckland referred again to the flint he had found in Paviland Cave in a letter to Harcourt:

> I found the teeth of rhinoceros in addition to hyenas, bears, and tigers, which have been noticed there [another hyena den in Devonshire] by Trevelyan, and found also a flint knife of the same kind as the one I have from Paviland, showing both these caves to have been inhabited by people who used such knives, *i.e.* aboriginal Britons.[27]

These aboriginal, ancient, or native Britons and the vestiges they had left behind in the form of stone tools were not of the same order of things as the rhinoceros, hyena, bear, and tiger. They represented a different epoch in the history of occupation of the British caves.

At the time of the Paviland excavation, the biblical chronology of human history was still in place, and thus there was as yet no broadly accepted concept of a great human antiquity correlated to knowledge about stratigraphy. Correspondingly, there was no excavation technique designed to register this antiquity, and no classification of implements beyond the general impression of a succession from stone to bronze to iron tools.[28] For all these reasons, the term *Ancient British* referred to anything pre-Roman. The fact that this was largely an empty signifier— there were no historical sources to fill the gap between Moses and the

classical writers—is evidenced by antiquarian fantasies from the Middle Ages onward that traced the first inhabitants of the British Isles to the Trojans, the Greeks, the sons of Japhet and Noah, the Lost Tribes of Israel, the Phoenicians, the Egyptians, and the Indians—origin stories that Buckland and others regarded as of particular appeal to the Welsh desire for noble pedigrees. After all, it was widely held that the territory of the earliest Europeans, or Celtae, had been reduced to "remote parts" by invading Germanic tribes, so that the Romans met with these ancient people at least in the south of Britain. Subsequently, the pagan Anglo-Saxon invaders of the British Isles drove the Christianized Romano-Britons into Wales, Ireland, and Brittany, where it was therefore most likely to find their descendants.[29]

Corroborating these notions was the observation that the Welsh, Cornish, Irish, Scots, Manx, and Bretons shared a language, which by the mid-eighteenth century had become commonly designated as *Celtic*. At the same time, the interpretation of Ancient British antiquities as Druidic monuments gained momentum, most ardently in connection with Stonehenge and Avebury. When Paviland was excavated, *Celtic* thus referred to a people, the descendants of whom could be found, among other places, in contemporary Wales, and to languages such as Welsh and others spoken in "peripheral" regions in the British Isles, as well as to religious and other cultural practices evidenced in regional oral traditions, ancient manuscripts, megalithic monuments, barrows, and pre-Roman artifacts. This so-called Celtic revival converged with the romantic appeal of rural landscapes, paganism, historical romances, and the celebration of local regions. Seen in this light, Buckland's creation of a romance about the Red Lady was inspired by the association of Wales with origin myths, from the King Arthur sagas to Druidical priests and philosophers. Mocking the romance of the peripheral, Buckland thus parodied the supposedly Welsh inclination for grandiose storytelling.[30]

However, the notion that the first inhabitants of Britain had been barbarians had not died out completely since classical times, and in the wake

Figure 3.1. Paviland Cave, *Reliquiae Diluvianae*, 1823, pl. 21; courtesy of The National Museum of Wales. William Buckland called this sketch "a Picture of Scrub [the dog] in his Hole at Paviland" (NMW, Bu P, Box 5, Folder 4, p. 168: Letter to Lady Mary Cole, 3 April 1823).

of the voyages of discovery, the Picts and occasionally even the Ancient British became aligned with the wild tribes of the New World. It was observed that the former and contemporary primitives showed similarities in material culture, such as ornaments, body paint, boats, huts, and, of course, stone tools. In a few illustrations of the final decades of the sixteenth century and the early seventeenth century, Ancient Britons appeared in accordance with classical sources as barbarous naked warriors covered with body paintings, but now this picture was reinforced by the sparely clad Red Indian warrior adorned by tattoos.[31] For Buckland, the interpretation of the Red Lady as Ancient British clearly referred her to "uncivilized races," to people considered savage in both appearance and behavior by the sources he relied on. Strabo had described the primitive shelters, ways of sustenance, and customs of the barbarians in the British Isles. He even related rumors of cannibalism, including the practice of eating deceased fathers, and of incest with mothers and sisters. Also in Giraldus's sympathetic picture, the remnants of the Britons in Wales were much given to prophecy and soothsaying, unsophisticated in the methods of husbandry, and living in wattled huts.[32]

Whether regarded as close to the early postdiluvians and thus not yet far degenerated, as primitives only surpassed in their savagery by the contemporary inhabitants of countries farther to the west, such as the Indians and Eskimos of the New World (who were farther away from the post-Flood center of dispersal and peopling of the earth [Mount Ararat] and by inference more degenerate), or indeed as noble savages, the Celts were believed to have been a postdiluvian people. An exception to this view was the Reverend Henry Browne, whose book on the stone circles appeared the same year as Buckland's *Reliquiae Diluvianae*. Browne rejected their Druidic origin by reference to classical sources that had depicted the Ancient Britons as far too primitive for such constructions. Nonetheless, he accepted the notion of the antiquary of Celtic Druid remains, William Stukeley (1687–1765), that Avebury represented a serpent, fancying that it had been built by the antediluvians, if not Adam himself, as a reminder of the Temptation. Browne expanded these speculations in *The Geology of Scripture* (1832), which he dedicated to Buckland.[33] As we will see, by this time, the presence of antediluvian humans in Britain ceased to be an impossibility in Buckland's mind because the meaning of this expression within his geological framework was undergoing considerable change.

On the sex of the skeleton, Buckland might again have drawn on the so-called pioneer of British archeology, although there is no comparison of male versus female burials in Ancient Britain in *Reliquiae Diluvianae*, and the reader has to infer that it was because of the ivory rings that Buckland assumed female sex.[34] Hoare already worked with the simple system that referred barrows containing "ornaments" to women and ones containing "weapons" to men. He judged that in the absence of metal and baked clay a barrow might be regarded as Ancient British. He furthermore identified different types of barrows, among these the Druid Barrow, "so called by Dr. Stukeley, and therefore let it retain its name; though I think the Druids had no claim to it, but rather the females, as we generally find beads, and other small ornaments in them appropriate to that class."[35]

A paper Buckland later presented to the British Archaeological Association, of which he was a member, supports the conjecture that he agreed with this way of sexing graves: "There never was, nor ever will be a period, when, even among uncivilized races, the female part of our species were not, and will not be, anxious to decorate themselves with beads. They are at this time highly prized by the negroes of Africa, and the Indians of America."[36] Thus, although self-adornment occurs in women and men of "uncivilized" tribes, and although Stukeley might have associated ornaments with male religious ritual, Buckland took the Paviland armlets as a marker of female sex.[37] Even though Buckland was here not speaking of the Red lady, Native Americans would not have been considered a far-fetched model for Ancient Britons and, as we will see in Part II, would remain with the Red Lady for a long time to come.

There is no evidence to suggest that Buckland interpreted the Red Lady and what he took to be associated objects along the lines of Browne and the more speculative antiquaries. In *Reliquiae Diluvianae*, he followed the lead of Hoare, Samuel Rush Meyrick's *History and Antiquities of the County of Cardiganshire* (1810), Giraldus, and Strabo, besides his personal network. In fact, Hoare, like Buckland and his geological peers, explicitly distanced himself from eighteenth-century grand theorizing. He is thus often seen as indicating the transition from a particular kind of antiquarianism to archeology characterized by systematic excavation and classification. In the preface to his volumes on Wiltshire barrows, Hoare made this program explicit and, ironically, added: "I shall not seek among the fanciful regions of romance an origin for our Wiltshire Britons."[38] Again, besides being an expression of Baconian ideals, the Englishman here might have distanced himself from "peripheral mythmaking," although this might rather have been a rhetorical ploy than a real break with the reliance on historical sources and linguistics.[39]

The tentative reconstruction of Buckland's making sense of Paviland Cave suggests that the burial of the Red Lady was interpreted in a classical, antiquarian, and archeological rather than an anthropological framework, and that the human bones were not given the same careful consideration as fossil animal bones (as discussed earlier for Kirkdale Cave; see Figure 3.1). Superficially, this suspicion seems substantiated by the fact that Buckland represented the Red Lady's skeleton as entire in the sketch of the cave. Correspondingly, he did not speculate about the missing head, but only later mentioned a tooth in a letter to Cole: "Mr. Knight has one of her Teeth, so that she must have had a Head, buried with Her & disturbed before my Arrival."[40]

More seriously, there is no discussion of the Red Lady's bones in *Reliquiae Diluvianae* (or anywhere else) that indicates an analysis with regard to her age and sex. In fact, Buckland received the bones of the Red Lady only after he had decided on the sex, since he reported their safe arrival in Oxford five days earlier in a letter to Cole dated 3 April 1823. In spite of this, he had been able to describe the witch of his February letter to Penrice Castle as old and very tall and thin, "as a Witch ought to be."[41]

This apparent lack of interest in the bones is surprising because at the time of Buckland's excavation at Paviland, the examination of human bones was used to determine the sex of a skeleton, as well as its kind and approximate relative age. Buckland was aware of the work of James Cowles Prichard (1786–1848), who would be his favorite candidate for the regius professorship of modern history at Oxford University in the 1840s. Buckland even suggested that the post be used to teach ethnology. As early as 1813, Prichard provided a synthesis of comparative philology, history (including sacred history), antiquarianism, "ethnology," and physical anthropology. He even integrated information from diluvial geology into his reconstruction of the history of humankind as a progression rather than a degeneration in anatomy, mind, and culture (with the exception of religion).

In discussing the relationship between science and religion, he, like Buckland, argued that the Bible was concerned with the interests of humankind rather than a literal account of a history of the physical phenomena. Prichard's was a picture of development from a single family and migrations from the East after the Flood. He was of Welsh origin and, when linking the Druids with the Brahmins, seems to have been inspired by the Celtic revival. Apart from his classical education at Cambridge and Oxford universities, recent Eastern studies, as well as physical anthropology, informed his description of the Celts as derived from an Asian nation. Although he was skeptical about racial classification on the basis of skeletal markers, he discussed the systems of Camper, Blumenbach, Cuvier, and others with regard to cranial as well as postcranial human bones.[42]

Although Prichard's early work dealt with the recent varieties of humankind and their unity as one species, there were also close analyses of potential fossil human bones that Buckland was aware of. In his paper on the human skeleton extracted from the rock on Guadeloupe in 1805, Charles König (1774–1851), keeper of the Natural History Section of the British Museum, began with an account of the significance of fossil bones and the history of their recognition and analysis. Blumenbach and Cuvier were among the authorities cited to position the current discovery in a progressive line from ignorance to enlightenment about the history of the earth. It might turn out that humans had existed in antediluvian times after all. The chemist Humphry Davy (1778–1829) chemically analyzed some of the bone from the Guadeloupe skeleton and found that it still contained animal matter, as well as "the whole of its phosphate of lime." This was not regarded as conclusive evidence for a recent date of the bones, however, since, as König stated in his paper, even bones from the Paris plaster quarries still contained gluten. König regretted the absence of the Guadeloupe skeleton's skull, "which by its form might have thrown some light on the period when it was deposited, or at least as to the nation to whom the individual belonged."[43] Blumenbach's excellent figures of skulls of Caribs in *Decas collectionis suae craniorum diversarum gentium illustrata* would have made it easy to see whether the frontal bone was similarly shaped because of compression. Analogously, Traherne, in his letter to Buckland partially reproduced earlier, referred to the bones rather than the artifacts in his inquiry about the Red Lady's sex.[44]

Not that Buckland would not have desired a bone of the scarlet woman, and he might have at least applied his famous tongue test to the Red Lady's bones, with the demonstration of which he had shocked audiences on several occasions. It simply consisted of touching a bone with one's tongue: if it stuck, the bone was likely to be fossilized.[45] Obviously, this was not a particularly "scientific" method, and we cannot know for certain whether Buckland ever considered the possibility that the Red Lady was in a fossil state. Either way, he wrote to Cole, "If you have not sent away all the Bones of the red Woman prey keep them for me I shd like to have the large thigh Bone which is broken through in the Middle unless you want it yourself."[46] The piece might have been intended as a special treasure for his blue bag, in which he carried fossils, and which became iconic for his legendary persona (Figure 3.2). One may imagine the stern appearance of the Oxford reader and canon-to-be in the course of delighting his students or a dinner party with the romance about a Welsh woman of enchanting powers. At a well-chosen point in his story, he would reach into his blue bag and, like a magician in performance, draw forth the ochre-stained bone of the Ancient British temptress.[47]

In Buckland's story, the Red Lady's bone would have carried meanings of both religious relic and fetish. (I abstain from delving into the issue of bone licking and the suggestion of Ancient British cannibalism in this context.) Its reddish coloring marked the Red Lady as enchantress in the double sense of witch and scarlet woman, as Cassandra and Mary Magdalene, both of whom are traditionally identified as red haired. The Red Lady's making a living out of entertaining a camp full of soldiers in any way desired further fit the fact that scarlet was the color of vice, and red-haired women had been the most expensive Roman slaves. Continuing in this vein, the red staining also prompted Buckland to jocularly trace her genealogy to Eve via Esau the redhead, if not to identify her with the first temptress, with whom the line of red-haired sinners had originated. Finally, it, too, was the color that allowed Buckland's "Red Woman" or "Witch of Paviland" to obtain the salon-proper title of Red Lady through its association with the most enchanting and ladylike trait, blushing, that Duncan in his poem granted her even in death (reproduced in the introductory chapter to this part), and which again alludes

Figure 3.2. William Buckland with blue bag around 1843, by R. Ansdell; courtesy of The Natural History Museum, London.

to shame. While thus both Duncan and Buckland speculated in jest about the meaning of the bones' coloring, scientifically, it remained a riddle. So did the huddled position of the skeleton in the grave.[48]

Although Buckland's humorous nature, in combination with the wide discrepancy between his letters to Cole and his account in

Reliquiae Diluvianae, makes his witch story appear like a joke, we are not mistaken in the supposition that the joke was once again not entirely private. Rather, the February letter to Cole about the Red Lady's romance seems to be the account of a lecture Buckland gave on the very day the letter was written; not surprisingly, he "puzzled them all as well as [him]self" with his story. Among his audience were the bishop of Oxford, four other college heads, and three canons of Christ Church.[49] The fact that the skeleton was female enabled Buckland to share a joke with his male audience, which often included lecturers and professors of various disciplines, about a taboo area, reducing the questions of its sex and age to the level of stereotypes. The use of humor in this setting would not only serve male bonding but would also build on the exclusion of women from this typically male space in society. Such a joke would not have been possible had women attended his lecture, and—in spite of the womanpower on which Buckland built during his career—he was not in favor of women's public participation in science.[50] Discussing the upcoming second meeting of the British Association in Oxford in 1832, he wrote to Roderick Impey Murchison (1792–1871):

> Everybody whom I spoke to on the subject agreed that, if the meeting is to be of scientific utility, ladies ought not to attend the reading of the papers—especially in a place like Oxford—as it would at once turn the thing into a sort of Albemarle-dilettanti-meeting [lectures at the Royal Institution], instead of a serious philosophical union of working men.[51]

Thus, as in the case of his interpretation of Kirkdale Cave, Buckland used humor to deal with the potentially uncomfortable in the evidence from Paviland Cave. Once again, as the reference to scapulomancy suggests, which, as we have seen, was based on the study of classical and antiquarian sources, the story was located between the realm of the jest and the serious. In this context, yet another meaning of the color red is invoked. As the color of power, it stands for the dominance of Buckland and his Oxford vis-à-vis Welsh superstition and female shallowness. But because it is the Red Lady's bones that are red, it also hints at the fear of empowerment of underprivileged groups through challenges to orthodox beliefs. After all, the evidence from Paviland

Cave was far from unambiguously pointing toward a postdiluvian age of the Red Lady.

One might summarize this evidence and Buckland's interpretation thereof as follows: First, on the one hand, the antediluvian fauna and the human bones and artifacts lay in the diluvial loam in the two-thirds of the cave toward the rear end, which did not appear ever to have been disturbed by the present sea. On the other hand, they had become mixed with recent bones and shells, and a great part of the larger bones were broken and dispersed. Buckland solved this paradox by suggesting that the cave floor had been mixed up by "ancient diggers" for ivory.[52] Second, the artifacts placed in the Red Lady's grave were of antediluvian tusk, and no pottery, metal, or glass was found, which argued for contemporaneity of the woman with the antediluvian animals. Buckland explained the first fact away by claiming that the human occupants of the cave used antediluvian ivory at a later stage, but while it was still workable. Third, the British camp on the hill immediately above the cave explained to Buckland the presence of a human being in the natural shelter, indicating contemporaneity of the camp with the occupation of the cave by the Red Lady.

We will never know what Buckland really thought about the Red Lady. Did he consider it possible that she was an antediluvian, but was afraid of the turmoil this interpretation might cause? Or did he wholeheartedly believe in his romance? To get at least closer to a probable answer to these questions, the further developments of the human-antiquity debates must be taken into account. But even at this early stage in the Red Lady's story, as entertaining and skillful as Buckland's diversion from the potential threat she represented might seem, there still remained one nagging question: If the Flood had deposited the diluvium in which the bones of antediluvian animals were found, why were there no human remains among them? In view of the ambiguous evidence just summarized, Buckland's good-natured indignation that "we cannot however admit our red Woman to have been Antediluvian" and his more formal judgment that "from all these circumstances there is reason to conclude, that the date of these human bones is coeval with that of the military occupation of the adjacent summits, and anterior to, or coeval with, the Roman invasion of this country" thus seem paradoxical.[53]

However, the inconsistency the negation of antediluvians caused in Buckland's diluvialism was counterbalanced by Cuvier's rejection of fossil human remains. In his *Recherches sur les ossemens fossiles* (1812), he had pronounced that all fossils so far claimed to be human were either those of animals, as in the case of Johann Jacob Scheuchzer's (1672–1733) *Homo diluvii testis*, or of a recent date. Nonetheless, Cuvier allowed for the possibility that human fossils might lie buried under the ocean, where there had once been dry land, while the new land might have been repopulated from spared regions. This was a possibility because within his scenario of exchange of dry land and sea, the flood was not universal.[54] Letters by Buckland show how much he cared for Cuvier's opinion. He was very pleased and probably relieved when, after he had disagreed with the great anatomist on certain points in his *Reliquiae Diluvianae* (1823), Cuvier took up his hyena-den theory in the second edition of *Essay on the Theory of the Earth* (1825; trans. 4th vol., 2nd ed., *Recherches sur les ossemens fossiles*, 1823).

Buckland basked in Cuvier's praise with great satisfaction and dedicated his memoir on the megalosaur from the Stonesfield quarries to Cuvier, who in return praised Buckland's way of doing science in his *Recherches sur les ossemens fossiles* (2nd ed., 1821–1824). In exchange for Buckland's courtesy, Cuvier made him a present of a *Mosasaurus* cast, a giant marine lizard, as a symbol of his high esteem. In accordance with their gift-giving friendship, professional collaboration, and general agreement in scientific matters, Buckland wrote in the postscript of *Reliquiae Diluvianae*: "M. Cuvier also expresses an opinion which coincides entirely with my own, 'that the human race had not established themselves in those countries where the animal remains under consideration have hitherto been found, in the period preceding the grand inundation by which they were destroyed.'"[55]

While thereby emphatically agreeing with Cuvier on the absence of any genuine human fossils among the finds so far made, Buckland believed that human fossils would eventually be found in the diluvial gravel of Asia, which was largely accepted as the cradle of humankind. This solution speaks to the fact that Buckland was painfully aware of the contradictory demands made on him concerning the question of antediluvian humans. In his lecture notes on the Mosaic Deluge, he pondered the two standpoints that humans either must have or must not have been in Britain before the Mosaic Flood and tried to find a "balance." The solution was simultaneously having antediluvian humans and not having them by arguing that though no humans had reached Europe or America before the Deluge, antediluvian humans had existed in Asia.[56] Shuffling the problem of antediluvian humans off to Asia seems as good a solution as one could have hoped for. On the one hand, it provided Buckland with hypothetical antediluvian humans and thereby made his theoretical compromise between biblical exegesis and geology consistent. On the other hand, it answered the danger of placing humans into his unorthodox reconstruction of antediluvian Britain and avoided disagreement with Cuvier and his British community.

Buckland used the presence of Pecora in the diluvial gravel of Europe as collateral evidence for antediluvian humans in Asia, since God had created the ox, deer, horse, and hog for human use and therefore at the same time. He reasoned that no genuinely fossil human bones were found in the British gravel since although Pecora and humans had been created at the same time (according to Genesis "the cattle" had been created by God as separate from "the beasts"), the latter had not spread as far from Asia as (northern) Europe before the Deluge. This was due to the fact that they had many more wants that functioned as checks on their propagation and thus migration. That no human remains had been discovered in Eastern regions, either, was explained by reference to the lack of true science, or knowledge, or even adequate roads. No gravel pits had been worked in Russia.[57] In accordance with Buckland's belief in divine benevolent providence, not only would antediluvian humans therefore have found a hospitable earth with houses and dinner ready at hand in the form of caves and Pecora, but also there was no reason to assume that the monstrous European fauna had been as dominant in antediluvian Asia.

BUCKLAND'S ENGAGEMENT WITH the Red Lady and the situation found at Paviland Cave has brought to light what might be called strategies of diversion. He created "romances" and gentlemen jokes around the skeleton, drawing on prejudices about Wales and on classical and antiquarian sources, but did not offer a description of its bones, a fact that, as we will see in Part II, intrigued anthropologists many decades later. His diluvialism depended on antediluvian humans, but he circumvented the controversial issue by transferring them to

Asia. However, although Buckland once again proved his skill in compromising, the Asia hypothesis could not settle the human-antiquity question for long because the evidence for human coexistence with the extinct antediluvian mammals was growing fast. As is discussed in the following chapter, Buckland himself was involved in the excavation of one of the most significant sites in Britain, where human-made flint tools were discovered en masse.

~ 4

Buckland Accused

> In considering Buckland's influence on cave studies a vital question is: to what extent was Buckland personally involved in the supression [sic] of the findings of McEnery, Godwin-Austen and Vivian? The key evidence appears to be lacking, but it does seem certain that Buckland convinced McEnery that the publication of his work would not be advisable.[1]

THE THIRD CAVE that will be of special concern is Kent's Hole, this "Object of superstitious wonder—Traditions and marvels."[2] As we have seen for Kirkdale Cave, Kent's Hole, too, was the site of rivalries: "The cavernous nature of the Limestone on the coast of Devonshire did not fail to attract Geologists from a distance, whilst those living in its vicinity pressed forward to anticipate the researches of strangers and to bear away from them the palm of original discovery."[3] Some visitors left their names on the cave walls, which invited an analogy to the metaphor of the cave as archive of history: "Its recent annals seem to be written on its walls as its ancient ones are preserved in its bosom—like medals in a tomb—documents in archives. Inscriptions in all languages."[4] This habit of claiming territory by inscribing one's name and the year of conquest had been practiced in many caves, and some inscriptions dated from the fourteenth century.

In 1824, Thomas Northmore (1766–1851), an antiquarian from Cleve in the neighborhood of Exeter and a hunter of Ancient British and Roman antiquities, was looking for evidence for a Druidic temple of the sun god (Mithras) when he encountered mammalian fossils in Kent's Hole. Buckland's *Reliquiae Diluvianae* (1823), which he found by chance on his brother-in-law's table, made him decide to extract as many organic remains from Kent's Cavern as the Oxonian had from Kirkdale Cave. To this purpose, he hired two assistants and an able draughtsman.

> Upon entering, then, the Cavern, and being at that time a novice in the art of exploring, I began to consider in what part it was most likely to find the expected treasures, and seeing a small recess in the Cavern, of a size sufficiently capacious to hold a large tiger, I began to dig therein through the stalagmitic covering, and in less than ten minutes I could not forbear exclaiming with joy—*Here it is;* and I pulled out an old worn-down tusk of *an Hyaena*, and soon afterwards a metatarsal bone of the *Cavern Bear*.[5]

Evidently, Northmore claimed to possess some of his rival Buckland's magic abilities and was his equal in zeal for new discoveries or in their imaginative interpretation. The biblical literalist Northmore interpreted the fossils from Kent's Cavern as the remains of animals that had been eaten by carnivores in the aftermath of the Flood.[6]

Buckland had visited Kent's Hole in the Devonian limestone of Torquay even before Northmore and W. C. Trevelyan discovered bones and teeth beneath the stalagmite floor. In 1825, Buckland again visited Kent's Cavern, as well as Pixies' Hole, Chudleigh, and wrote to Trevelyan: "In the Chudleigh Cave, I found fragments of old pottery in a hole like those at Kent's Hole. These I take to have been dug by the people who left old pots and flint knives, and mutton bones, and charcoal in them, and not by Bears, as you once fancied."[7] Buckland recognized what to him had by now become a typical Ancient British hearth assemblage. As in the case of Paviland Cave and Kent's Cavern, it was made up of a combination of traces, such as holes in the stalagmite, charcoal, pottery, remains of food, and the odd flint knife. Not only Trevelyan but also Northmore, who accused Buckland of a destructive excavation technique, reached another conclusion. The latter disagreed with Buckland and Cuvier, as well as the majority of the elite, on the "late formation of man."[8] That this cave was indeed a sight of struggle over the correct method and the right interpretation is documented in *The History of Chudleigh, in the County of Devon* (1852): "These learned gentlemen were highly pleased with their discoveries, however they differed widely in their geological theories, each considering that his own scientific deductions were illustrated and confirmed by what was here presented."[9]

In Kent's Cavern, Buckland unearthed a flint implement, as well as bones and teeth of the bear and rhinoceros, in what seemed to be another hyena den. Although the *Monthly Magazine* for March 1825 mentioned only "the celebrated Professor of Mineralogy and Geology in the University of Oxford," most of the work at Kent's Cavern was carried out by the geologist and Catholic priest of Torquay, John McEnery (1796–1841), between 1825 and 1829, albeit under Buckland's supervision.[10] McEnery's first visit to the cave had been as a member of an exploring party of about a dozen people, organized by Northmore, among them the commander of the coast guard and his people. On the journey that led through narrow passages, McEnery "could not divest [him]self of certain undefinable sensations, it being [his] first visit to a scene of this nature."[11] That the work was not easy for McEnery or the workmen is given frequent testimony in his manuscript, where he described how he lost his way or got stuck in the cave. Again and again, the reader is reminded of Greek myths of subterranean labyrinths and adventures of travelers to unknown countries. Even though McEnery's health and that of the workmen suffered from the bad air and dampness of the cave, he never ceased to be awed by his discoveries and could not help expressing his feelings.[12] In fact, he felt more than duly recompensed for the difficulties not only by the strong emotions he experienced but also by a glimpse at deep time:

> They were the first fossil teeth I had ever seen, and as I laid my hand on them, relics of extinct races and witnesses of an order of things which passed away with them, I shrank back involuntarily— Though not insensible to the excitement attending new discoveries, I am not ashamed to own that in the presence of these remains I felt more of awe than joy—but whatever may have been (sic) impressions or the speculations that naturally rushed into my mind, this is not the place to indulge them—My present business is with facts.[13]

That McEnery tried to force himself to stick to the "facts" in the writing of his manuscript that was intended for publication alerts one to the fallacy of inferring from the rather dry accounts of caving found in the scientific journals or books that the researchers did not also undergo strong feelings. McEnery's experience documents that Buckland was not the only one whose imagination was tempted to run away with him when exploring the underground windows to a former world.[14]

The spoils of the cave were rich, but as is well known, Buckland and McEnery did not agree on the age of the numerous stone tools found.

While the latter believed in their contemporaneity with the remains of the extinct animals, the former upheld his interpretation of a later introduction through ovens dug into the stalagmite layer by Ancient Britons. This explanation seemed impossible to McEnery because he could detect no traces of previous disturbance in the stalagmitic cover. Furthermore, he estimated that had the humans been in the cave after its deposition, they would not have been able to penetrate it. As we will see, the establishment of human antiquity inspired the reexamination of many excavation sites and the material gained from them, among which Kent's Cavern played a decisive role. As a result, once consensus had been reached in the scientific community on the contemporaneity of the extinct mammalian fauna and humans, Buckland was accused of having strategically suppressed the evidence gained by McEnery and those who continued his work at Kent's Cavern. The archeologist John Evans (1823–1908), for example, saw the reason for McEnery's doubts in the bias Buckland caused in geology. Among Buckland's critics also was Sollas, one of his successors to the chair of geology at Oxford and reinterpreter of the Red Lady: "Strange to say, this acute observer, who was naturally of the most open mind . . . received them [McEnery's results] with hostility and sought refuge in impossible and almost evasive objections. I can only attribute this to the power of preconceived ideas."[15]

As the epigraph at the beginning of this chapter suggests, this evaluation of Buckland's involvement at Kent's Cavern is still maintained.[16] However, Buckland did not reject the evidence for human antiquity McEnery brought forward against his better judgment, at least not immediately. Initially, he even seems to have welcomed it:

> Admitting that is [? it] was fully established that instruments of art occur in the midst of the bones in places to which we see not how they could have arrived by any postdiluvian operation, Dr. Buckland considers such a fact would go to supply a link in the chain of evidence which has been hitherto a desideratum of no small magnitude, namely that man in a savage state inhabited England before the Deluge.[17]

Quite to the contrary, where Buckland did issue critique, it turned out to be well founded, as in the case of the human bones from the cave. McEnery at first interpreted these as being contemporaneous with the antediluvian fauna in the same way he thought the tools were:

> Into this opinion [of human bones being contemporaneous with those of extinct animal species] I fell at first from the discovery of flint blades in contact with both in several parts of the cavern and the alternation of the stalagmite and I communicated my impressions to Dr. Buckland with all the earnestness of sincere conviction—It was to this doubt chiefly is owing the original delay occasioned in the publication of my labours—It is only from extended observation over the entire field of the cavern that I have come to the conclusion that human bones are long posterior to the sediment containing pebbles and bones and that they date no more than half way down from that period.[18]

This was therefore Buckland's cautioning for a good reason that delayed publication. It prevented hasty conclusions. The "extended observation over the entire field of the cavern" led to the insight that the flints found below the stalagmite were remarkably cruder (unpolished) than those associated with human bones, pottery, ornaments, or charcoal, which were indeed "sometimes in pits excavated in the stalagmite," in disturbed soil, or in association with a more recent burial. The sequence of stages of refinement in stone tools, as well as pottery, in combination with the appearance of other human artifacts at various levels, allowed McEnery to reconstruct the succession of human occupations in the cave.

Unfortunately, possibly because of the reasons discussed in previous chapters, Buckland at this point desisted from his initial enthusiasm with regard to the antediluvian origin of some of the flint implements and changed the direction of his influence. That there was a strong influence is evidenced in the fact that Buckland and Cuvier assisted in identifying the fossils McEnery discovered, to which purpose he sent them specimens and corresponded with them. He even paid Cuvier a visit in Paris. McEnery was also familiar with the two masters' writings and in the manuscript frequently referred to *Reliquiae Diluvianae* (1823), as well as to Cuvier's works. Despite his respect for the expertise of the two savants, however, McEnery by no means simply yielded to their views. Though reluctantly, and despite his acceptance that no fossil human bones had so far been discovered, he abided by his interpretation of the oldest flints: "It is painful to dissent from so high an authority [Buckland] and more particularly so from my concurrence generally in his views of the phenomena of these caves, which three years personal observation has in most every instance has [sic] enabled me to verify."[19]

Nevertheless, in the long run, Buckland's high position in science and society seems to have exerted its full power. Further toward the end of McEnery's manuscript, humans were no contemporaries anymore of the wild beasts such as hyenas and bears that had inhabited the cave. There, McEnery speculated that humans had entered the cave only shortly after the Deluge. He now explained that the flint tools had been left on the still-soft diluvial mud, from which they sank down to the fossil animal bones. Unlike the fossil bones, they therefore did not appear throughout the loam but rather at the top. According to this scenario, the early inhabitants of the cave had belonged to the primitive people who had entered Europe in the aftermath of the building of the Tower of Babel and their subsequent dispersion. They, as well as their Noachian predecessors, had once had the knowledge of metal but lost it, or the possibility to gain the raw material, on their journey. Gradually, as the discoveries in Kent's Cavern supported, they refined their technologies again. This is the orthodox story of the peoples having fallen from grace and having thus degenerated, reminiscent of Buckland's account in his notes on the dispersion of the human race.[20]

All in all, there are reasons to believe that not only McEnery but also Buckland changed his mind about the manufacturers of the oldest flints from an ante- to a postdiluvian people. Did Buckland distrust the undisturbed layer of stalagmite as evidence beyond doubt because of his discovery of what he thought of as postdiluvian human bones under this condition at Burrington (see Chapter 2)? Unfortunately, this is a circular argument. It seems more likely that Buckland's initial approval, as reproduced in McEnery's manuscript, represents an informal discussion between the two men that Buckland, possibly motivated by factors discussed in previous chapters, after further consideration of the consequences, would not have repeated in public.

There was no disagreement between McEnery and Buckland concerning the overall interpretation of the cave as a former hyena den. In the aftermath of the excavations at Kent's Cavern, the notion continued to be adapted in popular culture. From Octavian Blewitt's *Panorama of Torquay* we learn that Kent's Cavern had become a tourist attraction.

Blewitt guided his readers to this remarkable construction of nature, recommending proper dress, lights, and the guide who held the key to the cave. He warned that anyone who desired to dig the ground would need the permission of its owner. Blewitt conjured up Buckland's vision of antediluvian hyena dens when he poetically referred to the cave as a "resort of animals whose howlings shook the forests of the primaeval world,—and whose relics are still preserved as types of an epoch which is enshrouded in a veil of solemn mystery, relieved only by the sublime and pathetic narrative of the inspired historian."[21] This vision, which could only be aroused by the magic of the gifted geologist, became uncanny, for

> "In exhuming from their earthy beds or spar-bespangled vaults," says Dr. Ure [*New System of Geology*], "the relics of that primaeval world, we seem to evoke spirits of darkness, crime and perdition; we feel transported along with them to the judgement seat of the Eternal, and hear the voice of many waters coming to execute the sentence of just condemnation on an earth 'corrupt and filled with violence.' The powers of prophecy overshadow us. The bony fossil starts to life and conjures us in mysterious mutterings to flee from the wrath to come. How solemn, to walk through this valley of death!—Methinks the very stones cry out 'The Lord reigneth; righteousness and judgement are the habitation of His Throne!' "[22]

In this revelation scene, in which, similar to Conybeare's visual rendering of Kirkdale Cave, the antediluvian world is recalled to life, geological, mystical, and religious elements combine to create a dramatic vision with moral concern. Obviously, the Bucklandian antediluvian hyena den was attractive and malleable enough to be used for different purposes by different sections of British culture. As in the case of Eastmead's monster hyenas, it is not at all clear whether this adaptation of the antediluvian world as a place of darkness, crime, and perdition rendered it a more likely place for human beings, even though the pre-Flood scene at the same time calls out for their presence.

In fact, the human bones from Kent's Cavern had not been the only ones brought to Buckland's attention in the course of the 1820s.[23] In the aftermath of Paviland and Kent's Cavern, several groundbreaking finds of human bones in association with those of extinct animals in the diluvium were reported from the south of France that seemed to question Buckland's elegant solution of antediluvian humans restricted to Asia. Buckland was particularly intrigued by the finds made by Jules de Christol (1802–1861) in the caves of Pondres and Souvignargues, near Sommières (department of Gard). He observed that the caves of Pondres had been entirely filled with mud and gravel, containing the bones of hyenas, rhinoceros's, and humans: "the two other caves described by Mr. Christol, seem, from his account, to contain bones of human species, mixed with those of Hyaenas, in caves to which it seems there can have been no *Postdiluvian* approach; as they were entirely filled with gravel, so that no one could ever have gone in to bury them."[24] This appraisal of the situation reminds one of Buckland's initial interpretation of Kent's Cavern, where McEnery found flint tools mixed with fossil animal bones below an unimpaired stalagmitic cover. Once again, however, the facts that the inference of contemporaneity was not taken on by other authorities, most importantly by the commission appointed by the Académie des Sciences under Cuvier, and that crude pottery was found below the level of the faunal remains, made Buckland desist.[25]

In Buckland's case, the question regarding the evidence for the contemporaneity of humans with antediluvian animals is not "how many anomalies can a theory assimilate?" Rather, his diluvialism demanded exactly these, even if he so far restricted their presence to Asia. Why, in the face of this new evidence, did Buckland not return to the Red Lady and pronounce her the first antediluvian human skeleton found on the British Isles? Surely this would make for an even better "romance" than an enchantress of Roman-British times. Unfortunately, just as Buckland had done for the Red Lady, European experts explained the new finds from southern France by subsumption under the empty signifiers *Ancient British* or *Celtic*. The precedent Buckland had created with the Red Lady could even be brought in to strengthen the argument in the new cases.

This is what Jules Desnoyers (1800–1887) did in his address to the French Geological Society in Paris, of which he was a founder and at that point the secretary. However, he discussed the human remains in an anthropological as well as antiquarian context. He classified the skulls as Caucasian, which in agreement with Blumenbach he took to be the original European stock, and drew on sources such as Florus and Caesar, who had described the Celts and Gauls. Paviland Cave, too, figured in his list of caves that had functioned as their ancient dwelling places. He interpreted the artifacts from these caves in relation to the cultures of tumuli, dolmens, and stone monuments of Gaul, Britain, and Germany and concluded that "notwithstanding the exaggerated views of the supporters of the opinion of the advanced state of our early civilization," they had been in a state comparable with that of contemporary savages.[26] The ancient Gauls and Celts, to whom the Red Lady had belonged, had painted their bodies, sacrificed humans, lived in huts or caves, and used stone implements. Desnoyers juxtaposed the claim for the "geological" with his estimate of the "historical origin" of such human bones and artifacts. The Red Lady and her kin from southern France thus belonged to the realm of the antiquary and did not constitute proper objects for the geologist.

NONETHELESS, WITH THE growing number of reports about antediluvian humans in association with fossil animal remains, it became increasingly difficult to shuffle their bones and artifacts into another field of responsibility with its own empty signifiers to be filled according to inclination. Already with the new data from southern France as discussed in this chapter, the likelihood that other than scientific uncertainties weighed in on the question of human antiquity increases: Buckland's reliance on Cuvier and other authorities; his situation as exponent of the young science of geology at the University of Oxford, where religious concerns were prominent; his own double role as geologist and by now canon in the Anglican Church; and finally his promise to the prince regent that geology would not threaten the status quo.

Although biblical literalism, in its demand for verbatim adherence to the Scriptures, had been less of an issue in the more congenial atmosphere of the Oxford of the 1820s, the 1830s brought religious concerns back to the fore in the Oxford movement. This meant that the fact that geology was seen as "so liable to be perverted to Purposes of a tendency dangerous to the Interests of Revealed Religion,"[27] as Buckland had expressed it in his letter to the prince regent, was still a burning problem. Buckland not only had to show that geology did not harm religion, more than ever he had to prove its positive service to it. If the positive service of Buckland's sacred history and natural theology to the contrary consisted in preventing geology from such perversion, where did the real danger lie, and what were the "interests" Buckland referred to? These questions will be examined in the following chapter for the reasons their answers provide for Buckland's rejection, if not outright suppression, of the evidence brought forth by men lower down the social strata of caving, such as McEnery and de Christol. While in Chapter 5 the focus will be on the "perversions" and "dangers" of geology in the wrong hands, Chapter 6 will engage more closely with the "interests" of revealed religion, which of course are the interests of a particular group of people.

~ 5

Man as a Crocodile Superior?

As developed in this chapter, one incentive to reject fossil humans was Lamarck's and Etienne Geoffroy Saint-Hilaire's (1772–1844) theories of the transmutation of species and their revolutionary potential. In Britain, the first half of the nineteenth century saw an increasing number among the industrial working classes turning away from Christianity. In the aftermath of the French Revolution and exploited in an industrialization built on laissez-faire liberalism, science was sometimes perceived as a weapon against oppression by the ruling classes in church and state. Freethinkers tried to reach the working and middle classes through articles on science in the infidel press and public lecture series. Lamarck's notion of the inevitability of progress "from the bottom" through individual effort and improvement of circumstances inspired atheists and socialists in France and Britain alike. British libertarian freethinkers and French materialist radicals embraced Fossil Man as evidence for human descent from the animal world and organic transmutation in general as a way of rethinking the social body.[1]

Lamarck, professor of invertebrates, and Geoffroy Saint-Hilaire, the much younger professor of vertebrates, were both Cuvier's direct opponents at the Muséum National d'Histoire Naturelle in Paris. As the disputes around transmutationism in the Académie des Sciences and inside the museum walls were carried outside and taken up by materialist republicans and the socialist press, who attacked the authority of the church, as well as the privileges of the upper classes, it was also the social order that was under attack. As Adrian Desmond has shown, in Britain, anticlerical and antiaristocratic ideas were promoted on the basis of transmutationism in the illegal penny press, and transmutation mixed with political radicalism was discussed in secular anatomy rooms and Nonconformist colleges, first and foremost at the secular universities of London and Edinburgh. The threats to social hierarchies were therefore also aimed at the petty privileges of the Anglican gentlemen at the English centers of education.[2]

For Lamarck—no doubt influenced by his teacher and predecessor at the Jardin des Plantes, Buffon, particularly his *Histoire naturelle* (1749–1804)—animate nature was ordered according to a chain of organisms of increasing complexity from infusoria to humans. In his *Philosophie zoologique* (1809), ongoing spontaneous creation from inanimate into animate matter guaranteed the simultaneity of organisms at all grades of the chain of beings, since in analogy to ontogeny, Lamarck envisioned them to be driven up the ladder through an internal mechanism. If the series of increasingly higher beings was somewhat irregular, it was therefore due to a second mechanism of transmutation, the so-called use-inheritance, through which the organisms actively adjusted to gradually changing environments by developing new habits that eventually altered their structure inheritably. For organisms lower down the scale that did not show active adjustment to environments through new behaviors, the changed conditions in temperature, humidity, air, and light could still shape their form durably through the inheritance of acquired characteristics. Like Charles Darwin after him, Lamarck used domestication as an analogy to the workings of natural transmutation. Much faster than under natural conditions did the keeping of plants and animals in artificial environments and their hybridization transform their organs and structures. In contrast to Darwin, the emphasis was on new circumstances of life rather than on breeding. Humans could further tamper with the linear and complete chain of beings by causing the extinction of a species, for example, through overhunting. Generally speaking, however, Lamarck worked with the concept of plentitude and uninterrupted lines of transmutations rather than with extinction and novel creation.[3]

Cuvier's rejection of Lamarck's transmutationism depended on the claims that no humans had existed at the same time and in the same regions as the extinct animals, and that nowhere on earth were there any survivors of the fossil animals. He therefore relied on the concept of extinction. In his *Recherches sur les ossemens fossiles* (1812), he demonstrated that the fossil mammals and reptiles were not mere precatastrophic varieties of living species; they were different species that had become extinct in the exchange of dry land and sea. Furthermore, since there were no humans present at the time when the now-extinct animals lived, none of the large mammals could have become extinct in the process of human hunting life, as Lamarck suggested. Nor could any of these animal species have undergone transmutations induced by domestication. Cuvier further argued against domestication as a model for transmutation as adduced by Lamarck by claiming that while domestication caused an unnatural degree of variability within a species, in nature there existed bounds to flexibility in form. These limits to variability were fundamental, since each type was perfectly balanced and adjusted to its surroundings.[4]

The discovery of human fossils together with the fauna of former times—just like the discovery of any animal species still alive—would also have presented a conceptual problem for Cuvier's scenario of the exchange of land and sea. He thus argued that either there had been no humans before the last catastrophe, or they had only populated Europe after the last catastrophe from other parts of the world now drowned. Last but not least, Cuvier rejected the notion of a linear succession of organisms because he perceived unbridgeable anatomical gaps between what he called *embranchements*, the four main systematic branches of the animal world. Against Lamarck, Cuvier thus abandoned the eighteenth-century notion of a scale of beings while retaining its fixism. Geoffroy Saint-Hilaire, on the other hand, shared Lamarck's belief in the transmutation of organisms. His attempt to build a natural classification system on the basis of the analogy of organs and structures between groups of animals was inspired by the German *Naturphilosophie*. He explained the transmutation of the animals whose fossil remains were now discovered into their recent kin through the direct influence of the environment, building strongly on the principle of an underlying unity of plan of organic composition and on the analogy to individual embryonic development. He eventually tried to reduce the anatomy of all animal species to one plan. When he claimed to thus bridge the gap between vertebrate and invertebrate anatomy, the disputes between transmutationists and fixists in the Parisian Academy of Science flared up with unprecedented violence in the aftermath of Lamarck's death, only to subside with the death of Cuvier.[5]

Buckland again dealt with the problem by ridicule. Referring to Geoffroy Saint-Hilaire's work on reptiles, he taunted that the doctrine of the transmutation of species reduced man to a "Crocodile Superior." In the mid-1820s, Cuvier had identified a fossil from near Caen as an extinct species of gavial. Geoffroy Saint-Hilaire, who reexamined the specimen, instead announced that it was an intermediary between reptile and mammal, a teleosaur, supporting Lamarck's tree of vertebrate genealogy at the end of the *Philosophie zoologique* (1809). In the early 1830s, Geoffroy Saint-Hilaire expanded his work on the saurians, which led him to believe that the teleosaur, which had evolved from the ichthyosaur, was the ancestor of the Tertiary crocodile.[6]

In notes dating from the 1830s, Buckland objected that there were no intermediate states, and that a change in circumstances did not alter a species. The ichthyosaur had not existed before the Liassic and then had not changed up to the Chalk—the proportion of its bones had remained consistent. At the end of the Chalk, the saurians had suddenly gone extinct, while the crocodile, which had already lived with the ichthyosaur, was still crocodile in the present day. Buckland also rejected the ichthyosaur-plesiosaur-pterodactyl sequence, which was rendered impossible by stratigraphic knowledge. No saurian had thus changed into the plesiosaur or the crocodile, and it seemed even more illogical to claim that the saurians had become mammals. Rather, the mammals had simply been added, and the ichthyosaur, plesiosaur, pterodactyl, and crocodile had been contemporaries and part of an intricate system of interdependences.[7]

Buckland thus deconstructed the ontogeny-phylogeny analogy and the associated linear progressionism that he saw as prerequisite for Geoffroy Saint-Hilaire's and Lamarck's "Absurd Doctrines." Although he granted that both individual development and the succession of animal species throughout earth history showed an overall increase in complexity, there had been no transmutation of fish into reptile and into bird or mammal. Buckland quipped that if new organs were the effect of appetency, and fish had once turned into reptiles when they tired of swimming, just as reptiles had begun to fly through intensive wishing, then surely these days one would witness schoolboys producing wings to escape from school.[8]

On a more serious note, he reasoned that there were cases where the ancient animals were more complex in structure than the recent ones, as in the examples of ammonites and nautili. This was no major problem for Buckland, who saw no imperfection in any work of God. Although humankind was God's special creation, for whom he had prepared the earth through a series of dramatic events, as well as by steady changes under forces still at work, animals were proof of the Creator's existence and attributes in their own right. To the contrary, for Lamarck, perfection was achieved only in humans. Though both worked with a hierarchical chain of beings, for Buckland, every creature was absolutely perfect with regard to the sphere for which it was created; progress in his sense referred to the overall advance of the organic and inorganic systems toward modern conditions throughout the stratigraphic layers. Thus Buckland's "Chain of organized Beings" as designed by the Creator was flawless, hierarchical, and fixed: "I am not quarrelling or finding fault with a Crocodile / a Crocodile is a very respectable Person in his way / but I quarrel with the calling a Man a Crocodile Superior."[9]

Obviously, in agreement with Cuvier, Buckland identified the mechanisms of change not as those of transmutation, but as extinction and novel creation. Materialist transmutationism was a most unwelcome theory for Buckland, and Lamarck's conjecture of a race of ape developing into humans by inheritance of acquired characteristics was no more attractive to Buckland than accepting a crocodile as cousin: "We shall not be degraded from our high Estate to say to that Reptile Crocodile trust & greet you well right loving Cousin."[10] To prevent man's "degradation," Buckland, again similarly to Cuvier, maintained that breeding and nurture could modify the individuals within a species, but not transform one species into another: "all Monsters professing to be Unions of Diff[erent] Species [were] morbid unions of Parts of individuals of the same Species." Buckland's moral objection to Geoffroy Saint-Hilaire, in whose argument for organic transmutation through geological epochs deviations from the default path of embryonic development played a central role, was thus not restricted to the theoretical realm. Geoffroy Saint-Hilaire's experiments with the embryos of chickens were morbid actions that produced monstrous results: "all monstrosities [were] the effect of morbid action."[11]

By this time Buckland was working on his Bridgewater Treatise (*Geology and Mineralogy*, 1836) and had come to separate the last geological flood and the Deluge. He hid this check on the authority of the Scriptures in a footnote, explaining that the change was due to the preponderance of extinct animal species in the diluvium, as well as the absence of accepted human remains. In other words, while according to the Scriptures humans had been killed in the biblical Flood, but no species had suffered extinction, the opposite seemed to have been the case in the last geologically formative inundation. The necessity of allowing enormous stretches of time for geological changes, as well as the insight that the diluvium was not a universally uniform deposit, added to the problems Buckland encountered with the diluvial theory. In the adapted version of catastrophism, the flood that extinguished the animal species, which might have existed over more than one geological period, was thus one in a series of these geological cataclysms and did not correspond to the rather tranquil and recent event of the biblical Deluge.[12]

This nonbiblical catastrophism still satisfied his obligation to reconcile geology and the Bible at a place like Oxford University. At the same time, it allowed Buckland to define the prehuman world as the realm of science, while the human world, beginning after the last geologically significant inundation, but before the Mosaic Flood, would be left to the account of the Scriptures. As we have seen, already in his inaugural lecture, Buckland had argued that the word *beginning* in the first verse of the book of Genesis referred to an undefined primeval period antecedent to the last great change affecting the surface of the earth (then equivalent to the Flood) and to the creation of the present plants and animals. Geology might investigate the long series of geological changes that preceded that story of concern to the human race. Within the new scheme, the Bible had lost nothing of its moral authority, but the "primeval period" now included all of the history of the earth to its present state. The Bible contained no history of geological phenomena, which one might with justification expect to find in an encyclopedia of science, "but are foreign to the objects of a volume intended only to be a guide of religious belief and moral conduct."[13]

Within the new model in which the last geological cataclysm was dissociated from the Mosaic Flood, and which was associated with millions and millions of years of earth history, there was therefore no need to find human fossils associated with the fossil fauna anymore. One might even argue that given the new situation, it would have been highly disturbing to find human bones that preceded the last geological turmoil, since at that stage the earth was not yet ready for human arrival, and humans—as the Bible stated—were a recent creation. If anything, fossil humans would have supported the transmutationists' case. In *Geology and Mineralogy*, Buckland was thus quite firm on the point:

> No conclusion is more fully established, than the important fact of the total absence of any vestiges of the human species throughout the entire series of geological formations. Had the case been otherwise, there would indeed have been great difficulty in reconciling the early and extended periods which have been assigned to the extinct races of animals with our received chronology. On the other hand, the fact of no human remains having as yet been found in conjunction with those of extinct animals, may be alleged in confirmation of the hypothesis that these animals lived and died before the creation of man.[14]

This verdict was most controversial where the discoveries made by the physician and natural historian Philippe-Charles Schmerling (1790–1836) in Liège were concerned. Schmerling began to explore the caves on the banks of the Meuse and its tributaries in 1829. He discovered human bones of at least six individuals, worked bone, and flint flakes in the caves of Engis, Engihoul, Chokier, and Fond-de-Forêt (among them the Neanderthal child skull, Engis 2). Schmerling was confident that the sediments had not been disturbed, and that the human remains had been brought into their present situations at the same time and by the same means as the bones of the now-extinct animals with which they were associated. In his *Recherches sur les ossemens fossiles* (1833–1834), he therefore concluded that these were the traces of an antediluvian race:

> As I dare guarantee that none of these pieces [artifacts] has been introduced at a later time, I attach great importance to their presence in the caverns; since even if we had not found human bones, in conditions entirely favoring the consideration that they belong to the antediluvian epoch, these proofs would have been provided to us by the worked bones and flints.[15]

The decomposition, the often-broken and rounded state, the color, and the relatively slight weight of both the animal and human bones further strengthened Schmerling's conviction. Who might the antediluvian inhabitants of this region have been? The sutures and teeth informed him of the individuals' relative age: a youth and an adult or elderly person. The postcranial bone fragments seemed to indicate a height around five and a half feet. For racial classification of the crania, Schmerling drew on Blumenbach's descriptions of the forms and dimensions of skulls of various races.[16] Unfortunately, the faces of the human skulls, essential parts for classification (facial angle), were missing. Nonetheless, on the basis of the elongated form of the more complete skull from Engis Cave (Engis 1), the narrow and receding front, and the pronounced supraorbital ridges, he conjectured that the human being might have been of the Ethiopian type and of minor intelligence.[17] Well aware of the orthodox rejection of fossil human bones by such authorities as Buckland and the by-then-deceased Cuvier, whose verdict haunted science even after his death, Schmerling expounded the reasonableness of assuming antediluvians: It must have taken a long time to people the earth from a single original couple in India; the voyages of discovery had shown that humans indeed inhabited all the corners of the world; these geographical varieties exhibited huge differences in culture; to reach even the lowest rungs of civilization must have taken a very long time in view of the slow pace at which humans progressed.[18]

At the meeting of the natural historians in Bonn in 1835, Schmerling confronted the international audience with his spectacular discoveries. Among the gentleman naturalists was Buckland, who had inspected Schmerling's collection before the meeting. Buckland challenged the claim for contemporaneity of the human with the animal remains, humorously referring to Schmerling as "the doctor of the antediluvian hyenas."[19] He rejected Schmerling's estimate that the human bones were in the same state as those of extinct animals and claimed that they were less decayed. However, had that been the case, it would not have been regarded as finally decisive. As we have seen, even Buckland, although he had used the same argument against the fossil state of the human bones from Gailenreuth, was ambiguous about the status of the consistency of bones as a clue to their age. In 1844, at the Canterbury meeting of the British Archaeological Association, he cautioned his audience not to make inferences from the state of bones to their age: "The decay of bones was no *certain* proof of antiquity, nor their solidity a certain proof of recent interment."[20]

Nonetheless, at the Bonn meeting Buckland presented his argument very figuratively. It was in this context that he gave the most extravagant demonstration of the tongue test. Although the proceedings of the meeting only report Buckland's doubt about the contemporaneity of the human skull and the animal fossils, eyewitnesses told of other happenings. Buckland demonstrated to the assembly that although the human bones did not stick to his tongue, he could deliver his entire address with a bear bone from the same cave sticking to the inside of his lower lip or his tongue. The French geologist Jean-Baptiste Armand Louis Léonce Élie de Beaumont (1798–1874), who had replaced Cuvier at the Collège de France in 1832 and held the chair of geology at the École des Mines, remembered the occasion many years later in the context of a discussion on the means of determining the age of bones:

> Mr. Buckland immediately took a bear's bone, attached it to the end of his tongue, from which it remained suspended, which did not prevent the savant professor in the least from talking, and, while turning toward the different parties of the assembly, Mr. Buckland repeated at several occasions with a slightly guttural voice: *You say it does not stick to the tongue!*[21]

After what must have been a most hilarious spectacle, Schmerling had to accept the challenge. He could but try to achieve the same with one of the human bones—in vain.

Although already Blumenbach had observed that older bones stuck more easily to one's tongue than younger ones, in the 1830s this test was not accepted as producing reliable results.[22] To the contrary, on the basis of this test, the human bones from the caves in southern France (discussed in Chapter 4) would have had to be considered in a fossil state. Rather, Buckland once again used humor, this time in its practical form, to avoid a serious discussion of the issue at hand. According to Hermann Schaaffhausen (1816–1893), professor of anatomy at Bonn University, the audience at the Bonn meeting was presented with "amusing scenes between Buckland and Schmerling."[23] Playing the clown in front of an assembly of learned gentlemen in a country not his own, Buckland by his entertainment might have produced a relaxed atmosphere that made the men think of other than scientific matters. Buckland made sure he defined the terms of the game, and Schmerling was forced to enter the duel. Amused by the show, the audience might have forgotten to think about whether the competition was suited to decide the matter at hand, or whether it was not rather a mere question of who was the better performer.

Buckland left the arena victorious, and he published his verdict on the Schmerling discoveries in an authoritative style in the new theoretical context of his Bridgewater Treatise. Because the human bones found by Schmerling might stem from later interments, through which they would have been brought into association with the extinct fauna, the worked flints were those of the uncivilized tribes that inhabited the caves in more recent times. In confirmation of influential figures such as Lyell, Buckland supported the persistent skepticism toward the contemporaneity of humans and the extinct mammals found in Europe:

> M. Schmerling, in his Recherches sur les Ossemens Fossiles des Cavernes de Liège, expresses his opinion that these human bones are coeval with those of the quadrupeds, of extinct species, found with them; an opinion from which the Author, after a careful examination of M. Schmerling's collection, entirely dissents.[24]

This despite the fact that in the case of the caves explored by Schmerling, the standard arguments against the contemporaneity of human and animal remains from caves did not seem to apply: the human bones were not arranged in a burial, and no humans, animals, or water currents seemed to have disturbed the floor. Unfortunately, Buckland, as usual when writing about human bones, did not specify the nature of his "careful examination" or what it brought to light, so his verdict has to be taken at face value.

Clearly, when increasing knowledge suggested that the diluvium was far older than previously suspected, and the scientific authorities remained adamant about the absence of genuinely fossil finds, Buckland found a solution to replace his limiting antediluvian humans to Asia. In this new framework, there were certainly pre-Flood humans, but they were considerably younger than the diluvium, which had now been laid down by a last geologically subversive cataclysm not explicitly mentioned in the Bible. Pre-Flood humans had moved out of the sight of the geologist, adjusted as it was to long distances. At the same time, human coexistence with the fossil European fauna was expected less than ever before. Buckland was accordingly troubled by the discovery of fossil monkeys in the Tertiary of the Siwalik Hills in 1836. In a kind of self-fulfilling prophecy, he acknowledged, "there is no escape from the facts," but at this point he could still reassure himself: "But monkeys are not men, though the reverse is not always true."[25]

Although Buckland's nonbiblical catastrophism again seems an elegant solution to contradicting demands, we have seen in Chapter 1 that catastrophism was slowly running out of fashion when uniformitarian ideas gained ground. This did not necessarily affect the human-antiquity question, however, since exponents of uniformitarianism tended to be no less cautious in their treatment of human artifacts and skeletal remains. Lyell made it clear that the globe was of unconceivable antiquity, while Moses's chronology was strictly applicable to the creation of humans, which was more recent than the extinction of the mammoth.[26] Buckland and his uniformitarian fellow geologists did not disagree in this regard, and they felt justified in their cautious treatment of the human-antiquity question when confronted with support for fossil humans in association with the doctrine of transmutation from the politically radical stratum of society. Indeed, Fossil Man captured the imagination of popular writers, in the hands of whom he became a subversive tool for the "perversion" of geology, a danger to revealed religion and to the interests of the class to whose advantage the current system worked.[27]

The moral and social issues surrounding the question of fossil humans could not have disposed Buckland to favor human antiquity, let alone to personally contribute evidence for it. However, although conceptually linked, the questions of human antiquity and transmutationism were not identical; the issues could be kept apart. Even though, as we will see, the acceptance of human antiquity coincided with the publication of Darwin's *On the Origin of Species* (1859), and the two notions would indeed become inseparably interwoven, the younger scientists who won the case for human antiquity restricted their arguments to "the simple evidence" for human tools or bones in close association with those of fossil mammals of Quaternary Europe.[28] When these geologists and archeologists began to argue with increasingly stronger conviction, Buckland was once again forced to take a stance. Would he hand the Red Lady over to become the most powerful argument in the proantiquity case?

In 1841, the geologist Robert A. C. Godwin-Austen (1808–1884), who had reexamined Kent's Cavern, had an interesting public conversation with Buckland at the Plymouth meeting of the British Association. In Kent's Cavern and Ash Hole, Godwin-Austen had found human bones and artifacts in association with the bones of extinct animal species in undisturbed beds of clay underneath a stalagmite layer. For Godwin-Austen, this was proof of the coexistence of humans and the extinct Pleistocene fauna of Cornwall and Devon. Buckland, to the contrary, rejected all this evidence. Unfortunately, unlike McEnery, Godwin-Austen did not sufficiently distinguish between the human bones and the human artifacts from Kent's Cavern or among the tools themselves. He was, however, aware that the presence of a sepulture could have been used to argue for the intrusion of the human bones and artifacts into the older strata, and he stressed the fact that the human bones and artifacts at Kent's Cavern were scattered below the stalagmite in the same way as the animal bones. Most important, he then went on to compare the human bones he had discovered at Kent's Cavern with the state in which those at Paviland had been found.[29]

Buckland countered by reiterating his opinion that at Kent's Hole the human artifacts had been introduced into lower strata through holes dug into the stalagmite. "He also gave an amusing account of a cave in Wales [Paviland], in which he had found the skeleton of a female who had been buried among fossil remains, and who had evidently kept a sort of sutler's shop as appeared from the remains of Celtic implements of gambling, and other amusements of a camp."[30] Obviously, rather than entering into a serious discussion with Godwin-Austen, and rather than taking the new evidence from Kent's Cavern and other caves as an impetus to rethink his interpretation of Paviland, Buckland used one of his baffling stories between fact and fiction about the Red Lady to evade a scientific debate and to discredit the strong argument Paviland provided. In this diverting story, Paviland Cave appeared as an ancient version of a casino, where the Red Lady as easy woman entertained the soldiers of the nearby camp.

In the aftermath, the Torquay Natural History Society organized a systematic excavation of Kent's Cavern under the supervision of the geologist William Pengelly (1812–1894), among others. Buckland presided over the Geological Section at the meeting of the British

Association at Oxford, where McEnery's and Godwin-Austen's findings were confirmed by the new researches; the printed version of the letter presented merely mentioned implements found beneath the unbroken layer of stalagmite. Nevertheless, the repeated work at Kent's Cavern initiated the revival of the debate about human antiquity and the reexamination of many caves. McEnery's voluminous manuscript was finally recovered and published.[31]

I HAVE THUS far followed Buckland's work with a focus on his interpretation of human tools and skeletal remains as discovered by himself and others. It has become clear that the scientific controversy around the contemporaneity of humans with fossil mammals in Europe was linked to debates on transmutationism within the scientific communities and beyond. This, in combination with his nonbiblical catastrophism, has furthered an understanding of why Buckland abided by his verdict on the Red Lady's age, and by his stance toward human antiquity more generally, even when the proantiquity case grew in momentum. In his dealings with transmutationism, Buckland used humor as a form of attack, as he had done in the controversy around the glacial theory. Paradoxically, as a socially accepted and attenuated form of aggression, humor at the same time functioned as a stress releaser and unleashed its cathartic quality to satisfy Buckland's desire for harmony.[32]

Before looking at how the dam of resistance finally gave way to the force of evidence in the human-antiquity controversy toward the end of the next chapter, I step back one last time to contemplate Buckland as a person, as inseparably a scientist and a clergyman. One particular instant in Buckland's life as a priest will serve to illustrate his understanding of the social and the natural spheres as structured by the same holy principles and governed by the same divine laws. In Buckland's cosmology, the social or predominantly human realm was deeply affected by changes in the way one thought about the natural or predominantly mineral, plant, and animal realm, both of which were of the same making. Of course, this does not come as a surprise if one considers Buckland's natural theology background. But the consequences this perception had for his thinking about the order of society still need spelling out in the attempt to uncover the "interests" under threat through the "perversion" of geology discussed earlier.

6

The Holy Order of Nature and Society

THE EVENTS DESCRIBED in chapter 5 show that Buckland increasingly belonged to the older generation of geologists, and his catastrophism had grown out-of-date before he had taken on the glacial theory. Although he had fought for the inclusion of science at Oxford in the earlier years of his career, and for a less exclusive participation on the basis of religion, politics, and social privilege through the formation of the British Association for the Advancement of Science in the early 1830s, in the main he was a typical representative of the privileged English gentleman of his time. As Jack Morrell and Arnold Thackray reveal in their history of the young British Association for the Advancement of Science, there had been opposition to its latitudinarian stance from the Tractarian or Oxford movement and the scriptural geologists or biblical literalists. They had feared not only the admission of dissenters to the educational institutions of the Anglican clergy, but also a reversal of authority between religion and science, with the first being of secondary importance in the search for truth as defined by the BAAS, as had increasingly become the case in Germany and France. The BAAS was certainly not the only society established that had as its aim to forge new bonds between different geographical and social locations. A hunger for knowledge brought about literary and philosophical societies in expanding industrial centers and provincial towns. The growing intelligentsia encompassed plantation owners from the empire, bankers, entrepreneurs, engineers, manufacturers, wealthy university teachers and clergymen, physicians, aristocrats, and gentlemen with the necessary money and leisure to engage in science and the fine arts. With it came the growth of scientific publications that profited from steam printing just as much as the expanding periodical press and newspaper industries.[1]

Buckland lived and worked during great changes in British and international life, characterized by the growth of towns with London as the undisputed economic and political center, the improvement of travel and communication by land and water, and the mechanization and rationalization of such industries as textiles, iron, coal, and increasingly agriculture. Social groups were reformatted and sought organization and a right to a say in matters of concern. The outlook of Buckland and his circle had no doubt partly been shaped by the continued strained social peace of their times. They realized the need for moderate reform, but wanted to maintain the existing institutions of church and state. As we have seen for Buckland individually, the BAAS thus tried to create a space for science against constraining reactionary forces, but also without subscribing to radical social reforms. Politically belonging to the more conservative part of the core of gentleman scientists of the BAAS rather than the Whig majority of moderate reformist ideals, Buckland shared with the latter a latitudinarian, Broad Church Anglicanism. Although the association proclaimed that it was nonexclusive and apolitical, socialists, working-class members, Catholics, Jews, ultra-Tories, and Utilitarians were absent from its ranks.[2]

At the other end of the spectrum from the Tractarians and literalists, France, beginning with the threat of the French Revolution and continuing with transmutationist, antireligious, socialist doctrines, was often seen as the source of many evils. When France experienced the July Revolution in Paris in 1830, British radicals were attempting to democratize and secularize Britain.[3] As we have seen, the transmutation combats between Cuvier and Geoffroy Saint-Hilaire flared up once more in the 1830s and also incited English minds. Authorities in church and state felt menaced. England had been shaken awake by a series of agricultural riots and violent protests in the new industrial centers, where the conditions of the poor had become unbearable. The large divide from the enormous wealth of the gentility and the inherited privileges of the aristocracy were resented also by a rising middle class. The Poor

Law Amendment of 1834 aggravated the situation of the destitute, and the First Reform Bill of 1832, which gave more seats to industrial towns and included a greater number of households within the voting public, had not gone far enough. The fact that the Reform Bill included too small a percentage of British adult males incited Chartist reaction.

The production and reception of Buckland's Bridgewater Treatise is illustrative of the area of tension occupied by the gentleman scientists of Buckland's type. It represented a manifesto of the BAAS's stance toward science and its relation to religion. As such, it was very successful because it was widely read and discussed. However, it was also wildly attacked by biblical literalists, on the one hand, and appropriated by radical circles, on the other. Buckland's demonstration of the overall increase in complexity in the fossil record especially lent itself to an argument in favor of the transmutation of species in the infidel press. Among the gentlemen of science and in fashionable society, the book could be read as a specialist tract and a synthesis of science and religion; in the public arena of the BAAS, it was welcomed as representative of the liberal Anglican conciliation; in middle-class homes, it could function as a morally uplifting text suited to reading aloud; in newspapers, factional periodicals, and pamphlets and from the pulpit, it was also attacked as blasphemy; and, at the other end, in radical artisan circles, it could be appropriated for materialist purposes.[4]

In the hungry 1840s, the real social unease was again linked to the heresies of transmutationism when the hideous subject of the transmutation of species was brought into the very parlors of the English with the anonymous *Vestiges of the Natural History of Creation* (1844). The development of organic forms on the basis of a uniformitarian natural law introduced by the author, who forty years later was officially pronounced to be the Scottish publisher and journalist Robert Chambers (1802–1871), was understood as a challenge to the existing social order, suggesting the potential of each individual organism for improvement. As James Secord develops in his cultural history of the production and reception of *Vestiges*, like Buckland's Bridgewater Treatise, it was read in different local and social settings, interpreted in various ways, and adapted to diverse purposes. With its avoidance of direct divine intervention and the rather marginal and apologetic allusions to the Divinity in general, it could certainly be seen as undermining the authority of the church, and although with delay, Adam Sedgwick finally protested in revolt that "the world cannot bear to be turned upside down; and we are ready to wage an internecine war with any violation of our modest principles and social manners."[5]

Chambers's combination of transmutationism with the argument from design was particularly unpalatable for natural theologians who had relied on divine providence as an alternative exactly to the principle of transmutation to explain the progressive complexity exemplified in the fossil record throughout the rock series. In addition, natural theology as advocated by the members of the BAAS in the 1830s and 1840s was increasingly aiming at a secularization of science while maintaining the claim that religion and science would not be at odds. *Vestiges* now seemed to prove that the Tractarian and scriptural geologists' accusation that instead, natural science would lead to materialism and atheism—a fear addressed in Buckland's early letter to the prince regent—had been well founded. As we have seen, Buckland had deconstructed an exact progressionism throughout the fossil record in connection with the theories of transmutation advocated by Lamarck and Geoffroy Saint-Hilaire. He felt the impetus to do so even more strongly with the appearance of *Vestiges*, and in lectures he warned his students against the dangerous theory promoted by the book.[6]

Later in the decade, the Continent was unsettled by revolutions, and in England the discontent grew. On 10 April 1848, London was in great alarm, with 150,000 volunteers sworn in as constables, because of the great Chartist meeting. The Chartists, who were socialist but not irreligious like the more radical freethinkers, demanded universal suffrage for males, equal electoral districts, annual parliaments, the payment of members of Parliament, vote by secret ballot, and abolition of property qualifications for the franchise. Even though the People's Charter had been presented to the House of Commons in 1839, 1842, and 1848, it had merely been ridiculed, despite a large number of bourgeois signatures. Now, to hold back the tides of unrest, the Duke of Wellington had placed his troops in the houses and gardens of Bridge and Parliament streets to be ready in case of emergency.[7]

On Easter Day, 23 April 1848, in the choir of Westminster Abbey, newly reopened after a restoration, Buckland gave a sermon that was clearly meant to remind his congregation of the true order of the world. This order, installed by the only God, was expanding its dominion with the British Empire. The congregation had reason to be proud since "colonial bishops . . . have gone forth under the national sanction of the Government of this country to preach the gospel in many of the extreme regions of the world . . . Never before did the compass of Christianity circumscribe so vast a circle." The hierarchy ordained by God structured the natural as well as the social realms also in "the dark and savage regions of the world," and here, as there, it was a fixed order. Just as there was no transmutation from one species into another in nature, every man and woman had his and her proper place in society:

> The God of Nature has determined that moral and physical inequalities shall not only be inseparable from our humanity, but coextensive with His whole creation. He has also given compensations co-ordinate with these inequalities, working together for the conversation of all orders and degrees in that graduated scale of being which is the great law of God's providence on earth. From the mammoth to the mouse, from the eagle to the humming-bird, from the minnow to the whale, from the monarch to the man, the inhabitants of the earth and air and water form but one vast series of infinite gradations in an endless chain of inequalities of organic structure and of physical perfections . . . Equality of mind or body, or of worldly condition, is as inconsistent with the order of Nature as with the moral laws of God. . . . There may be equality in poverty: equality of riches is impossible. Equality of poverty is the condition of the negro, the bushman, and the Esquimaux. Equality of wealth and property never has and never can exist, except in the imagination of wild transcendental theorists.

The chain of beings appears as a tripartite structure ordering the inorganic cosmos, the natural world of animal species and human "races," and finally the social world of human classes. It was this inflexible and all-pervasive order that gave security and stability to society. The wild imagination of transcendental theorists had to be kept at bay if peace were to be maintained. Thus Buckland warned the people of Britain against Continental threats and reminded them of their due and happy place in a society organized according to the sacred principles of crown and church: "In the years of peril and perturbation which agitated Europe half a century ago, it was the personal character of the king of this country (King George III.) which, under Providence, was mainly instrumental to preserve us from the sanguinary revolutions which then overran the fairest part of the Continent." The queen would once again, with God's blessing, preserve Britain from "the returning hurricanes of European political revolution." Each British subject was called upon to do his service to the beloved queen, to protect "that great united kingdom and justly balanced constitution at the head of which a gracious Providence has placed her."[8]

Within a system where the natural and the social were of one and the same making, terms from one realm, such as *hurricanes*, smoothly entered the other. Britain did not and was not going to have a history marked by *revolutions*, a term used for both geological and social upheavals. Where politics was concerned, Buckland sided with "the uniformitarians" or even "transmutationists," recognizing the necessity of gradual adjustment to changing circumstances, the traditional recipe of British stability and success. Whereas many radical transmutationists had no need of cataclysms in their science, they regarded them as a social necessity. Buckland, to the contrary, preferred smooth and little-by-little progress in political matters. However, he saw no such way leading either from peasant to monarch or from mouse to mammoth (who would speak of elephants knowing of mammoths?). What we have seen with regard to his belief in bounded variability within the natural entity of species was true also for social entities; change of station could only happen within certain bounds without disrupting the entire system.

The utilitarianism of early industrial Britain taught taking advantage of nature, while the duty of the gentleman scientist was also to employ science for the general good, as well as to educate the public and to spread and maintain the belief in a social and natural order that represented God's will. According to the same logic, the British Empire had a responsibility to distribute Victorian ethics beyond the British Isles. Buckland lived up to the expectation both in his role as scientist, as illustrated, for example, in the Bridgewater Treatise, and in his role as clergyman. For that purpose, a sermon was an effective means in Victorian England since, as George Young puts it, "the sermon was the standard vehicle of serious truth, and to the expositions and injunctions of their writers and statesmen the Victorian public brought the same hopeful determination to be instructed, and to be elevated, which held them attentive to the pleadings, denunciations, and commonplaces of their preachers."[9]

Especially in the advanced stage of his career, Buckland increasingly regarded science as something also meant for practical application. He worked on the draining of marshes and bogs, and he was a member of the Royal Agricultural Society, as well as an honorary member of the Institution of Civil Engineers. He even bought some clay land near Oxford and turned it into a model sanitary village. He personally instructed farmers and circulated papers on agriculture, such as one on the problem of the potato disease. He further experimented on different sorts of wheat and barley and pioneered the use of coprolites as fertilizer. As a firm believer in progress, he trusted in the ability of science to provide remedies for hunger and disease, which he recognized as side effects of industrialization and population growth.[10]

A foreshadowing of Buckland's taste for the utilitarian potential of mineralogy and geology had been given in the Bridgewater Treatise and when members of the British Association for the Advancement of Science, accompanied by a large crowd, visited Dudley Caverns in 1839. The owner of the limestone quarries had made a fortune on iron smelting. The visitors traveled in boats on the canal; rockets of red and blue light were lit, "the effect of which was striking and magnificent in the extreme" and opened to the eye "scenes of *indescribable* grandeur" of "the immeasurable sea of light and flame," of "the fairy-like spectacle."[11] Buckland lectured for over an hour, pulling all the strings of his great oratorical talent and humor. He declared to his large audience in this artificial subterranean dome that the richness of Britain's soil in iron, coal, and limestone was no coincidence, but a sure sign of divine providence and of God's intention to make Britain the most powerful and the richest nation on earth. After all, Buckland argued, it was due to iron that humans rose to the dominion of the earth and to the high culture they currently enjoyed. He reminded his audience of the inexhaustible source of power iron possessed through its magnetism, and of how it guided the pioneers' way across the waters to explore new land. Continuing on this note, the crowd left the caverns singing "God Save the Queen."[12]

This is the image Buckland wanted for his geology and his scientific persona, not the "perversions" and dangers to the status quo dormant in the concepts of human antiquity and transmutationism. Symbolized by the candle he brought into Kirkdale Cave, and in a more advanced stage of his career by the fireworks at Dudley Caverns, the knowledge his researches brought forward would be used for the best of empire, crown, and church. Though in their prime not lacking in surprises, his visions in the enlightened darkness of the cave were ultimately of a conservative nature. He had been a knight, not a revolutionary—he was a dean, not a Chartist. Geology meant welfare tied to a social stability based on a hierarchy that was increasingly in need of protection through education and entertainment. As we have seen throughout the preceding chapters, Buckland was no exception in that his science was affected by his interrelated professional, religious, and social positions.

Around the time of the Dudley Caverns excursion, the last major threat to Buckland's negation of human antiquity took shape. In France, Jacques Boucher de Crèvecoeur de Perthes (1788–1868) published the first of his three volumes of *Antiquités celtiques et antédiluviennes* (1847–1849). The life of the romantic writer of novels, travelogues, political pamphlets, tragedies, and satires as told by Claudine Cohen and Jean-Jacques Hublin itself is the stuff of a romance. Most important in this context, Boucher de Perthes was also an "amateur" scientist with a particular interest in archeology. In 1837, he had begun his work in the quarries of the Somme Valley, where he discovered different kinds of flint instruments that he classified as antediluvian and Celtic, pottery, and fossil animal bones. The combination of typology, stratigraphy, and paleontology allowed the reconstruction of a sequence of human inhabitants of the valley from the antediluvians to the (pre-)Celts, Gallo-Celts, Gallo-Romans, and Romans to the people of the Middle Ages and modern times.

However, far from postulating such a long pedigree for modern humans, Boucher de Perthes regarded his antediluvian stone axes as the products of an extinct human race, exterminated by a geological catastrophe (although Boucher de Perthes assumed the passing on of the souls). Following this as well as earlier cataclysms of total annihilation, there had been a special creation of modified and well-adapted animal species and humans alike. Yet from the catastrophes that followed there had always been survivors, as was recorded in historical documents. Boucher de Perthes linked the antediluvians with Adam's clan, the postdiluvians with that of Noah (or the antediluvians with a pre-Adamic creation and the postdiluvians with Adamic creation). Noah's three sons Sem, Cham, and Japhet had given rise to the three main human races, Mongolian (yellow), African (black), and Caucasian (white).[13]

In 1844, the Académie des Sciences in Paris reacted negatively to his finds: Some of the discoveries were mere plays of nature, while the genuine ones were not as old as Boucher de Perthes claimed. In addition to his outdated geological views and the inconsistencies within the text, Boucher de Perthes had also identified unworked flints as tools, supposedly human-made animals and humans cut out of stone, as well as allegedly religious totems. He even claimed to have found evidence for early hieroglyphs. He had furthermore been the victim of frauds by his workers, who fabricated some of the flint tools.[14] Therefore, if Buckland was aware of the new discoveries made by Boucher de Perthes, he was reassured in his stance toward the question of human antiquity one last time by the scientific community's reaction. Buckland was effectively removed from the scientific arena by the close of the 1840s and died in August 1856, three years before consensus was reached on the contemporaneity of humans and the fossil mammals of Quaternary Europe, in the process of which, as discussed in the remainder of this chapter, Boucher de Perthes's discoveries in fact played a key role.

However, the very year *Vestiges* appeared, and the majority of the Parisian Académie des Sciences rebuffed Boucher de Perthes's discoveries, Buckland applied his full intuition and experience to human bones—the only such instance, at least to my knowledge the only documented one. In front of the British Archaeological Association, he bragged about his twenty years of experience with bones long buried in the earth, examining state of petrifaction, impregnation, decay, friability, and dampness. He inferred what the Ancient British ate from the ways in which the teeth were abraded and even solved the old puzzle of the origin of a peculiar kind of soil in tumuli when he noticed the presence of earthworms. As he had elsewhere reconstructed the behavior of extinct animals by inference from their living cousins, he now drew parallels to "modern uncivilized races of men." But in this context there was no question of the human bones and grave goods pertaining to antediluvian times. They were debated in an archeological context as being the traces left by the Ancient British, Celts, Romans, Saxons, and Angles.[15]

To the contrary, motivated by the accumulating evidence in favor of the coexistence of humans with the extinct Pleistocene fauna, Pengelly and Falconer joined forces in 1858 to excavate the newly discovered Brixham Cave near Torquay, supported by the local Torquay Natural History Society, the Geological Society, and the Royal Society. To this purpose, a committee of leading experts was put together, including, besides Falconer and Pengelly, Ramsay, lecturer in geology at the Government School of Mines and member of the Geological Survey, the geologist Joseph Prestwich (1812–1896), Lyell, Richard Owen (1804–1892), the great comparative anatomist or "British Cuvier," and Godwin-Austen. The same year, Falconer, president of the excavating commission, Pengelly, and Ramsay presented their first results to the British Association. They claimed the contemporaneity of the flint tools discovered with extinct animal species.[16]

Pengelly had used a meticulous method of excavation, removing the cave floor layer by layer and carefully noting the horizontal and vertical location of each fossil before its removal. Compared with the then-predominant procedure of digging one or more vertical shafts, this excavation technique provided more reliable information on the association of fossils and strata. Each fossil was carefully numbered, described in notebooks, and stored in boxes. The names of the members of the committee and the systematic method applied, as well as the fact that Brixham Cave had not been previously ransacked, seemed to guarantee the quality of the work. In spite of this, Owen, as well as Prestwich, treasurer of the committee, still doubted the implements' antiquity. Nonetheless, Pengelly felt that the dam of resistance was giving way.[17]

Later the same year, Falconer visited the Continent because of poor health and became acquainted with Boucher de Perthes's collection. In a letter to Prestwich, Falconer urged the persistent skeptic to come and see for himself. Falconer described the difficult situation that "Abbeville is an out-of-the-way place, very little visited, and the French savants who meet him in Paris laugh at Monsieur de Perthes and his researches."[18] Nonetheless, although he disagreed on several points with the French "amateur" (hieroglyphs and industrial interpretation of various other objects), Falconer was impressed by the collection, which strengthened his determination to carry on the work at Brixham Cave. His letter was successful, and in April 1859 Prestwich, Britain's most renowned Quaternary geologist, later joined by John Evans, expert on stone implements, visited the valley of the Somme.[19]

Here they found clear sections of undisturbed strata in gravel pits, not in caves but in open land, and met with flint implements in situ. On his return to London, Prestwich brought his conclusion concerning the age and genuineness of the Boucher de Perthes collection before the Royal Society, as well as the Geological Society, while Evans presented a paper to the Society of Antiquaries. Lyell, after he had followed the trail of the curious to the Somme Valley, declared his acceptance of human antiquity at the Twenty-ninth Congress of the British Association for the Advancement of Science in Aberdeen in September 1859, where he presided over the Geological Section and Prince Albert was general president. Prestwich also forwarded a paper to the French Académie des Sciences. This led Albert Jean Gaudry (1827–1908), paleontologist at the Muséum d'Histoire Naturelle, the anthropologist Jean Louis Armand de Quatrefages (1810–1892), Édouard Lartet, and others to visit Abbeville and Amiens.[20] Consensus was reached, and at this point Falconer's reminiscence fits into the chronological narrative:

> Viewed in the light of a strictly physical inquiry, the chief rational argument in support of the opinion that the advent of man upon the earth dates from a very modern epoch, was first, the negative evidence in the non-occurrence of human relics [i.e., uncontroversial fossil human bones], and next the fact, that taking him in conjunction with the Mammals, with whom he is now associated, they appeared, as a group, to belong to a new order of things strikingly different from that of the immediately preceding period.[21]

However, the matter of human antiquity had not been viewed in "a strictly physical light" in the first place, but had rather "provok[ed] ridicule, or shock[ed] the strong prejudices on the subject which were then dominant among educated men."[22]

In the paper Prestwich read to the Royal Society on 26 May 1859, he, like Falconer in this quote, observed that the rejection of the evidence for human antiquity had mainly been due to negative evidence (the absence of uncontroversial fossil human bones) and preconceived opinion, apart from the ill-observed and dubious cases. He reopened the discussion on the work of Schmerling at Liège, as well as that of McEnery, Godwin-Austen, and others at Kent's Cavern, indicating that the acceptance of a great human antiquity resulted in a wave of cave explorations and reexaminations throughout Europe: "He [Prestwich] is therefore [because of the acceptance of Boucher de Perthes's discovery] of the opinion that these and many similar cases require reconsideration; and that not only may some of these prove true, but that many others, kept back by doubt or supposed error, will be forthcoming."[23] Paviland Cave did not figure among the sites proposed for

The Holy Order of Nature and Society 115

reexcavation, which forestalled the chance to recognize the skeleton as the lacking piece of evidence. In fact, Prestwich at the same time reaffirmed that no human fossils beyond doubt had so far been discovered, nor did human coexistence with the extinct fauna as evidenced by the human-made tools necessarily imply a very great antiquity. As we will see, it would take the interference of French prehistorians to bring the Red Lady, shielded by Buckland's authoritative interpretation and half-serious "romances," back into the spotlight.

This points to the fact that the question of human antiquity in Europe and in general was in fact not yet closed and, as we will see, in new guises is still under debate.[24] It was a happy ending with a vengeance, since there was no consensus on what exactly *antiquity* meant beyond the existence of humans with the Pleistocene fauna. Furthermore, human antiquity was not established on the basis of human bones or teeth, but on human-made tools. The question whether the absence of fossil human bones could be interpreted as negative evidence, that is, the question whether the absence of evidence in this case was equal to the evidence of absence, which had troubled geologists and archeologists in the course of the establishment of human antiquity, is one that will reappear prominently in the debates on human antiquity of a century later that will be examined in Part II.

BUCKLAND DID NOT live to hear Lyell give his speech at the BAAS in Aberdeen, accepting the Somme Valley evidence for prehistoric humans. How would he have felt when at the same meeting Lyell, whose *Principles of Geology* (1830–1833) had been a great inspiration to Darwin, announced the forthcoming *On the Origin of Species* (1859)? With evolutionism added to uniformitarianism as the new frameworks for geology and biology, Buckland would have turned into a living fossil. However, he might have been positively surprised to learn that even despite the initial tide of indignation, once taken up by members of the scientific elite who were themselves prime examples of the Victorian lifestyle, evolutionism turned out to be less socially explosive than anticipated.[25] Buckland's system of parallel natural and social hierarchies did not simply crumble, although it was increasingly grounded on other than divine authority. In fact, as we will see in Part II and even Part III, the *scala naturae* principle had its reverberations in human phylogenies, in which the Red Lady would also find its place.

∼ Conclusion:
The Red Lady Is No Fossil Man

$\mathbf{B}$UCKLAND'S SPECTACULAR GLIMPSES into long-gone scenes of animal and plant life added wonderful knowledge to the new science of geology. His merit is increased by his skill at communicating his visions to his audiences. His oratorical talent, humor, and lively imagination, facilitated by a classical education, a rather eccentric personality, and a natural theological approach, were instrumental in the promotion of his beloved science at Oxford and beyond. The time's romantic fancy for the imaginative, magical, mystical, and poetic added to the appeal of the scenes resurrected by a historical geology, the practice of which was emphatically empirical and inductive. Buckland's expertise in paleontology, combined with his acute reasoning from different kinds of traces of the past and from present-day analogues, enabled him to reconstruct comprehensive lost worlds. In this, as well as in his insistence on reading the book of nature as a historical romance, his intimacy with ecclesiastical and classical texts and iconography was hardly a hindrance.[1] But where the Red Lady and potentially antediluvian bones in general were concerned, these same skills could be put to work for diversion and obfuscation. His humor amused in both the sense of entertainment and distraction from the real issues at hand. It kept the Red Lady and "her" kind in limbo between the serious and the fanciful. His acquaintance with the Scriptures and classical sources allowed his sharp imagination to create romances to the taste of the time that also built on deep-rooted stereotypes about what were perceived as primitive and peripheral regions of Great Britain and on fantasies about primitive peoples in the New World. Where human bones and artifacts were concerned, his mastery of the evidential paradigm was applied as historical, not geological, analysis, in his role as antiquary, not as paleontologist. If all his charms and quixotic chivalry threatened to fail, he could also resort to unadorned authority.

There were reasons for this beyond the serious geological difficulties associated with human finds, particularly in caves. As an Anglican clergyman and natural scientist at Oxford University, Buckland had to reconcile the new geology with the ecclesiastical and classical traditions. He seems to have achieved a certain harmony through the application of the Bible in the interpretation of geological data, even though this necessitated a nonliteral reading. According to Buckland's exegesis, the vast history of the formation of the earth's crust to its present shape was mostly contained in the single word *beginning* in the book of Genesis. However, as long as he considered the biblical Flood as having been responsible for the deposition of the superficial gravel and the excavation of valleys, the territory covered by the Scriptures strongly overlapped with Quaternary geology, and humankind's place in nature was a particularly delicate topic within this trading zone.

It thus might well be that although Buckland's diluvialism theoretically depended on antediluvian humans, he regarded it as more prudent to postulate antediluvian humans in far-away Asia than to bring the theme to the fore by supporting a high antiquity in Europe. After all, the insights that geological times exceeded the biblical chronology by great measures and that prehistoric worlds full of monsters had preceded the present state of things may have seemed enough rattling of conventional concepts for the time being. It would have been problematic to claim human coexistence with the huge and ravenous beasts that belonged to a premodern time, as once again stressed in the quote from Falconer in the previous chapter. The progressive view of a world designed by God dictated that humans had been created at the very end, after the earth had been carefully prepared for their appearance. But even without this conceptual backdrop, a consideration of the diluvial fauna suggested a different "ecosystem" than the one of the existing species of which humans formed a part.

With his diluvialism, typical of the practice-oriented English geologists organized around Oxford and Cambridge, Buckland was sometimes in opposition to the more theoretical Scottish school of uniformitarians at Edinburgh University. (The theoretical differences between the Wernerian and the Huttonian traditions within the Scottish university were greater, however.) Edinburgh University wished to remove religious concerns from the scientific enterprise and regarded with scorn the institutional connection between the two realms as practiced at Oxford. The exclusion policy of Oxford University was a thorn in the eye of Whig politicians. Nevertheless, as we have seen, the Scottish exponents of uniformitarianism were no less cautious in their treatment of human remains. In Lyell's system, the appearance of humans signified a new order of things—an obvious break in uniformitarian reasoning.

In this respect, Buckland's later nondiluvial catastrophism as advocated in the Bridgewater Treatise (1836), which clearly demarcated the territory into which geology might safely inquire from the exclusive domain of the biblical record, was not altogether different. Here, the biblical Flood had not been of geological importance, and sacred history started where geological interest ended, at the present state of the earth. In this scheme, humans belonged entirely to the realm of the Scriptures and were later creations than the mammoth, hyena, and cave bear, which had lived before the last geologically relevant cataclysm. Like the uniformitarians, the catastrophists tended to shy away from speculating about human origins on scientific grounds. That this stance toward human antiquity transcended the nuanced diluvialism-catastrophism-uniformitarianism debates among the elites becomes clear from Sedgwick's presidential address to the Geological Society in 1831, in which he recanted diluvialism (although not catastrophism). He declared that where humans were concerned, there was no geological continuity whatsoever, nor was it possible (or morally sound) to account for their appearance on earth by any kind of natural law.[2]

It was mainly yet another community that came to trust first in the association of fossil bones of extinct animal species with those of human artifacts or remains. These were the provincial clergymen and other educated men outside the scientific elite, such as Frere, McEnery, Schmerling, and Boucher de Perthes. Significantly, the excavation of Brixham Cave that played such a central role in the establishment of human antiquity was a joint effort of local and London societies. It also marked the beginning of human-origins research as a cooperation between geology, paleontology, and archeology.[3]

Lyell's private journals reveal that the source of his suspicion against catastrophism was, among other things, the fact that the notion of progressive advancement in the state of the earth and the complexity of its creatures could support transmutationism.[4] So could the concept of a human prehistory. Buckland, too, was an enemy of the idea of the transmutation of species as it was advocated inside scientific circles by Lamarck and his friend Geoffroy Saint-Hilaire, and he felt threatened by its politically radical guises. In Buckland's natural system characterized by plan and direction, mankind occupied a special place. On a lower scale, each animal in nature, as well as each individual within a society, was adapted to a particular "ecological" space or station. Education of the public could help maintain this order, and the privileged position of the gentleman of science was also tied to the responsibility to employ science in the service of society. According to the same logic, it was the duty of the British Empire to spread its brand of civilization. The boundaries at stake here were those between social classes, men and women, lower and higher races, and human and animal. For Buckland's vision of Britain and empire, it was essential that these boundaries remain in place, that the parallel orders of nature and society could be claimed as given by God. Because of the interconnectedness between science, politics, and religion, the competing concepts of humankind and its place in nature were central pillars supporting the respective worldviews. A materialist and/or transmutationist model lent itself to different inferences for a social system than a fixed, religiously infused one.

With his Broad Church orientation and moderate Toryism, Buckland was between the extremes of the Tractarians and the social reformers. His geology likewise was under attack from religious conservatives, on the one hand, and more progressive scientists, on the other hand. Among the latter were those adherents to human coexistence with the antediluvian fauna and organic transmutation with whom Buckland never came to agree. However, despite his need for harmony, which heightened his sensibility to those who occupied similar social locations, and which, apart from exterior circumstances, might have been at the root of his search for balances, it cannot be said of Buckland that he generally blindly followed faith.[5] Nonetheless, Buckland's readiness to compromise and perhaps his opportunism won in the long run and made him renounce the glacial theory, as he had given up convincing evidence for antediluvian humans when not supported by his peer network.

Buckland did not live to witness the most crucial events, such as the official acceptance of human antiquity in 1859 or the publication of Darwin's and Alfred Russel Wallace's (1823–1913) theories of evolution by natural selection, but he was well aware of the things at stake and the territories that needed to be defended. Where the Red Lady in particular was concerned, Buckland's judgment of an origin in historical times was not definitively abandoned until after the turn of the twentieth century. As we shall see, at that time, also the sex of the skeleton was once again altered. Obviously, many changes took place between this part about the Red Lady's romance, which revolved around Buckland and the establishment of human antiquity, and the first major turning point in the skeleton's postmortem story, which will be focused on Sollas and the racialist theories of human evolution around the turn of the twentieth century. In fact, the period also marks a turning point in the history of the young science of paleoanthropology. Buckland had never heard of a race called Cro-Magnon, but his very Red Lady would share the honor of being part of the first efforts to systematize what by then had become a confusingly diverse collection of specimens of Paleolithic humans in Europe.

~ II

William Sollas, a Cro-Magnon Man, and Issues of Human Evolution

IN THE EARLY twentieth century, William Buckland's interpretation of the Red Lady as Ancient British was definitively overthrown. William Sollas (1849–1936), one of Buckland's successors to the chair of geology at Oxford University, pronounced it a Cro-Magnon man. It is this reanalysis of the skeleton from Paviland Cave within the new framework of paleoanthropology that will be discussed in what follows. Obviously, this redefinition took place in a very different context with regard to the theoretical and methodological approach, as well as to the institutional and cultural setting. Most important, a new kind of object was under study: fossil human bones. Although already during Buckland's lifetime what in retrospect are genuinely fossil skeletal remains had been discovered, they were not generally recognized as such. In this opening chapter, some of the developments that brought about the paleoanthropological framework into which the Red Lady was integrated will have to be introduced, even if by necessity in a fragmentary way, to understand how categories such as *Cro-Magnon* and *Paleolithic* became available for the Red Lady's reinterpretation.

Part I ended the first major episode in the Red Lady's afterlife at the close of the 1850s, when human antiquity had just been established on the basis of prehistoric tools, and evolutionary theory had made its very convincing entry. Clearly, in addition to the belief in a great age of humankind, an evolutionary framework was needed before paleoanthropology as we know it could take hold. Furthermore, in the course of the second half of the nineteenth century, anthropology began to be institutionalized in Europe and to a lesser extent in the United States. Anthropological congresses were held, and societies, chairs, institutes, and museums, with their associated organs of publication, were founded. As discussed at the beginning of Part I, physical anthropology had as its main object the existing human races, which it described and classified through a variety of instruments for comparative measurement. Its institutions allowed the accumulation and distribution of skeletal remains and ideas on human diversity and its origin.[1]

Many of the anthropologists who were instrumental in the institutionalization of physical anthropology were polygenists, and associated ideas persisted in paleoanthropology. In the course of the first half of the nineteenth century, polygenism gained ground over monogenism through the analysis of large collections of human skulls and their systematic measurement, which seemed to indicate that the different racial types could be traced far back in time. This notion was supported by increasing insights into prehistoric cultures, which were perceived as similar to those of modern savages. Within a nonevolutionary framework, primitive peoples were thought to be the descendants of a pre-Adamic creation, while the ancestors of the civilized races represented a second creation—a view that shows some similarity to the one Jacques Boucher de Perthes had promoted in the first volume of his *Antiquités*. Others, such as John McEnery, dealt with the primitiveness suggested by the tools dug up from the soil by reasoning that the first humans had been civilized but had degenerated later. Still others, such as James Cowles Prichard, saw the development as progressive (see Part I)—a framework that, as we will see in what follows, allowed them to adjust to the new understanding of self, related to apes and only recently emerged from savagery, envisioning progress without limits. However, also after the year 1859, there continued to coexist a welter of different accounts of humankind's place in nature, comprising nuanced stances between the various opposite "isms" of evolutionism and antievolutionism, progressionism and antiprogressionism, monogenism and polygenism, and materialism and spiritualism, which highlights how little self-evident the question had become.[2] Nonetheless, as Stephen Jay Gould has observed for this crucial time in the history of anthropology, the second half of the nineteenth century was not only the era of evolution in anthropology. Another trend, equally irresistible, swept through the human sciences—the allure of numbers, the faith that rigorous measurement could guarantee irrefutable precision, and might mark the transition between subjective speculation and a true science as worthy as Newtonian physics. Evolution and quantification formed an unholy alliance; in a sense, their union forged the first powerful theory of "scientific" racism.[3]

Most important in this context, apart from the eventual convergence of the theoretical framework of evolutionism and a great age of humankind with physical anthropology and its tools of racial classification, fossil human bones—including the Red Lady—entered the "unholy alliance." In the no-longer-Cuvierian framework, *fossil* was not synonymous with *extinct* anymore, and the animal and human bones found in Pleistocene strata might be ancestral to contemporary forms.[4] During the second half of the nineteenth century, these discoveries of skeletal and cultural remains were made most notably in France. As described by Gould, the physical anthropologists who achieved a typology of prehistoric humans from Europe were well trained in the classification of modern humans as distinct races and in their hierarchical systematization by means of a wide array of anthropometric, particularly craniometric, instruments and practices. However, the growing understanding of the diversity of the skeletal remains and even more of prehistoric tools also demanded an integrating chronological system that would draw on knowledge from geology, paleontology, and ethnology, besides archeology and physical anthropology.

A milestone for prehistoric archeology was the work of the French lawyer and paleontologist Édouard Lartet (1801–1871) and the English banker and prehistorian Henry Christy (1810–1865) in the 1860s. They systematically excavated caves situated along the banks of the Vézère, in the Dordogne (Périgord, southwestern France; La Gorge d'Enfer, Laugerie-Haute, La Madeleine, Le Moustier, and Les Eyzies, among others), where they discovered rich industrial beds, including engravings of animals on stone and bone. Édouard Lartet had earlier discovered fossil human bones belonging to seventeen individuals of an anatomically modern type who had coexisted with the cave bear at Aurignac. Because his interpretation of the human bones from Aurignac was doubted, Édouard Lartet had one of them chemically analyzed together with some animal specimens from the same layer. All the bones under consideration turned out to contain similar proportions of nitrogen, which meant that they had lost about the same fraction of gelatin. The argument from the amount of "animal matter" in bones to their relative age has surfaced in Part I, but this seems to have been the first instance in which what has come to be known as the nitrogen test (a more sophisticated method than the tongue test) was applied to samples of different fossil bones from the same site for relative dating. However, not enough was yet known about the loss of gelatin in aging bones to regard this outcome as conclusive evidence for contemporaneity of the bones.[5]

Édouard Lartet also proposed a system of relative dating based on paleontology that correlated the human skeletal and cultural remains to fauna from French cave sites predominant during a certain prehistoric period (Hippopotamus Age, Mammoth/Cave Bear Age, Reindeer Age). Subsequently, the banker, politician, and natural historian John Lubbock (1834–1913) refined the standard archeological system, the Scandinavian tripartite division of prehistory into Stone, Bronze, and Iron ages, by subdividing the Stone Age into the Paleolithic, the Old Stone Age of chipped or flaked stone tools, and the Neolithic, the New Stone Age marked by polished stone tools. In the aftermath, this division came to be used synonymously with the geological designations of Pleistocene and Holocene. Lubbock's *Pre-historic Times* (1865) was an attempt to synthesize knowledge from archeology, ethnology, geology, and anthropology, and to a lesser extent history and philology, in a new "Prehistoric Archaeology."[6]

Further specimens of the modern-looking type that Édouard Lartet had found were brought to light at Solutré in 1866 and at Cro-Magnon (near Les Eyzies) in 1868. At the rear of the Cro-Magnon rock shelter, Louis Lartet (1840–1899), Édouard Lartet's son, unearthed four adults and a fetus that were accompanied by red ochre and seashell necklaces. Not surprisingly, this is where the Red Lady comes in. After their return from the Périgord in 1863, Édouard Lartet and Christy had visited the Oxford University Museum to inspect the remains of the Red Lady. Hugh Falconer, who seems to have been the first to have referred to the Red Lady after Buckland, accompanied them. In his 1860 communication to the Geological Society about the ossiferous caves of the Gower

Peninsula, Falconer had reiterated Buckland's judgment: "Paviland Cave was next referred to; but the author restricted his remarks to the remains of *Elephas primigenius* and human bones that were found in it, and argues that the latter (*i.e.* the skeleton of the 'Red Lady') are of more recent date than the former."[7] Falconer had even suggested that the ivory artifacts were more likely to have been imported from France than to have been produced from the mammoth tusks found in the cave.

To the contrary, not only had Édouard Lartet's and Christy's measurement of the Red Lady's long bones at the Oxford Museum shown that this individual had been of "very great stature," but in hindsight they realized that the character of the Paviland burial corresponded with what they found at Cro-Magnon:

> According to Dr. Buckland's account these Littorina-shells were associated with some remains of a human skeleton, which was regarded by him, we know not why, as that of a woman, and has since become famous as "the Red Lady of Paviland." . . . we assured ourselves, by measurement of some of the long bones of this so-called "Red Lady," that they belonged to an individual of very great stature. Some of these bones are reddened with oxide of iron; and we saw that the remains of ivory implements are in the same state of alteration as the fragments of Elephants' tusks from which, as Dr. Buckland says, the implements had been made. These circumstances are strikingly analogous to what we observe in the Cave at Cro-Magnon, where also were found, besides Periwinkle-shells, the stump of an Elephant's tusk, with implements and ornaments made possibly from the ivory of the same tooth. Moreover, at Cro-Magnon among the human skeletons were found fragments of haematite or red ochre, unctuous to the touch, the contact of which, after the decomposition of the corpses, has left superficial red stains on a skull and femur referred to an old man (see page 74), just as we see on the bones of the so-called "Red Lady of Paviland."[8]

Louis Lartet brought the fossil human bones from the cave of Cro-Magnon to Paris, where the anatomist and anthropologist Paul Broca (1824–1880), the zoologist and anthropologist Armand de Quatrefages (1810–1892), and the anthropologist and ethnographer Jules Ernest Théodore Hamy (1842–1908) examined them. All these contributors, including the Franco-German Franz Pruner-Bey (1808–1882), were trained physicians. When the last skeptics, such as the French prehistorians Gabriel de Mortillet (1821–1898) and Émile Cartailhac (1845–1921), had to give in because of the discovery of further bones of a similar type in the cave of Laugerie-Basse in 1872, the craftspeople and artists of the Reindeer Age were established as a fossil human race.

The descriptions of the industries and fossil bones from the south of France were published collectively in *Reliquiae Aquitanicae; Being Contributions to the Archaeology and Palaeontology of Périgord and the Adjoining Provinces of Southern France* (1875), where Pruner-Bey pronounced the following judgment on the Cro-Magnons, or the Red Lady's people: "Of somber aspect, with an imposing stature, and conscious of his strength, ignorant of moderating his passions by a cultivated *morale*, he could be violent, and turn against the weaker sex the weapon intended to kill his prey."[9] Broca defined the Cro-Magnon race as a curious mixture of superior (big brain, dolichocephalic skull, highly developed frontal region, orthognathism of upper face) and inferior physical traits (great breadth of face, alveolar prognathism, great ascending ramus of lower jaw, pronounced muscular insertion, flattened tibia). He concluded that the Paleolithic troglodytes had in some respects approached the living inferior human races and even the anthropoids. On the other hand, they surpassed the most civilized existing humans in cranial capacity. In parallel to their morphological makeup, Broca imagined them to have been cultural chimeras of high intelligence and artistry, with sculptures testifying to a refined aesthetic sense, living a life of violence and savagery.[10]

De Quatrefages, on the other hand, warned against the characterization of certain morphological traits as primitive. In accordance with his suspicion toward evolutionary interpretations of the fossils, he emphasized that "we ought to be very careful in assigning significance to characters of which in reality we know neither the cause nor the value; and I should always with regret employ expressions liable to deceive others not acquainted with their real value, and likely to lead them to believe in agreements between Man and Ape, of which nobody is thinking."[11] This makes clear that while *Reliquiae Aquitanicae* might be seen as a precursor of contemporary scientific publications on the excavation of and data from particular Paleolithic sites, such as have actually been produced on Paviland Cave and will be discussed in Part III, there was as yet no consensus on an evolutionary framework. Nonetheless, as a comprehensive and, we might say, interdisciplinary work, with contributions from some of the leading experts in prehistoric archeology, physical anthropology, ethnology, paleontology, and geology, *Reliquiae Aquitanicae* stands at the beginning of the systematic scientific endeavor to unravel human prehistory.

Édouard Lartet's and Christy's tentative integration of the Paviland material into the newly established canon did not immediately command agreement. John Evans (1823–1908), in *The Ancient Stone Implements, Weapons, and Ornaments of Great Britain* (1872), in the main supported the report about Paviland Cave Buckland had given in *Reliquiae Diluvianae* (1823). The extensive volume on the forms, possible uses, methods of manufacture, and circumstances of discovery of prehistoric implements, weapons, and ornaments from the remote antiquity of Great Britain was another milestone in the establishment of an archeological framework. For his interpretation and classification of the artifacts, Evans carried out comparisons with Continental Paleolithic and contemporary savage cultures, but he also drew on geological knowledge where the formation of caves and the dating of the archeological material were concerned. With the estimate that the Red Lady belonged to the Neolithic period, Evans followed another British authority, the geologist William Boyd Dawkins (1837–1929). Nonetheless, the way in which Evans expressed his doubts with regard to Buckland's determination of sex is reminiscent of Lartet's and Christy's observations: "In size it equals the largest male skeleton in the Oxford Museum, so that the name of 'red woman' appears misplaced."[12] In the famous *Cave Hunting* (1874), Boyd Dawkins still included Paviland in the chapter "Caves Containing Human Remains of Doubtful Age," suggesting that the Red Lady had been interred in a preexisting Pleistocene deposit. He followed Buckland's interpretation of the cave finds, accepting the disturbed state of the cave, as well as the notorious sheep bone among those of extinct mammals, as evidence against a Pleistocene age of the human bones.[13]

It was only in the groundbreaking *Crania ethnica: Les crânes des races humaines* (1882) that de Quatrefages and Hamy confirmed Lartet's and Christy's earlier suspicion and classified the Red Lady as the first Cro-Magnon ever discovered:

> With some doubt we begin [our discussion of the Cro-Magnon] with the one from Goat's Hole at Paviland, announced by Buckland in 1823. Under archaeological circumstances to some extent comparable to those met with at the by now classic stations of Cro-Magnon, Bruniquel, etc., this naturalist has found in Goat's Hole the left side of the skeleton of a woman, of which he has given no description. These human remains did not appear to him to fulfill the conditions at that time required to be considered as fossil.[14]

This compendium of human races, augmented by 100 plates with lithographs from drawings by Henri Formant and accompanied by close to 500 illustrations throughout the texts, was compiled at the Muséum d'Histoire Naturelle, where de Quatrefages held the chair of anthropology and Hamy was his assistant. Although not in favor of human evolution from simian origins, de Quatrefages was a defender of human antiquity and monogenism, which motivated this comprehensive study of the collections of his museum, the anthropological society in Paris, and other major collections at home and abroad. *Crania ethnica* thus not only cemented the Red Lady's place among the Cro-Magnons, but also added to the work on Paleolithic human races in Europe initiated by *Reliquiae Aquitanicae*. Significantly, the work was subdivided into two parts, the first consisting of a classification of the fossil human races and the second of a classification of the extant human races. De Quatrefages and Hamy allocated the fossil human material from Europe to three races, including those of Canstadt and of Cro-Magnon.[15]

The Canstadt race, which was the oldest and most primitive, was named after a skullcap of uncertain origin—possibly discovered as early as 1700 near Cannstatt (near Stuttgart in Germany)—that showed close similarities to the Feldhofer Neanderthal. The latter skull and postcranial bone fragments had been discovered in the Feldhofer Grotto in the Neander Valley (Germany) in 1856. Thomas Henry Huxley, who, inspired by Charles Darwin, was one of the most prominent early proponents of evolutionism, analyzed a cast of the skull and concluded that it was "the most pithecoid of known human skulls."[16] Nonetheless, since the skullcap was of normal cranial capacity, he classified the Feldhofer Neanderthal as a mere variant of the modern human type. It was closer to the Australian Aborigine, whom Huxley thought occupied the lowest rung in the hierarchy of extant human types, than to the ape. The Neanderthals thus might have evolved into modern human races.

To the contrary, possibly to prevent an evolutionary interpretation, the Berlin pathologist-anthropologist Rudolf Virchow (1821–1902) described the Feldhofer Neanderthal as a pathological aberration of known humans rather than a newly found type and by his great authority silenced many of his opponents. In time, similar finds were made at La Naulette (near Dinant) and Spy (region of Namur) in Belgium in 1866 and 1886, at Krapina (Croatia, 1899), and eventually at La Chapelle-aux-Saints, Le Moustier, and La Quina (1908) and La Ferrassie (1909) in France. Earlier discoveries, such as Philippe-Charles Schmerling's child cranium from Engis (Belgium, 1829–1830) and the female cranium from Forbes Quarry (Gibraltar, 1848), were reevaluated, and it became clear that the Neanderthals were indeed a distinct and fossil human type.[17]

Once different Paleolithic human races were recognized, they became caught up in theories of racial origins. In this respect, de Quatrefages and Hamy made another important observation in *Crania ethnica*. Their analysis of the fossil human bones led them to conclude that the Aryan theory was wrong. Philologists had traced Sanskrit and Indo-European languages back to a hypothetical Aryan tongue spoken in the distant past, which had led to the idea that Aryan peoples had invaded Europe from Asia, marginalizing the original European inhabitants and language that had been close to the Finnish and Basque tongues. The belief seemed substantiated by the biblical stories of the Garden of Eden and the Tower of Babel, as well as by ancient Eastern civilizations (the term *Aryan* is derived from the Sanskrit word for "noble"). New impact was added when the Swede Anders Adolf Retzius (1796–1860) brought the cephalic index into anthropology in the 1840s. He distinguished the longheaded (dolichocephalic) from the shortheaded (brachycephalic) skull type on the basis of the ratio of the width to the length. Shortheads (also called round- or broadheads) came to be associated with a lower position in the hierarchy of human races. In analogy to the language scenario, the idea grew that an autochthonous brachycephalic race had inhabited Europe before the invasion of the Indo-European dolichocephalic race, and that the ancestors of the Basques, Finns, and Lapps had been marginalized by succeeding invasions of the Aryan race.[18]

However, the examination of Paleolithic skulls by de Quatrefages and others did not confirm this, since all the brachycephalic finds they described were younger than the dolichocephalic race of Canstadt. Also, the race of Cro-Magnons, with the so-called Old Man from the Cro-Magnon rock shelter as type specimen, showed that a highly developed dolichocephalic race had been living in Europe already in late Paleolithic times, or long before the great Neolithic migrations.[19] Broca had realized as early as the 1860s that the new finds of fossil humans did not support the Aryan theory. He consoled himself with the fact that the prehistoric typological diversity in Europe was a confirmation of polygenism. He thus emphasized that the Quaternary human remains from Les Eyzies ("Cro-Magnon") belonged to a different race than those from the Belgium caves ("Neanderthal"). Others, to the contrary, tried to maintain the Aryan theory in the face of new evidence. Pruner-Bey, for example, allowed for only one Paleolithic European race, a brachycephalic Mongoloid type, and tried to subsume all dolichocephals under the Neolithic Aryan type. He reconciled this preconception with the analysis of the Cro-Magnons by the suggestion that even though they were dolichocephalic, they resembled recent Estonians (and were therefore Mongoloid). He even judged from the bony palates of the Cro-Magnon specimens that they had spoken neither Aryan nor Semitic but a weak phonology such as Finnish.[20]

The Aryan-invasion theory had political implications. According to Retzius's scheme, the French were typically brachycephalic, while the Germans were dolichocephalic, which adds meaning to the fact that the theory was deconstructed by the French savants, particularly as the Franco-Prussian War (1870–1871) seemed to point to a racial inferiority of the French. However, this was not the end of the Aryan myth. The Indo-European cluster was split into Teutonic/Saxon (dolichocephalic; predominantly England and Germany), Alpine (brachycephalic; Celts, predominantly Ireland, Wales, and France), and Mediterranean (dolichocephalic; Italy, Spain, and so on) toward 1900. The more general Indo-European or Caucasian as Aryan came to be narrowed down to Nordic-Teutonic, with the cephalic index, as in all craniometric determinations of race, as the most important characteristic, although some referred to light skin and hair, blue eyes, and tall stature as additional markers.[21]

In Britain, this marked the apex of a history of Anglo-Saxonism that can be traced back at least to the Reformation, when English scholars had turned to their Germanic roots in the Angles, Saxons, and Jutes to found the English Church on pre–Roman Catholic institutions. Similarly, in the seventeenth century, the defenders of parliamentary rights vis-à-vis the king looked to Anglo-Saxon "original" freedom. This interest in the language and customs of Saxon times grew into the national myth of a specially endowed people of predestined greatness that fed into the expanding imperial pretensions. The Enlightenment flair for history and its belief in a directed progress moving Western societies to ever-higher forms of civilization seemed confirmed by the achievements of industrialization. From the Napoleonic Wars, Britain emerged as the center of Western world hegemony, and in the course of the nineteenth century comparative philology and physical anthropology lent support to the ideology of a superior English race. This kind of biological Anglo-Saxonism, with the English as of a Nordic/Teutonic/Germanic pedigree, not only worked against the rights of the peoples suppressed throughout the distant empire, but also had something to say against the Celtic peoples at home, particularly during the final years troubled by the Irish Home Rule question.[22]

By the time fossil races were integrated into the Aryan theory, the theory, as well as its association with dolicho- and brachycephalic types, existed in many different versions, and there was as yet no consensus on the correlation of geological strata, archeological industries, and human remains. Boyd Dawkins's scenario of the peopling of Europe as explained in *Early Man in Britain* (1880) is indicative of this. It also represents an attempt at synthesizing the practices and then-visible effects of European expansion with the human evolutionary past that will characterize many evolutionary scenarios discussed in the following chapters. Apart from transporting hierarchical relations among the races of Europe through his view of its prehistory, he compared the successive displacements of one race by another in the Paleolithic with, for example, the marginalization of the American Indians under way at his time.[23]

Boyd Dawkins thought that during the European Paleolithic the people who inhabited open sites (drift period) had been succeeded by an invasion of the caveman (cave period). Although the skeletal remains were too fragmentary to allow racial classification (they were both dolichocephalic), the changing animal fauna, climate, and human culture had left abundant traces that suggested that these two had constituted distinct races and had differed from the later Neolithic invaders and from any of the racial types presently found in Europe.[24] Boyd Dawkins considered the Paleolithic caveman the ancestor of the Eskimo, chased out of Europe by the Neolithic Cro-Magnons. The Cro-Magnons, themselves in turn driven from European soil by the later Neolithic invasion of a small, dark, non-Aryan people into Europe, were seen as now represented by the Red Indians. Subsequently, a brachycephalic Celtic race invaded Europe, to which Boyd Dawkins referred as "the van of the Aryan race."[25] This adaptation of the Aryan theory is reminiscent of Pruner-Bey's Mongoloid theory, and Boyd Dawkins, like Broca, opposed the view that the Paleolithic European races had been absorbed by later invaders, and that traces of their heritage could still be found in current European nations.

Boyd Dawkins's rejection of the dominant interpretation of the Red Lady and the Cro-Magnon race, which he defined as Neolithic, was due to his choice of ethnological analogue. Because he identified the European Paleolithic caveman with the recent Eskimo on the basis of similarity in culture, he denied the existence of Paleolithic interments. After all, one of the most striking features of Eskimo culture was their complete disregard for the bodies of the dead, which tended to be eaten by animals or to end up piecemeal on refuse heaps. Consequently, in his race-succession scenario, Boyd Dawkins did not even mention the Red Lady and Paviland Cave as a site of burial, and the Cro-Magnon race in general was defined as Neolithic.[26]

What I have called the race-succession scenario—an idea of progress as geographical replacements of what were perceived as lower by higher races—presented an alternative to the grand model of an overall linear evolutionary sequence from the simplest and most savage to the most elaborate and civilized with regard to both cultural and morphological development, which was favored by many nineteenth-century evolutionists. Like the migratory model, the roots of the "evolutionary" model can be traced back beyond the nineteenth century, but with the advent of prehistoric archeology it was applied to Paleolithic industries, as well as human fossils. The British cultural anthropologist Edward Tylor (1832–1917) is one of the names closely associated with this view. In his last book, *Anthropology: An Introduction to the Study of Man and Civilization* (1881), he gave an excellent summary of the young field to readers who had enjoyed or were enjoying the ordinary higher English education. It provided an overview of the various fields the new anthropology subsumed, such as comparative anatomy of apes and humans, physical anthropology, historical and comparative philology, ethnology, and Paleolithic archeology.[27]

For the twenty-first-century reader, the most striking aspect of this advocacy for anthropology as essential to an integrated education is the fact that Tylor discussed the development of human culture from savage via barbaric to civilized without regard to geography or time. He led the reader through the wonderful progress from the most primitive weapons and tools consisting of pebbles simply picked up from the ground to the achievement of the steam engine by jumping in his illustrative examples between Paleolithic and Neolithic cultures in Europe and the numerous cultures of savages and barbaric tribes all across the world, and between the ancient Eastern and the modern Western civilizations. The story line was not chronological or geographical, but was outlined along the overall gradual improvement in material culture.[28]

Lubbock was also among those British followers of Darwin who held such an "evolutionary" view of human history. To the more strictly archeological part of *Pre-historic Times* (1865) he added chapters on the customs and manners of modern savages because ethnological studies would "throw some light on the remains of savage life in ages long gone by."[29] Indeed, although his general approach to the Paleolithic record was diametrically opposed to Boyd Dawkins's, he agreed that "so far as the present evidence [from the Dordogne] is concerned, it appears to indicate a race of men living almost as some of the Esquimaux do now, and as the Laplanders did a few hundred years ago."[30] Geology was brought in to provide estimates about the age of the Paleolithic human tools, and the anthropological contribution consisted of illustrations and a short discussion of fossil human bones. Lubbock also approached the question of the applicability of the Darwinian theory to human evolution and, because he was something of a pioneer in experimental study of animal behavior, even drew analogies between ape behavior and human culture, such as in tool use and the building of shelters. Overall, Lubbock's aim was to illustrate the gradual progress in culture from apes via the different Paleolithic and Neolithic industries up to the highest civilizations of his time, a seemingly inevitable march on which the human races had forged ahead independently and to unequal extents.[31]

Lubbock's speculations highlight the limiting factor that the fossils so far discovered all seemed rather modern looking. They could not serve as true links between the anthropoid apes and the lowest savages—both the Cro-Magnons and the Neanderthals were considered to be well within the range of human variability. This problem was handled creatively by de Mortillet, who also introduced the archeological system that came to dominate the subdivision of Lubbock's Paleolithic. Like Boyd Dawkins, de Mortillet initially rejected the Cro-Magnons as a Paleolithic race. Out of antireligious sentiments, he did not believe in Paleolithic burials. However, his interpretive framework was closer to that of Lubbock than to that of Boyd Dawkins. De Mortillet, too, was a believer in the universal law of progress, a belief he incorporated into the prehistoric section of the Universal Exhibition in Paris in 1867. The exhibits on the history of industry (*l'histoire du travail*) were arranged chronologically and according to nations. On the subject of prehistory, the visitors were presented with artifacts from the Paleolithic, Neolithic, dolmen, lake-dweller, Celtic, Gaul, and Gallo-Roman periods. In the visitor's guide, de Mortillet included other galleries considered of interest to the subject, such as those on the colonies of France and of other European nations that illustrated the material cultures of savages. He emphasized that the exhibits were intended to acquaint the visitor with the laws of human evolution, which were the law of human progress, the law of similar developments in all human races, and the great antiquity of humankind.[32]

As discussed earlier for the compilation of *Crania ethnica*, also where the classification of Paleolithic artifacts was concerned, prehistoric archeologists analyzed museum and private collections and often visited the sites. De Mortillet was in a particularly favorable situation through his immediate experience with large collections of Paleolithic artifacts in his work for the Universal Exhibition and his involvement in the organization of the Musée des Antiquités Nationales in Saint-Germain-en-Laye. He established a classification system that in part is still used today. He subdivided the Paleolithic period into cultures named after French cave sites: the oldest Paleolithic culture, the Chellean (also Acheulean), consisted of tools made by chipping off small flakes to put a sharp edge on the original core of the flint; the Mousterian referred to tools made from the flakes rather than the core; the Solutrean showed beautifully shaped blades of chipped stone; and the last Paleolithic culture, the Magdalenian, which corresponded to Lartet's Reindeer Period, was characterized by sophisticated artistry in bone.

In *Le préhistorique: Antiquité de l'homme* (1883), de Mortillet consciously imported geological practices into prehistoric archeology: "Following an excellent method applied in geology—one is not to forget that paleoethnology is directly derived from geology—I have given each period the name of a very typical site."[33] The designation *paleoethnology* for the new science had been introduced as an alternative to the somewhat clumsy *prehistoric archaeology* at the foundation meeting of the International Congress of Anthropology and Prehistoric Archaeology (La Spezzia, 1865). The term was understood in a very broad sense as the study of human origins and evolution antedating written documents. In *Le Préhistorique*, de Mortillet thus gave an outline of the gradual and progressive evolution of human culture and morphology from the Eolithic up to the Neolithic, a development he thought had taken place in Europe. De Mortillet introduced the term *Eolithique* to refer to the Tertiary and the traces of the most primitive tool culture it was supposed to contain (so-called eoliths as opposed to paleoliths).

In his scenario, a hypothetical missing link between the highest anthropoid ape and the lowest savage, to which de Mortillet gave the genus name *Anthropopithecus*, had been the shaper of these eoliths. This genus had evolved into Neanderthal Man and eventually the Cro-Magnons—a morphological progress paralleled by the development through the material cultures outlined earlier and accompanied by changes in geology, fauna, and flora. De Mortillet allowed for only one invasion from the East into Europe, that of the Neolithic agriculturalists, who had brought with them the seeds of religion—and thus the practice of burying their dead. For de Mortillet, a radical socialist and staunch opponent of the church, which signified antiprogress, this invasion had a negative connotation. It meant the corruption of the last true Paleolithic race of autochthonous Cro-Magnons and their dilution to the degree found in the morphology of the recent European population.[34]

The German biologist Ernst Haeckel (1834–1919), too, had quite precise ideas about what was needed to finally put the truth of human evolution from an apelike precursor on a solid empirical basis. Although he considered himself a staunch Darwinian, Haeckel's strong progressionism led him to develop a tree of life that had a main stem with mankind at its apex, and he offered predominantly "Lamarckian" explanations for this progress to take place. In contrast to de Mortillet, however, he thought that the first hominids had once originated in southern Asia, on a hypothetical continent he called Lemuria, because he considered the orangutan brain to be closest to that of modern humans. In *Generelle Morphologie der Organismen* (1866) and *Natürliche Schöpfungs-Geschichte* (1868), he created a hypothetical missing link between Asian ape and human savage, the *Pithecanthropi alali*. In his science fiction, *Pithecanthropus* had evolved into *Homo primigenius*, the first humans on Lemuria, who in turn gave rise to the human subgroups of the Lissotrichi (straight-haired) and Ulotrichi (woolly haired). These spread across the earth, splitting into several branches, with the Indo-Germanic as a Lissotrichi offshoot.

Later Haeckel realized that the Lemuria hypothesis had been rendered improbable by the new geological insights and postulated India as the birthplace of humankind, particularly the Himalaya region. He claimed that the inhabitants of Ceylon (Weddas), today's Sri Lanka, were the most primitive living humans and that they provided a missing link to the highest apes. They could stand in as models for the earliest humanlike forms. Although this suggests a common human origin far back in prehistory, Haeckel in fact regarded the recent human types as representing different species that could be grouped into different genera. The various human species not only had reached the characteristics that differentiate humans from animals at different points in the course of evolutionary time but had also perfected them to different degrees up to the present. In his scenario of human racial evolution, Haeckel strongly relied on the analogy to philology (which he called *vergleichende Anatomie der Sprache* or "comparative anatomy of language" and *Ontogenie und Phylogenie der Sprache* or "ontogeny and phylogeny of language"). It was from this analogy that he claimed a speechless human ancestor, *Pithecanthropus alalus*. Also on the basis of the language analogy, Haeckel went so far as to suggest that the living human types had originated from different *Pithecanthropi* precursors. Comparative philological studies that claimed that the current languages could not be reduced to one origin functioned as Haeckel's "proof" for polygenism.[35]

In an idiosyncratic interpretation of Darwin's intertribal selection (restricted to prehistoric times), Haeckel justified contemporary imperialism and genocide as natural processes that had driven human evolution since primordial times. Within this framework, the progress of human anatomy and culture had depended and still did depend on the replacement of lower by higher human species. Haeckel claimed that "the woolly haired species" were unable to develop high culture, and it was therefore left to "the midlandic species," and among it the Indo-Germanic race in particular, to spread its dominion across the globe by means of its highly developed brain and civilization. Haeckel prophesied that the species of the temperate zones would exterminate the less developed human species everywhere except in the inhospitable tropical and polar regions, a process he saw at work against the American Indian, the Australian Aborigine, the Hottentot, the Papuas, and others, who he thought were approaching their complete extermination.[36]

The reason why it is worth looking at Haeckel's prophecies is that they were self-fulfilling in more than one respect. His highly influential mixture of a linear evolution of hominids with parallel evolution in the human racial branches, and a teleological concept of progress in the direction of higher intelligence through interracial competition, will come to mind in the discussion of Sollas's model. Haeckel and Huxley, under whose influence Sollas came at the Royal School of Mines, were friends, and *Natürliche Schöpfungs-Geschichte* had been translated into English as *The History of Creation* under the supervision of Huxley's disciple Edwin Ray Lankester (1847–1929), who for a short interval was Sollas's colleague at Oxford University.[37] Furthermore, shortly before Sollas's reanalysis of the Red Lady, Haeckel's prediction of an Asian missing link between ape and humans came true. With this discovery, the point was reached where the notion of a human prehistory, evolutionary theory, the quantitative tools of physical anthropology, and fossil human bones converged in paleoanthropology, even though physical anthropology and paleoanthropology were still mainly the prerogative of men trained in the medical sciences and to a lesser extent in zoology and geology. However, the questions and controversies caused by the interest in human prehistory in the latter decades of the nineteenth century were carried into the twentieth: Had progress been brought about by continuous development of local types or through replacement of stagnating by more advanced types? How much diversity was evidenced in the fossil record? Where did hominids originate? Had there been hominids in Europe already in the Tertiary, or had they arrived in the Quaternary? What was the status of comparative ethnology in the interpretation of prehistoric artifacts and human bones? The answers given to these questions had a bearing on how fossil human bones—newly discovered or reexamined—were read, including the Red Lady.

☙ As will be explored in what follows, Sollas's work in anthropology, and on the Red Lady in particular, took place during a time of transition with regard to these questions. Paleoanthropology had barely gained the necessary institutional, theoretical, and methodological background and, most of all, prehistoric human bones (including pre–*Homo sapiens*) to work with when it experienced a shift in outlook toward the hominid phylogeny that this evidence seemed to support. Although a linear view of evolution from fossil ape to modern humans gained wide acceptance in the aftermath of the discovery of a possible missing link in Asia, the unilinear phylogeny was soon challenged and generally rejected in favor of dendritical structures. Sollas's work on the Red Lady and its place in his larger view of human phylogeny and evolution were caught up in these concerns. While in Part I the Red Lady came to stand in for fossil human remains in the controversies surrounding the question of human antiquity, in which "her" discoverer was a central player, Part II will see the Red Lady as representative of "his" Paleolithic race of Cro-Magnons. What did it mean to be Cro-Magnon according to Sollas and more generally? What was the place of this Paleolithic race in hominid phylogenies, and what role did it play in evolutionary scenarios? And how was the meaning of *Cro-Magnon* affected by the developments alluded to?

Like Part I, this second major episode in the Red Lady's biography will begin with an introduction to her interpreter (Chapter 7). As a successor to Buckland as professor of geology at Oxford University, Sollas was in the midst of imperial concerns and experienced Oxford's war effort. His position at Oxford also brought him close to the remains of the Red Lady. In fact, Sollas made the acquaintance of the Red Lady early in his scientific education, when the skeleton was still considered of doubtful identity. He could not have known at that point that years later, in the aftermath of Henri Breuil's (1877–1961) visit to the Oxford Museum, he would lead a reexcavation of the cave and carry out a reevaluation of its contents (Chapter 9).

The Red Lady was defined by the characteristics of the Cro-Magnons. The primitive physical traits and savage behaviors that had been attributed to the race by anthropologists treated in this chapter were no longer emphasized in the early twentieth century. As the category *Cro-Magnon* came to be defined through differentiation from the Neanderthals, the more bestial aspects were predominantly reserved for the latter type. The new image of the Neanderthals and by inference the Cro-Magnons was crucially driven by the work of Marcellin Boule on the La Chapelle-aux-Saints Neanderthal. His distancing of the Neanderthals from modern humans, represented by the Cro-Magnons, initiated a change in hominid phylogeny from unilinear to branched. Sollas was clearly among those who were influenced by Boule's views, and he grappled with the consequences for his interpretation of the fossil record (Chapter 10).

On a larger scale, Sollas assigned the Red Lady a place in his system of hominid evolution. Sollas's ideas, like Buckland's, were far from static. Rather, he went through considerable changes in his theoretical outlook on human diversity. Chapter 8 will introduce the quantitative, technology- and number-driven approach later applied to the Red Lady's bones on which his early unilinear phylogeny was based. This work was followed by *Ancient Hunters and Their Modern Representatives* (1911), a grand synthesis of geology, ethnology, and prehistoric archeology in which the Red Lady was assigned its specific place in the wider evolutionary framework. It expounded Sollas's race-succession scenario for human racial evolution and geographical distribution, which, despite the precedents discussed in this chapter, was not the dominant interpretation of either the European anthropological or archeological Paleolithic record at the time.

The differences between consecutive editions of *Ancient Hunters* (1911, 1915, 1924) will bring back to the fore the linearity-diversity/continuity-replacement debates discussed earlier and highlight that in this respect views on hominid evolution in general must be treated separately from human racial evolution. Thus, although Sollas's notion that migration had played a paramount role in the long independent histories of the human races was in harmony with the trend toward a diversification of the pre-*sapiens* hominid phylogeny, Sollas resisted the latter up to the third edition of *Ancient Hunters* (1924) (Chapter 10). That a race-succession scenario as advanced by Sollas could accommodate not only a new model of cultural evolution that emphasized migration and diffusion of inventions, as introduced by the Australian-born Grafton Elliot Smith (1871–1937), but also the dendritical hominid phylogeny and the associated central Asia hypothesis, will be discussed in Chapter 11.

The Red Lady's genealogy, racial characteristics, and affiliations were defined by these notions and the changes they underwent. In fact, in an outgrowth of what came to be referred to as the Praesapiens theory, the Red Lady played the part of an ancestor in the noble pedigree of "the British" reaching back to the Pliocene. Arthur Keith (1866–1955) promoted the idea that modern human anatomy and the parallel evolutionary lines of living human races were of very great antiquity. Consequently, the known hominid fossils could no longer serve as ancestors of modern humans because they were either too young or too primitive. In phylogenies such as that of Keith, these were therefore relegated to side branches of the hominid family tree. For the theories of human evolution at the time, the presumed antiquity of the first hominids in Europe and ideas on their "modernity" or "primitiveness" were thus central. Those who accepted the supposedly human-made tools or bones of Tertiary Man in Europe in contrast to de Mortillet now tended to see them as supportive of a more diverse hominid phylogeny and in particular the wide gap between the Cro-Magnons and the Neanderthals (Chapter 12).

The theorizing on the antiquity of anatomically modern humans and of the human racial lines and on hominid and racial evolution was linked to imperialism and war of that time. Throughout the following chapters, beginning with Sollas's experiences at Oxford University, the context of imperialism and World War I will enter the discussion. Sollas's theory of human racial evolution, in which the Cro-Magnons at times appeared as winners in the interracial struggle for European territory, as well as those of his medically trained British peers, in the main Elliot Smith and Keith, must be viewed in this context. Their blending of anthropology and politics will bring to the fore the question of the ethical implications of their science (Chapters 11 and 12). Both of them, as well as the other main British scientists occupied with the analysis of fossil human bones and therefore discussed in what follows, such as Lankester and Arthur Smith Woodward (1864–1944), were powerful figures of science at the time and were distinguished by knighthoods.

7

William Sollas

WILLIAM JOHNSON SOLLAS (Figure 7.1) was born in Birmingham on 30 May 1849, so he was Buckland's contemporary for about seven years. After the City of London School, he went to the Royal College of Chemistry in 1865, where his father was registrar. When he had worked for two years in the laboratory of Edward Frankland (1825–1899), he gained a scholarship to the Royal School of Mines, where he attended the courses of the illustrious Andrew Crombie Ramsay and Huxley. In Ramsay's geology lectures, Sollas heard of Buckland and the Red Lady for the first time. In the course of a general outline of the history of geology, Ramsay, who, as we have seen in Part I, had been one of the geologists immediately involved in the processes that led to the establishment of human antiquity, talked about the fossil human remains from caves. Sollas and his fellow students learned of Schmerling and his work in Liège, the Neanderthal finds from Gibraltar, the excavation of Brixham Cave, and McEnery's and succeeding work at Kent's Hole. Ramsay gave Buckland credit for having been the first to note the mammalian remains in limestone caves. About Paviland Cave and the red woman Sollas entered in his notebook that

> a few implements were found with bones of extinct mammalia, and the skeleton of a woman. (this was some time ago.) wh: in consequence of its red color obtained for her the name of the "Red

Figure 7.1. Photographic portrait of William Johnson Sollas, 22 February 1921?; courtesy of The American Philosophical Society Archives.

Lady of Paviland." it is now many years since this discovery was made, and the general opinion was, that the cave was a place of sepulture, whether or not this is the case it is impossible to say since no person of sufficient authority was present when the bones were exhumed.[1]

It seems not only as though Buckland would eventually have suffered from the same arrogance he had sometimes shown his less accredited contemporaries, but also that Ramsay was wrong in his verdict, since his own student would confirm Buckland's opinion that Paviland Cave had been a place of burial. Ramsay also lectured on the later prehistoric occupants of Europe, who had left their traces in the Dordogne in France, at Swiss lake dwellings, and in tumuli and kitchen middens of northern Europe. Ten years after Charles Lyell's pronouncement of the former coexistence of post-Pliocene mammals with humans in Europe and the publication of Darwin's *On the Origin of Species* (1859), Ramsay's lectures presented the contemporaneity of humans and the extinct European Pleistocene fauna as uncontroversial. From Ramsay, Sollas further learned of the Aryan theory, which in his teacher's version meant that the Aryans had arrived in Europe from Asia some 4,000 years ago and had replaced and driven north and south the people from whom the current Laps and Basques descended.[2]

At the same school, Huxley taught Sollas and his classmates the basics of the physiology, morphology, distribution, evolution, and etiology of humans. From Huxley, Sollas heard that because of their interbreeding capacity, all humans belonged to a single species, which could be subdivided into Ulotrichi and Lichotrichi, the crisp and woolly haired versus the smooth, straight or wavy, and fine haired. Skin and eye color, body size, facial angle (orthognathism versus prognathism), steatopygia, form and size of sexual organs, coarseness of jaw and teeth, cephalic index (brachy- versus dolichocephaly), and brain capacity could be used for finer distinctions. Thus, among the Ulotrichi, the Negroes and Negritoes could be distinguished from the smaller Bushmen, whose women showed steatopygia and protruding sexual organs. As Sollas understood it, Huxley subdivided the Lichotrichi into the Australoid type ("chocolate" color, coarse jaw and teeth, prognathous), the Mongoloid type (yellow skin), the Xanthochroi (the fair-colored Europeans), and the Melanochroi (the Europeans of darker color).

The human male differed from the female in pelvis size, and humans from animals in their possession of a chin, with extraordinarily great men such as Goethe and Napoleon sporting particularly impressive specimens. In Huxley's view, it was the Australian Aborigine, and not the Negro, as was sometimes held, and certainly not the fine and handsome Asian, who most closely approached the ape. The fossil race of the south of France (Cro-Magnon), who with their compressed tibia resembled the modern Negroes, and the Neanderthals had already been very similar to us, and although the Neanderthals clearly represented the lowest human form, their large crania suggested that they might have been very intelligent.[3]

Huxley, who thought of himself as Darwinian, did not consider climate alone responsible for the differentiation of the human races, since similar forms were found under widely different circumstances and vice versa; nor was language a helpful criterion for the classification of humanity according to race. Although the sun tanned the skin, exercise toned the body, and diet changed one's appearance, such extrinsic factors would not be handed on to the next generation. It was the intrinsic factors that created variance between parents and offspring and among offspring that was passed on. In contrast to Darwin, however, Huxley was of the opinion that *natura fuit saltus*, that evolution took place by discontinuous large changes that transformed organisms at once. Whether because of extrinsic or intrinsic factors, the races also showed differences in temperament. The Australians and Negroes were passive by nature, while the Xanthochroi and the Mongoloids had so far won the competition between the racial groups.[4]

As we shall see, the teachings of both Ramsay and Huxley had a strong influence on Sollas's thinking and work, so much so that the former keeper of the geological section of the British Museum, Smith Woodward, observed in his obituary: "The broad basis of the scientific training he received here, and its eminently practical nature, had a life-long influence on the philosophical character of his original work and on the experimental ingenuity which he applied to the solution of his more difficult problems."[5] Smith Woodward paid tribute to the practical training Sollas enjoyed at the Royal School of Mines. However, it was not only the content of the lessons and hands-on experience with fossils but also the attitude toward education and science in general that turned Huxley into an inspiration for future scientists.[6]

In Part I we have seen that Buckland and his gentleman peers created an image of themselves as endowed with special faculties, or genius, that offered them insight into the secrets of nature. In contrast, in a talk titled "On the Educational Value of the Natural History Sciences" Huxley had given in St. Martin's Hall the year he became lecturer in natural history and paleontology at the Government School of Mines (1854), he claimed that the times in which one believed that science could only be carried out by a priestly caste of specially endowed minds rightly belonged to the past:

> Science is, I believe, nothing but trained and organised common sense, differing from the latter only as a veteran may differ from a raw recruit: and its methods differ from those of common sense only so far as the guardsman's cut and thrust differ from the

manner in which a savage wields his club ... So, the vast results obtained by Science are won by no mystical faculties, by no mental processes, other than those which are practised by every one of us, in the humblest and meanest affairs of life.[7]

To return once again to Huxley's later treatise on the method of Zadig, retrospective prophecy as a function of the historical sciences was defined as nothing but the method of all humankind. Huxley placed paleontology and geology in general firmly within this realm, for which he denounced the notion of special "magical" capacities that he relegated to a former age of unreason. He attributed paleontological prophecy to a sharpened common sense trained to reason from analogy.[8]

To the contrary, while such early paleontologists as Cuvier and Buckland had been masters of this kind of retrospective prophecy, and Buckland certainly esteemed the Baconian ideals of empiricism and induction, he made use of the trope of magic to style himself as a prophet in the old sense (apart from granting theory and religion a guiding role in his geology while denouncing grand theorizing). However, Paul White argues that Huxley, who saw himself as a "man of science" rather than a "scientist," was also a transition figure. Although he advocated an everyman's science, he drew on models of genius from romantic literature in the shaping of his persona and felt that his high culture was superior to that of commercial and professional men. In addition, by turning science into the new religion, he has been seen as assuming the role of nature's priest.[9]

Lorraine Daston and Peter Galison also observe a transition from the idea that genius offers privileged insight into the secrets of nature to the notion of the scientist as ordinary diligent person. They further observe that this transition coincided with the rise of objectivity as an epistemic and moral category around 1800. This new sense of the scientific self and the new understanding of objectivity were accompanied by technologies of automaticity ("mechanical objectivity"). As expressed in the passage quoted from Gould in the introduction to this part, the elaborate systems of instruments and measurements through which anthropologists approached (fossil) human bones became an integral part of rational science and distanced the scientist from his objects in the endeavor to produce objective knowledge. However, although paleoanthropologists of the late nineteenth and twentieth centuries referred to intuition in

Huxley's sense of an acuteness based on a training of the common senses, rather than in the sense of "mythical intuition," the status of instruments for comparative measurement and, above all, statistics remained controversial. As we will see, Sollas, for one, was a firm believer in the new mechanical objectivity and was at times ridiculed for his "overreliance" on machinery, measurements, and numbers.[10]

Finally, it seems that Huxley argued for a democratization of access to education not so much out of egalitarian ideals as because of his belief in the natural supremacy of the white male. At the end of the American Civil War, Huxley reassured that one need not fear the freeing of black slaves or the admittance of women to education. On the basis of the law of natural selection, he thought that neither black people nor women would be able to surpass the better among the white men, whose place at the top of society was therefore not threatened by the smaller-brained, prognathous, and on average intellectually inferior black, or by the comparably defective white woman. It appears, and this notion is supported by his pejorative allusion to the savage's craftsmanship in the quote earlier, that some were more apt for science than others after all.[11]

Strongly impressed also by his teacher's reliance on the moral authority of seemingly unmediated nature, Sollas continued his career as an apprentice at the School of Mines and subsequently moved to St. John's College, Cambridge, where he came under another important influence. Thomas George Bonney (1833–1923), his tutor in geology, was the main impetus for Sollas's choice of geology as principal subject and for taking the natural science tripos in 1873. The following year, he married Helen Coryn of Redruth, Cornwall, with whom he had two daughters, Hertha and Igerna, who would become helpmates to his science. As a lecturer at Cambridge, Sollas began to work on the Cambridge Green Sand and its fossil content, as well as on what would develop into one of his main interests, sponges. He published an account of the sponge fauna of Norway and a monograph on the Tetractinellidae collected by the oceanographic *Challenger* expedition (1872–1876). His pioneering microscopic work on sponges earned him the Wollaston Fund of the Royal Society in 1878.

The following year, Sollas was appointed lecturer in geology and zoology at University College, Bristol, where he was also curator of the Bristol Museum. He was promoted to professor in 1880 and two years later resigned from the curatorship to devote himself exclusively to the college. Again he engaged in the local geology of the Bristol district and described a new species of plesiosaurs, among other paleontological work. Successively, he became professor of geology and mineralogy at Trinity College, Dublin (1883), where he worked as a petrologist for the Geological Survey (1893–1897). In 1889, he was elected a fellow of the Royal Society, from which he received the Royal Medal in 1914. At Dublin, he was especially interested in the Leinster granites, but also in crystals, glaciers, and coral reefs. He was a member of a committee to investigate the structure of coral islands by means of borings and later led the Royal Society expedition to Funafuti. He also finished his work on sponges and studied foraminifera, brachiopods, echinids, and starfish.[12]

The part of Sollas's scientific career of most interest here is his time at Oxford University, where he held the chair of paleontology and mineralogy from 1897 onward. This is due to the fact that the move from Dublin to Oxford roughly coincided with a change in research emphasis from geology to anthropology. It was also at Oxford that Sollas contributed an important invention to the paleontologist's material culture, in which he was greatly aided by the university's reader in mechanics. Since it was impossible to cut fossils into slices thin enough for microscopic analysis, the specimen had to be ground down after slicing, whereby it was not only lost, but important parts were also ground away and thus made inaccessible to the eye of the observer. Sollas solved the problem by the invention of a slicing machine.[13]

The experiments with the newly designed machine, in which his daughter Igerna, a zoologist who held teaching and research posts at Newham (1903–1913), was involved, provided him with sections of specimens less than one millimeter thick. He kept a record of each slice in the form of an image drawn with the help of a camera lucida or a photograph. At the end of the process, he could restore a model of the original specimen through the production of copies of each slice on sheets of gelatin or glass. The method was a great advantage over the old procedure, although the original specimen was still lost to the collector or curator. However, the possibility of microscopic analysis of each layer, as well as the fact that in the end there existed a substitute for the original specimen, was compensation enough for the anatomist/paleontologist, so that Smith Woodward responded: "The life the Artist-Sportsman took/The Sportsman-Artist could restore/As full of life in every look/And far more lasting than before."[14] William and Igerna Sollas made extensive use of the new technology in their paleontological work.[15]

The move from geology to anthropology was made very deliberately, but not without anxiety. Indeed, Bonney felt prompted to reassure Sollas: "Now ... don't worry yourself about the Paleolithic Adam (Eve would be more appropriate) having tempted you from the study proper of geology into the realm of primeval man. Geology cooperates [?] with archeology."[16] The temptress from prehistory is most likely an allusion to the Red Lady, on which Sollas began to work around the time of the letter, and which would perfectly fit the gender confusion. Furthermore, Bonney's characterization of the Red Lady, both as Eve and as temptress, had been used by Buckland before him. Sollas gave in to the temptation and concentrated on Paleolithic archeology and anthropology, even though not all of his new colleagues allowed this intrusion into their disciplinary territory without resistance. Sollas was also unhappy about the fact that his greatest work in anthropology, *Ancient Hunters and Their Modern Representatives* (1911), was listed in *Nature* under geology rather than under anthropology, even though "I am anxious to keep it as far as possible from geology."[17]

The change of subject was not complete, and Sollas continued to engage in local geology in the district and assisted his daughter Hertha in her translation of the Anglo-Austrian geologist Eduard Suess's (1831–1914) *Das Antlitz der Erde* (4 vols., 1883–1909) into English. From 1908 to 1910, he was president of the Geological Society, which had given him the Bigsby Medal in 1893 and the Wollaston Medal in 1907. He was made a senior member the year before he died. He held honorary doctorates from Dublin, Bristol, Oslo, and Adelaide and foreign memberships of the geological societies of America and Belgium. In 1934, he became a fellow of the Imperial College of Science and Technology.[18]

The Oxford of Sollas was radically different from the one Buckland had known. Soon after Sollas's arrival, Oxford was shaken by the war in South Africa (1899–1902). It also experienced an acceleration of the women's suffrage movement and was thrilled by the exoticism of its Rhodes scholars. Although the interconnections between the Oxford University and the empire were as old as the empire itself, which is not surprising for an institution that contributed essentially to the production of Britain's leadership and elite, their climax was relatively short and coincided with the period from the last quarter of the nineteenth century to the outbreak of World War I. Modern history, the classics, and philosophy provided the imperialist cause with theory, rhetoric, and valiant models of the past. The comparison to the Roman Empire was drawn with the observation that in contrast to the Roman hierarchy, which had been based on class, the British Empire was built on racial distinctions. Oxonians from Thomas Arnold (1795–1842) to John Ruskin (1819–1900) preached the moral imperative of the superior English gentlemen to civilize the dark corners of the world. Richard Symonds observes that the imperial zeal for the civilizing mission had partly functioned as a substitute for the earlier religious zeal, which had declined after the traditionalism of the Tractarian movement. The new fervor was inspired by an evolutionary outlook as presented in Darwin's *On the Origin of Species* (1859) and *The Descent of Man* (1871).[19]

All in all, at Oxford the years leading to World War I were marked by an enthusiasm toward the empire, and between 1896 and 1914, Oxford won more than half the places in the Indian Civil Service. Also the Colonial Service and the Egypt and African services recruited young men directly from the university. Oxonians were working as teachers, administrators, and evangelical missionaries throughout the empire, and in the Boer War Oxford had its volunteers. Of course, the empire brought other opportunities, such as travels as ethnologists, naturalists, geologists, or surgeons and expanded networking with Canadian, South American, and Australian colleagues, with the British Association for the Advancement of Science sometimes meeting in its former colonies. There was also ample opportunity to enrich collections, as exemplified by the great skull series from all over the world that the craniologists George Rolleston (1829–1881) and Henry V. Moseley (1844–1891), who succeeded each other in the chair of comparative anatomy, acquired for the Oxford Museum. While Rolleston had served as an army surgeon in Crimea, Moseley had traveled for four years as a naturalist with the *Challenger* expedition.

Besides geography, a readership that was first held by Halford John Mackinder (1861–1947) from 1887 onward as a result of the Royal Geographical Society's concern for the geographical literacy of those in charge of imperial policies, anthropology was the discipline that profited most from the interest in matters of empire. The installation of a readership in anthropology in 1884 had been a condition on which Augustus Henry Pitt Rivers (1827–1900) had donated his ethnographical and archeological collection for the foundation of the Pitt Rivers Museum in 1883. The readership was first held by Tylor, who was made professor of anthropology in 1896. Robert Randolph Marett (1866–1943), Tylor's successor as reader of social anthropology, instituted a one-year diploma course in anthropology for Colonial Office, Indian Civil Service, and Sudan Civil Service probationers. This resulted in a network of exchange of information and specimens that had its center in the Pitt Rivers Museum under the curator Henry Balfour (1862–1939).[20]

Other branches also profited. Under the Darwinian Edward Poulton (1856–1944), Hope Professor of Zoology (1893–1933), the Hope Department in the Oxford Museum became the imperial center for entomology at a time when this research and fieldwork were crucial in finding solutions to human and animal diseases in new territories brought under British rule. From all corners of the empire, correspondents sent Poulton specimens and articles to correct. Both Poulton and the notable zoologist Lankester, who held the Linacre Chair of Comparative Anatomy after Moseley (Lankester became deputy professor in 1890 and full professor in 1891), wanted Oxford to become a great imperial university, the imperial center of university education and research. Lankester, who had been an assistant to Huxley, whom he admired as a scientist, as well as an administrator, as mentioned earlier, supervised the translation of some of Haeckel's work into English and would become a well-known popular science writer.[21] Sollas's work in geology also benefited from the imperial network when, among other things, he classified the rocks the archeologist Marc Aurel Stein (1862–1943) brought back from the third British central Asia expedition and collaborated on the geological parts of Stein's *Innermost Asia* (4 vols., 1928). During this period, Sollas also received an invitation to classify the collection in the Museum of the Royal Garden in Ceylon.[22]

On the other hand, Sollas, like many of his colleagues, felt that the sciences at Oxford and in the nation's curricula in general did not receive their due share. In his anniversary address as president of the Geological Society in 1910, he complained about the teaching system. It still put too great a weight on the classics, while the scientific disciplines were becoming more and more specialized, producing a great number of papers and books that scientists could hardly keep up with in their own fields. In addition, science had become more international, and a practitioner depended on the mastery of several modern languages to gain access to the knowledge produced outside Britain. It therefore seemed a great waste of time to study Latin and Greek in the case of scientists, who would have no need of these in their daily work. Despite these complaints, at Oxford compulsory Greek was abolished only in the interwar years, and Latin remained compulsory up to the 1960s and 1970s.[23]

Obstacles to scientific research and teaching were exacerbated during the war years. When World War I broke out, Sollas was sixty-five and not part of the contribution of Oxford senior geologists who were subjected to an order compelling all officers with a knowledge of geology or mining to be used in underground warfare in France (mining, tunnel building). Instead, there is a rumor about him lying full-length on the floor of the University Museum and firing a .22 rifle at a target attached to a bookcase at the far end of the room, sometimes hitting books instead. Nonetheless, Sollas's world changed dramatically. Most current undergraduates and many fellows and staff signed up. They were replaced by soldiers, British and American, who transformed Oxford into a military camp. Then there arrived the refugees from Belgium and Serbia, for whom lodgings, a toy and a linen-goods factory, and a school were installed. The examination schools were converted into hospitals, and schools for military training and aeronautics were opened. Port Meadow was used as an airfield, and quarters for the Royal Flying Corps were set up. The fact that the lights had to be dimmed because of the fear of air raids added to the warlike atmosphere.

Sollas and his colleagues lost many students, and the majority of their class attenders were now women. He complained in a letter to Bonney on the loss of staff, in response to the latter's refusal to travel to Oxford because of "the horrors of war." Bonney deplored the decline of studentships to a dozen, more than half of whom were women, while the college was filled by cadets and soldiers.[24] World War I was crucial for the integration of women into what had been the male colleges. With the university nearly emptied of its male undergraduates, women's fees became an important financial source, and women were needed as tutors. As a culmination of this process, the Sex Disqualification (Removal) Act of 1919 led to the full admittance of women to university membership. However, a suspicion of female education persisted throughout the

1930s and 1940s, and the women's colleges did not succeed in gaining the same esteem as their traditionally male counterparts.[25]

While the integration of women into the university moved forward during the war, the scientists faced other obstacles, such as the increase in prices for publication and the paper shortage. On the other hand, the fact that the sciences were employed for war purposes gave them a boost in prestige. Oxygen provision in high altitudes, diagnosis of enteric fever, and a respirator against chlorine gas were among the objects of research. In 1919, the degree of doctor of philosophy became available, and in the 1930s, the number of postgraduate students in science tripled. However, geology, botany, zoology, and other sciences were almost totally outside the colleges but had a readership, conservatorship, or chair of the university, with researchers working in stalls annexed to the museum building. In fact, in geology, finalists sometimes sank close to zero.[26] Sollas, being by nature no diplomat, often despaired of the situation, and he complained to his friends in correspondence, among them Robert Broom (1866–1951) in South Africa, who kept him informed on his insights concerning the natives. Thus, relying on his extensive network across the empire and beyond and drawing on the great collections of the Oxford Museum, Sollas managed to make important contributions to the field of anthropology despite adverse circumstances.[27]

Like Buckland before him, Sollas was a legendary figure, and there are anecdotes about his eccentricities. He not only called his two daughters Hertha and Igerna, but also, not entirely untouched by the spiritualist movement of his time, claimed after the death of his first wife in 1911 that she had only translated herself into another sphere, whereupon he married the widow of his colleague Moseley.[28] Sollas's idiosyncrasies also showed in the professional sphere. He is said to have once tried to pay a demonstrator with a microscope instead of money, and another time to have spent all the money for the coming year on a bison model. Toward the end of his long life, things naturally became more difficult to handle alone. At one occasion, he was found with his face cut at home, claiming to have been bitten by a salmon when swimming in the Bristol Channel. Although it was clear that the cuts were from glass, their exact origin remains unknown. At any event, the university took the rare step of appointing a deputy professor to the department, James Archibald Douglas (1884–1978).

Douglas had been to Bolivia and Peru in 1910–1912 and had done war work in charge of mines and corps of miners in France and peacetime work in Persia. In 1924 Douglas had become a long-serving paleontological consultant to the Anglo-Persian Oil Company and an expert on Persian paleontology. The appointment of Douglas crossed Sollas's intention to turn the Geology Department into a center for Jurassic geology by winning the world expert William J. Arkell (1904–1958) as his successor. Douglas and Sollas were on strained terms, and it was Douglas who recorded Sollas's eccentricities, or "Sollesiana," on tape. The last incident that has been passed on to posterity took place two years before Sollas's death, when Wilfrid Edward Le Gros Clark (1895–1971) had gained the chair of comparative anatomy and the zoologist Solly Zuckerman (1904–1993) had installed a baboon colony in its building opposite Sollas's lab. By that time, science had become considerably modernized and Sollas nearly deaf. He is reported to have responded to a complaint by Douglas on the baboons' nerve-wracking noise: "Balloons! They are probably something to do with the Air Force."[29] Even if under increasingly awkward circumstances, Sollas remained in office at Oxford until his death on 20 October 1936.

⁓ THE INSIGHTS GAINED into Sollas's personality, scientific training, and institutional setting in this chapter provide important background for the discussion of his work in anthropology, and on the Red Lady in particular. Ramsay introduced Sollas not only to the Red Lady, but to the European fossil record in general. Huxley's introduction to the tools of physical anthropology applied to the anatomical markers of supposedly heritable racial differences, such as brain capacity, facial angle, coarseness of jaw and teeth, cephalic index, and body size, and his racial hierarchy with the Australian Aborigine at the lowest rung influenced Sollas's approach considerably. Sollas's infatuation with instruments and measurements that would guide his work in anthropology has been foreshadowed by the slicing machine. This technology-mediated and number-driven approach, by which Sollas would also access the Red Lady, as well as the strong influence of Huxley's thinking, will be further explored in the following chapter on Sollas's initial forays into anthropology.

Sollas wanted to enter the territory of anthropology, for which his anthropological training and his knowledge of geology and paleontology prepared him well. The step was nonetheless accompanied by a certain anxiety, in particular where the old temptress Red Lady was concerned. At the same time, we will see in Chapter 9 how his giving in to the temptation resulted in a cooperation with the leading experts in Paleolithic archeology on the Continent, as well as at his own institution, thereby lending authority to his position in the new field. In shaping his relations to other scientists working on human evolution, his irascible personality will be seen to further entrench, rather than ameliorate, differences between him and some of his colleagues with regard to the analysis of particular hominid fossils or overall frameworks of interpretation throughout the succeeding chapters. Finally, that Sollas was educated under Ramsay and Huxley when the Aryan theory, with its emphasis on migrations of races, was taught, in combination with the atmosphere at Oxford University in times of imperialism and war, will provide one possible inspiration for Sollas's model of racial distribution and the moral lessons he derived from it (or the moral lessons from which he derived it).

~ 8

Ancient Hunters and Their Modern Representatives

> The Neanderthal and Pithecanthropus skulls stand like the piers of a ruined bridge which once continuously connected the kingdom of man with the rest of the animal world.[1]

As DISCUSSED IN the introductory chapter to this part, a linear evolutionism that conceptualized the extant savage races as living fossils, shortening the gap between apes and humans, was widely adopted in the last third of the nineteenth century. We have also seen that among the holders of this view was Huxley, who integrated the Feldhofer Neanderthal into the hierarchy of variations in the human species. Sollas, who was influenced by his former teacher, began his forays into the territory of anthropology at an advantageous time. Unlike Haeckel and de Mortillet, he had no need of a hypothetical missing link. He was still professor of geology and mineralogy at Trinity College, Dublin, when he had the chance to study the two piers that were left of what he considered to have once been a continuous bridge between apes and humans—*Pithecanthropus erectus* and the Neanderthal. The Dutch physician Eugène Dubois (1858–1940) brought the femur, calvarium, and some teeth of *Pithecanthropus* (today *Homo erectus*) to Europe in 1895. In Java he had made the discovery that seemed to fulfill Haeckel's prophecy of an Asian ape-man when visiting the Dutch East Indies as a military surgeon. Sollas immediately began examining the precious remains. He published a report on his findings in *Nature*, taking his place among the first to accept the Javanese fossil as a hominid ancestor of upright posture and comparatively small brain.[2]

Early in the twentieth century, the German anatomist Gustav Schwalbe (1844–1916), who followed Haeckel in his conception of evolution as progressive, incorporated the newly found missing link into a line of descent that now went from *Pithecanthropus erectus* via the Neanderthals to modern humans. However, Schwalbe defined the Neanderthals as a separate species (*Homo primigenius*), rather than as a fossil human race, as done by Huxley, thereby enlisting it unambiguously for an evolutionary scenario of hominid development. Sollas responded to Schwalbe's work by a craniometric study of the Neanderthal skull that had been discovered at Forbes Quarry, Gibraltar, in 1848, comparing the specimen to Australian and great-ape skulls from the Oxford University Museum. In this comparison, he experimented with different developmental stages and different baselines for aligning the skulls to refute Schwalbe's classification of the Neanderthals. He claimed that Schwalbe's measurements were misleading, that they indicated a greater difference between the Neanderthals and human skulls than there actually was, because the line Schwalbe used for adjustment was unreliable. Also, Sollas showed that cranial and facial angles could vary from skull to skull not only depending on evolutionary stage, but also on developmental stage. A chimpanzee infant was in certain respects closer to an adult European than an adult chimpanzee, so that a bias might result from unknowingly comparing two skulls of individuals at different developmental stages.[3] Although in general not easily accessible because the argument relies heavily on intricate measurements and is thus highly numerical and jargon laden, the main point becomes immediately obvious from the visualizations (Figure 8.1): The article was to prove the close resemblance between Australian and Neanderthal skulls, when aligned correctly, and with all variables taken into account.

Sollas used his own method to arrive at "a true profile or meridian section of the cranium" because he found fault with Broca's craniograph, which in his opinion produced results that were affected by the person who carried out the drawing. In addition to this drawback, the other instrument then much used by anthropologists, the orthopter, did not provide an exactly median section. Sollas therefore constructed his own instrument in his laboratory, proud of the fact that the expense was kept below £2. He called it a hyptograph because it functioned on the principle of physical contact. Put simply, the skull was surrounded with

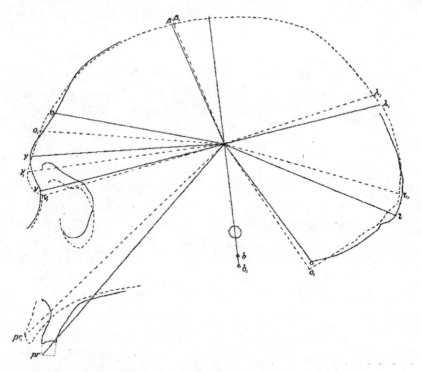

Figure 8.1. Diagrammatic superposition by the initial line of the Gibraltar Neanderthal (continuous line) and an Australian skull (broken line) in profile (*v*: nasion; *γ*: glabella; *o*: ophryon; *β*: bregma; *λ*: lambda; *ι*: inion; *pr*: prosthenion; *b*: basion); from William Sollas, 1908 (1907), fig. 18, p. 324; courtesy of The Royal Society of London.

a number of needles touching it along the meridian curve. These had to be numerous enough to provide a precise outline. This was done while the skull rested on a rectangular wooden board and the needles could be stuck between two additional wooden boards on each side of the skull. A drawing board with a piece of paper was then brought into contact with the undersurface of the needles so that the curve of the needle points could be translated into a graph. To test the accuracy of the instrument, the drawings obtained were measured and the numbers compared with those gained from the actual skull with a craniometer. Sollas was pleased that his method proved "extremely accurate."[4]

The paper skull produced with the hyptograph could now function as a substitute for the real thing. Sollas thought that there was no constant baseline as a standard line of reference; he did not consider even the Frankfort line (ear-eye) stable enough as a base for the comparison of skulls. However, he believed that reliable angular measurements could be obtained on the basis of an anatomical point as center. To this purpose, he determined "the geometric center" by cutting the skull out of the paper and piercing a pin through it close to the margin. From the pin a thin thread with a weight attached was suspended, functioning as a plumb line. The position of the thread was marked on the lower margin of the paper skull. The hole made by the pin was now connected to the marked point. This procedure was repeated from a hole at a ninety-degree angle to the first line. Where the resulting lines intersected, the center of gravity was located, which according to Sollas was also the geometric and morphological center. From this center, radii might now be drawn to each of the anatomical points (considered important in craniometry and indicated in Figure 8.1 at the periphery of the skull). The radii lengths, as well as the angular distances between them, served for comparisons between skulls. For superposition, paper skulls were best adjusted along the line connecting the geometric center and the basion (foramen magnum, where the skull meets the cerebral curve), which served as point zero for the determination of angles.[5]

The outcome of the labor-intensive procedure was in support of Sollas's former teacher Huxley in classifying the Neanderthal as a mere variant of the modern human type, a race of *Homo sapiens*. As is visualized in Figure 8.1, it was especially the native Australian who was supposed to link the Neanderthals to modern humans. However, rather than bring the Neanderthals closer to modern humans, this seems to have distanced the Australians from the Europeans. Sollas here introduced a concept that would prove central for his scenario of hominid evolution, that remoteness in space stood for remoteness in time and could thus indicate great distance in phylogenetic relation:

> The Neanderthal race, the most remote from us in time of which we have any anatomical knowledge, and the Australian, the most remote from us in space, probably represent divergent branches of the same original stock. In that most important of all characters, cranial capacity, the two races are almost identical.[6]

This quote also once again points to the central role of a hierarchy of brain sizes from which an evolutionary lineage could be inferred. "In that most important of all characters, cranial capacity," not only the Neanderthals and Australians but also the Neanderthals and *Pithecanthropi* overlapped. If one imagines the human lineage as a bridge across the river of time, the Neanderthals skirted the human bank, while *Pithecanthropus* stood at the edge of the pier from the ape-human ancestor's side. Indeed, although the exact position of *Pithecanthropus* on the line between ape and human might have varied, at the turn of the twentieth century, most paleoanthropologists worked with a similar linear model.[7]

The exhaustive and meticulous craniological analysis, in which Sollas established new standards for comparison and a new technique for drawing profiles of skulls, earned him the respect of the academic community, and he became president of the Geological Society from 1908 to 1910. The degree to which Sollas wanted to enter the field of anthropology can be inferred from the fact that he delivered the presidential address of 1910 on the subject of human evolution.[8] In this lecture and three preceding ones he gave to the Royal Institution in 1906, which were published in 1909 in *Science Progress* under the title "Palaeolithic Races and Their Modern Representatives," Sollas laid the basis for his ideas on the evolution of the human races.[9] These took definitive shape in a book issued in 1911, 1915, and 1924.

Ancient Hunters and Their Modern Representatives (1911) was Sollas's great synthesis of geology, ethnology, Paleolithic archeology, and paleoanthropology. For this achievement, he visited a number of sites and strongly relied on his international network. In France alone, he studied the painted caves of the Ariège and the Hautes Pyrénées assisted by Cartailhac and accompanied by his friend Marett; he visited those of the Dordogne with Breuil, the French expert on Paleolithic art, among others; and Boule introduced him to the La Chapelle-aux-Saints Neanderthal specimen. Similarly extensive investigations of sites and collections of Paleolithic industries and human fossils were undertaken with the aid of the national experts in Germany and Belgium and of course in Britain where he could rely on Tylor, the Cambridge anthropologist-ethnologist Alfred C. Haddon (1855–1940), and Balfour, among others. The book is amply illustrated, and Sollas obtained permission to reproduce images of Paleolithic artifacts, savages, fossils, and geological maps from other researchers. He was not humble about his achievement: "I believe this is the first time that a general survey has been attempted—at least in the English tongue—of the vast store of facts which have rewarded the labours of investigators into the early history of Man during the last half-century."[10]

The reader gains an understanding of Sollas's intent already from the contents. After chapters on the geology of the Ice Age and human antiquity in general, the list of chapters announces discussions of Paleolithic periods alternating with ones of recent savages. The treatment of the eoliths therefore precedes the Tasmanians, the Middle Paleolithic the Australian Aborigines, the Aurignacian Age the Bushmen, and the Magdalenian the Eskimo. The reader is thereby warned that the comparative method of ethnology may here be applied to Paleolithic archeology and paleoanthropology. Indeed, the human races in past and present were Sollas's main concern, as indicated by the title. Following Huxley's teaching, Sollas subdivided the human species according to the structure of hair. Thus the Lissotrichi were the straight-haired (Mongoloid), the Ulotrichi the very curly haired (Negro and Negroid), and the Cymotrichi the wavy-haired (including Europeans). Further criteria to describe races were pigmentation, stature, and eye, nose, and mouth shape. Of paramount importance, because they were observable on skeletal remains, were the cephalic index, cranial capacity, degree of prognathism, development of brow ridges, and dentition.

In *Ancient Hunters*, Sollas explained the geographical distribution of the living and fossil races according to a model of racial succession in the tradition of Boyd Dawkins, which suggested that some human races had always migrated and expanded, marginalizing and extinguishing forms less advanced in morphology and culture.[11] Sollas envisioned that the ancestors of modern savages had successively been forced out of Europe to the peripheries of the earth by more advanced invaders. According to this scheme, inferior Neanderthal tribes had been driven out of Europe when they had to compete with more highly developed tribes of their own race. The lower tribes migrated as far as the Australian region. There, these Mousterian newcomers replaced the forefathers of the Tasmanians in all of Australia except in the south. Correspondingly, the modern Australian Aborigines not only possessed the cultural equivalent of the Mousterian lithic industry but also resembled the Neanderthals physically.[12] The Paleolithic Neanderthaloid invaders of Australia had therefore been the ancestors of the modern Australian Aborigines, to whom Sollas referred as "Mousterians of the Antipodes."[13]

Because it had been the lower Neanderthal tribes that had migrated to Australia, the "Australians [were] a lower race than Neanderthal,"[14] that is, classic European Neanderthals. The next invaders of Australia, the British, caused the final extinction of the Tasmanians, on which Sollas commented, "It is a sad story, and we can only hope that the replacement of a people with a cranial capacity of only about 1200 c.c. by one with a capacity nearly one-third greater may prove ultimately of advantage in the evolution of mankind."[15] Sollas thought that the last Tasmanian had died in 1877.[16]

In Paleolithic Europe, the remaining Neanderthal tribes had been replaced by the more advanced peoples of Aurignacian industry. These had consisted of two races. On the one hand, there had been a Negroid race called the Grimaldi race (Grottes des Enfants, Mentone). When the Grimaldi race was driven out of Europe, they supposedly fled all the way to southern Africa, leaving behind their characteristic cave-wall paintings between the Dordogne and the Cape. Sollas identified the Bushmen as the descendants of the Grimaldi race, whom they resembled in culture, as well as in bodily form. Both, for example, painted cave walls and made statuettes of women who were clearly steatopygous. Just like the Tasmanian and the Australian Aborigine, in Sollas's days the Bushmen faced extinction through the invasion of their territory by a yet more advanced race, the agricultural Boer. On the other hand, the second race present in Europe during Aurignacian times had been tall and was described by some anthropologists as Mongoloid. However, the statuettes they had left behind resembled more the modern Europeans, whose ancestors, Sollas reasoned, they may have been.

The people of Magdalenian culture, who had driven the Grimaldi race out of Europe, had again consisted of at least two races: the Cro-Magnons (Périgord) and the small Eskimo-like Chancelade (Léo Testut, Périgueux, 1888). Not only the Magdalenian culture but again also the morphology of the Chancelade race reminded Sollas of the Eskimo to a degree that he suggested blood ties, as Boyd Dawkins and the French anatomist Testut (1849–1925) had done before him.[17] Toward the end of the Pleistocene, the representatives of the Chancelade race had been driven out of their European territory to North America by the Neolithic agriculturalists; so were probably the tall Cro-Magnons, who resembled the American natives, who followed the Eskimo to the interior of America. The American Indians spread over the entire American continent, only to be expelled from a great part of their territory by the arrival of highly civilized Europeans. Sollas summarized his imperialist model of human racial distribution and evolution as follows:

> If the views we have expressed in this and preceding chapters are well founded, it would appear that the surviving races which represent the vanished Palaeolithic hunters have succeeded one another over Europe in the order of their intelligence: each has yielded in turn to a more highly developed and more highly gifted form of man. From what is now the focus of civilisation they have one by one been expelled and driven to the uttermost parts of the earth: the Mousterians survive in the remotely related Australians at the Antipodes, the Solutrians are represented by the Bushmen of the southern extremity of Africa, the Magdalenians by the Eskimo on the frozen margin of the North American continent and as well, perhaps, by the Red Indians.[18]

Obviously, in Sollas's encompassing picture of human prehistory, history, and present, racial violence and imperialistic behavior were central mechanisms of progress. There seemed to be a certain orderliness, if not moral authority, in the view that higher races inevitably replace lower races, especially in regions considered more desirable, such as Europe. Sollas viewed human races as having long separate (pre-) histories and associated them closely with different fossil human races and even species or genera (Australian with Neanderthal, Tasmanian with *Pithecanthropus*). This had the effect of distancing and dehumanizing non-European races—a process that depended on an easy nature-culture link when Sollas inferred blood ties on the basis of cultural similarity. He also confused Paleolithic cultures with biological races when he used terms such as *the Mousterians* or *the Aurignacians*. The same kind of monolithic thinking is obviously at work in his treatment of recent groups such as "the Australian Aborigines" as representing an ethnological entity. In general, the racial categories Sollas described for the Pleistocene were based on very few fossil fragments indeed, and not even all of Sollas's Paleolithic groups had been substantiated by fossil remains. Sollas explained the cultural progress observed through successive archeological beds by each higher level being the product of a higher human race or races. This did not leave much room for variation or progress within one race, which makes his model appear somewhat antievolutionary.[19]

The concept of a geographical gradient from higher human types in Europe to lower types at the peripheries could also be applied to the evolution of humans as a species. It allowed Sollas to reconcile the possible contemporaneity of *Pithecanthropus* and *Homo heidelbergensis* in the early Pleistocene with his linear view of hominid evolution. *Homo heidelbergensis* was represented by a fossil jaw from Mauer in Germany that showed affinities to the Neanderthal. Sollas imagined the lower form, *Pithecanthropus*, to have survived as a kind of anachronism into later times in marginal regions such as Java, since, as we have already seen in the case of the Australian Aborigine, "in geology, time is equivalent to space, being represented in the stratigraphical series by a vertical line, and in geographical distribution by a horizontal line, the distance which intervenes between remote races measuring the time required for their differentiation from a common stock or their migration from a common centre."[20] The geographical distance between the European and the Asian fossil thus indicated their phylogenetic distance: "No doubt it [*H. heidelbergensis*] was preceded by still more primitive ancestral forms, and one of these, surviving in Java after its fellows had become extinct elsewhere, is possibly represented by *Pithecanthropus erectus*."[21]

Sollas therefore saw no a priori contradiction between the contemporaneity of higher and lower forms and a linear view of evolution, since even in modern times lower and higher races, such as the native Australian and the modern European, coexisted. In other words, just as Sollas regarded the Australian as an anachronistically primitive human, a relic of a Neanderthal-like race, so had been the Javanese *Pithecanthropus*, which in more favorable, and in Sollas's ethnocentric view more central, regions had long since evolved into higher forms:

> It is worth noticing that the lower form (Pithecanthropus) occurs in that hemisphere where the most primitive of known races of men (the Australians and Tasmanians) continued to hold exclusive possession of a large isolated territory into comparatively recent times, while the higher form, *Homo heidelbergensis*, lived in Europe,

which has nurtured for a very long period the most highly endowed races of the world.[22]

One gains the impression that Sollas conceptualized Europe as a kind of evolutionary vortex into which ever-higher human types were sucked from afar, so that those already stuck in it were spat out and whirled to the edges of the world, where the forces of the vortex were no longer sensible, and time, or development, came to a standstill. Sollas left open the question of the whereabouts of the center from which the races that invaded Europe originated in the first place. However, even though he seemed to speculate on an African origin of hominids in his presidential address of 1910, it does not seem unlikely that he agreed with Boyd Dawkins on (later) invasions from the East, on whose ideas he drew strongly for his racial-succession model.

Sollas's first edition of *Ancient Hunters* (1911) was a success, and within weeks he had sold 468 copies in England and 100 in New York. Nonetheless, Sollas feared reviews from some of his renowned colleagues, not without reason. We have already seen that Sollas and Boyd Dawkins disagreed on the interpretation of the Cro-Magnon remains and (Paleolithic) burials in general, and they would clash again after the appearance of the third edition of *Ancient Hunters* with regard to the question of Paleolithic art in Britain. Boyd Dawkins regarded an engraved horse from the Cresswell Caves as genuine, while Sollas pronounced it a forgery.[23] Chapter 12 deals with Sollas's at-times-vociferous differences with Lankester on the eoliths question, and his fears were at least justified in this regard, as Lankester attacked *Ancient Hunters* in the *Saturday Review*.[24] Keith was another central persona with whom we will witness Sollas bickering. He referred to the powerful Keith as the British oracle of paleoanthropology in despite of his old-fashioned methods and great influence. At one point he poured out:

> Sir Arthur Keith . . . is indeed the most arrant humbug and artful climber in the anthropological world . . . He makes the rashest statements in the face of evidence. Never quotes an author but to misrepresent him, generalises on single observations, and indeed there is scarcely a single crime in which he is not adept. Journalism, my dear boy, journalism pure and simple, and backed by

all the journalists, poor dears. He has gone up like a rocket, and will come down like the stick.[25]

Regardless of these personal differences, the Anglophone press coverage was very favorable, while some of the foreign newspapers and especially the more scholarly reviewers, though generally enthusiastic, felt that Sollas had gone too far in his identification of Paleolithic with modern races.[26] Among these was the archeologist Samuel Hazzledine Warren (1873–1958), who especially applauded Sollas for his cautious treatment of such controversial matters as Tertiary Man and his tools in Europe (see Chapter 12). He was less convinced by Sollas's inference of lines of descent between the Paleolithic and modern human races. Apart from finding fault with the application of the term *Paleolithic* to recent humans, Hazzledine Warren cautioned that "how far we may be justified in construing comparison into identification is a subject that may long remain a matter of controversy."[27] In contrast, other voices excused Sollas's approach until many years after the last edition had been issued. In 1955, William C. Brice commented that "even granted that the Trobriand Islanders and the rest are not true modern representatives of ancient hunters, how much without them would we know about 'the social purposes which items of prehistoric gear were designed to subserve'."[28] Nonetheless, the times had changed, and Brice sadly observed that Sollas's prominent tomes no longer figured in the library of the new Disney Professor of Prehistory at Cambridge.

It was with the ideas on hominid and human racial evolution as outlined in *Ancient Hunters* (1911) in mind that Sollas entered Paviland Cave. Sollas's early anthropology situates him in the tradition of his teacher Huxley. The hierarchy of the races of *Homo sapiens*, which emphatically included the Neanderthals, was temporalized in an evolutionary scheme that conceptualized extant nonwhite races as living fossils. Thanks to Dubois's discovery, Sollas could connect that evolutionary series to fossil apes. Also in the establishment of this once-continuous bridge, he relied on craniometric measurements, most prominently cranial capacity. His participation in what Gould has identified as a preoccupation with numbers as a strategy for "scientification" would also strongly mark his work on the Red Lady. Ironically, though, the Red Lady partly disappointed his desires in this respect, since, as we know, "she" is headless.

However, Sollas's linear view of evolution is complicated by his notion of anatomically modern humans. In his racial hierarchy, existing races could figure lower down the scale than Paleolithic ones. The Australian Aborigine was below the higher type of Neanderthal. Furthermore, the living human races had evolved from different Paleolithic types: the Australian Aborigine from the Neanderthals, the Native American from the Cro-Magnons, and so on. It therefore seems that once an original stock of *Homo sapiens* had evolved, it diverged into different lines that led to the various existing human races. In the following chapter, we will see how his work on the Red Lady would once again set in motion this system of ancient hunters and their modern representatives, unsettling the relationships Sollas had established between them.

The year after *Ancient Hunters* (1911) appeared, Sollas was tempted away from his laboratory by the scarlet lady of the Oxford Museum and her French accomplices to carry out his own excavation at Paviland Cave, trespassing on the sacred ground of the anthropologist/archeologist. He was well prepared, equipped with the methodological tools of physical anthropology and a system of Paleolithic archeology into which to incorporate the data. However, the Paviland material would change his conception of the Cro-Magnons from Magdalenian ancestor of the Native American to Aurignacian ancestor of modern Europeans.

~ 9

The Red Lady Is a Cro-Magnon Man

THE SUCCESS OF the first edition of *Ancient Hunters* can be measured by the short time within which it ran out of stock. Macmillan asked Sollas for a second edition only months after its first appearance. There had been many new discoveries in France, and Sollas had spent his Easter vacation in Paris in an attempt to catch up on the new insights. Breuil helped him correct the blunders of the first edition, and Sollas visited his French friends and colleagues again over Christmas. However, the work on the new edition was slowed by the reinvestigation of Paviland Cave. In addition, Sollas, immensely proud though under strain, was preparing the Huxley Memorial Lecture of 1913 for the Royal Anthropological Institute of Great Britain and Ireland on Paviland Cave and the Red Lady.[1]

We have last heard that when Édouard Lartet and Christy visited the Oxford Museum in 1863 and saw the human bones and artifacts from Paviland, they realized the similarity to what they had encountered in the Périgord. Furthermore, the bones later discovered at Cro-Magnon had also been ritually buried and were red from ochre. In both cases, the burial had been associated with mammoth tusks, ivory ornaments, and marine shells. Even though this interpretation had been supported by the authors of the influential *Crania ethnica* (1882), it was not universally taken up. However, because of a visit by Cartailhac to the Red Lady at Oxford in 1911, Sollas cautiously reproduced the French prehistorian's opinion that the skeleton and paraphernalia were of Aurignacian age in the first edition of *Ancient Hunters* (1911).[2] Cartailhac's visit set in motion an avalanche of events, at the beginning of which stood the interference of Boule and Breuil in the autumn of the following year. The great French experts in paleoanthropology and Paleolithic archeology, respectively, traveled to England to inspect allegedly human-made tools from a Pliocene deposit below the Red Crag in Ipswich. But they not only confirmed Sollas's opinion that these flints were mere plays of nature, they also examined the Paviland collection. While Boule recognized the bones of the Red Lady as Cro-Magnon, Breuil did not hesitate to classify the ivory implements as Aurignacian.

Breuil and Sollas then proceeded to the Swansea Museum in Wales and examined the flint and bone implements from Paviland there. Breuil identified Upper Aurignacian and early Solutrean tools. Thence they traveled on to the cave itself.[3] Because Breuil soon discovered several Aurignacian artifacts, he urged Sollas to continue the work and write a full account of it. Breuil, of course, would examine and classify the implements. This was a favorable turn for Sollas, since his work on the notorious Paviland material, once excavated by the legendary Buckland, aided him in gaining ground in the field of anthropology. The detailed study of the bones and artifacts from the cave helped him clarify his thoughts on the European Paleolithic races and match the Cro-Magnon race with the Aurignacian, besides the Magdalenian culture—a change that found entry into the second edition of *Ancient Hunters* (1915). Credit must no doubt be given to Breuil, who had been instrumental in bringing the very category of the Aurignacian as an extensive and important Upper Paleolithic cultural epoch to life. The change was welcomed by a reviewer of the second and third editions of *Ancient Hunters* (1915, 1924), whereas others seem to have lost trace of Sollas's intricate web of relationships between the Paleolithic hunters and their presumed modern representatives.[4]

When Sollas began his work at Paviland, he had secured the permission of the owner, "Miss Talbot," the help of a member of the Cardiff Museum, and the collaboration of his colleagues Marett and Balfour, his assistant, who functioned as photographer and illustrator, and finally two workmen.[5] Since Sollas assumed that the inhabitants of Paviland had obtained the flint from the adjacent glacial drift that had accumulated from the movement of the ice, he inferred that the Aurignacian people had lived in the cave after the peak of the last glacial period. There were no discernible industrial layers, so the dating of the tools had to be done solely on typological grounds. Sollas therefore cautioned that "an implement which is assigned say to the Mousterian class is not necessarily of Mousterian age."[6] The excavations yielded some eight hundred implements, which renders Buckland's mention of only one flint in *Reliquiae Diluvianae* even stranger. These could be classified into Mousterian, pseudo-Mousterian (with traces of Aurignacian workmanship), earliest Aurignacian, Middle Aurignacian, Upper Aurignacian, and Solutrean.

The discovery of special minerals seemed to suggest that the Red Lady's people had not only known the use of ochre as pigment but had also collected "precious stones as curiosities." Sollas interpreted the quartz crystal as having been associated with magic powers, as it was by medicine men "of most existing primitive tribes."[7] Implements of bone and ivory, some similar to those unearthed by Buckland and some new kinds, were discovered. The rods Buckland had referred to tokens in games Sollas speculated to have been used as bag handles, sinew twisters, netting pins, or bow drills. Most astonishingly, Sollas and his team found the "ivory pendant" that formed the complement to Buckland's mammoth tusk with an irregular cavity, probably due to an outgrowth in the pulp cavity after the animal had been wounded. As if symbolic for the progress of science the history of excavations at Paviland Cave was meant to illustrate, Sollas's egg-shaped ivory pendant fit nicely into the cavity of the injured tusk. Reminiscent of Buckland's witch story, Sollas inferred the practice of magic from this rare object: "Magic powers were probably attributed to so rare and remarkable an object and it might have been suspended in the cave or slung round the neck of the hunter to bring him good luck."[8]

Sollas recognized the similarity of Buckland's description of the burial to the interments of Mentone as pointed out by Cartailhac, where the bodies were laid out approximately parallel to the cave axis, with red ochre spread over and about them. In both caves, large stones had been placed at the head and the feet of the dead bodies, and ivory rods and ornaments, as well as seashells, had been buried with them. Accordingly, the bones of the extremities were compared with the Cro-Magnon finds from the Mentone/Grimaldi sites (Grotte du Cavillon, Grotte des Enfants, Grotte de la Barma Grande 1 and 2).[9] In the measurements and classification of the human bones, Sollas was significantly assisted by a colleague of Oriel College. Although the humerus was comparatively slender—as in fact were all the bones—and the muscular attachments were not well marked, and although there was only half a pelvis, Sollas and his collaborator considered the evidence sufficient to turn the Red Lady into Paviland Man because of the considerable size of the head of the humerus, femur, and tibia. This was supported by a comparison with measurements of Red Indians, which showed that men and women differed in the range of head diameters. Sollas added that "for civilized and European races the results were similar."[10]

The estimated age of the Red Lady was "not much over twenty-seven" on the basis of the ossification of the epiphyses and the fact that the line of demarcation (between epiphysis and shaft) was still visible. Cro-Magnon racial characteristics as already defined by Broca and others that were shared by the skeleton were a prominent ridge (*linea aspera*) on the femur, associated with a shallow depression outside its upper and outer bifurcation, platymery (transverse flattening) in femur and tibia, and the presence of only two trochanters (femur). Sollas reconstructed the stature of the Red Lady by applying two methods to the long bones, one of the anthropometrist Léonce-Pierre Manouvrier (1850–1927) and the other of the biometrician Karl Pearson (1857–1936).[11] Taking femur and tibia together in accordance with Pearson's method showed that the Red Lady must have been about 1.70 meters tall. The stature of the Red Lady corresponded with what René Verneau (1852–1938), director of the Ethnological Museum in Paris, had found for Cro-Magnon skeletons, namely, that the femur was shorter relative to the tibia, and the humerus relative to the radius, than in modern Europeans (the latter was inferred since the radius of the Red Lady is damaged), and that the arm-leg ratio was less in the Cro-Magnons than in modern Europeans. In these respects, Verneau had shown, the Cro-Magnon race approached the modern Negro.

Very unfortunately for Sollas, who, as we have seen, was in love with craniology, had attempted to find adequate standards for making measurements comparable, and had developed a new machine for obtaining sagittal sections, the Red Lady's skull was still missing. The missing skull was actually part of the Red Lady's legend, and everyone taking on the post of professor of geology at Oxford University was asked whether he had found it. The story may have originated in the fact that Buckland had represented the skeleton as entire in his section of the cave, suggesting both Buckland's lack of interest in the bones and the power of images as evidence, since the legend assumed that there had initially been a skull. Sollas removed the blemish when he reproduced Buckland's sketch in *Ancient Hunters*.[12] Although he could not provide the head, Sollas rather parenthetically mentioned the discovery of another human fragment that was only later identified as such in the material yielded by the excavation. This distal end of a left humerus, however, "differs greatly in character from that of the 'Red Lady.'"[13]

Where the reconstruction of life in the cave is concerned, it is hard to imagine a greater change than from Buckland's dreary and lonely abode to the scene Sollas revived for Paviland Cave:

> As a temporary habitation it would be difficult to find a more excellent cave than Paviland; situated on the face of the steep limestone cliffs of Gower, it looks out over the changeful waters of the Bristol Canal; behind it is a fertile land which must have provided a rich hunting ground in early times; it is roomy, well lighted and dry, with a natural chimney to promote ventilation—serving also to carry off the smoke of a fire kindled beneath; in front of the entrance is a rocky platform with natural seats where the hunter can sun himself in the open air. Add to this that it is concealed from the landward view and difficult of access to those unfamiliar with the way. Evidently in every respect a highly "desirable" hunting lodge! How its advantages appealed to the palaeolithic man of Glamorgan during the Aurignacian age is shown by the great kitchen midden which forms its floor. Here, it is plain, he fabricated his implements and weapons, here he roasted his meat, flesh of the horse, the bison, the mammoth, and the bear, and here on one solemn occasion he entombed his dead.[14]

This is a strikingly colorful, lively, and happy picture of the days of the Paleolithic inhabitants of Paviland.

Sollas contradicted Buckland's interpretation of the broken animal bones as due to diggings by later explorers. It had been the human inhabitants in Paleolithic times who had broken the bones of the mammoth, woolly rhinoceros, reindeer, bison, horse, and cave bear to get to the marrow. Many bones showed scratches such as would be produced in scraping off the flesh with a scraper. As Buckland had already realized, none of the animal bones had been gnawed by hyenas, and this led Sollas to draw a conclusion that at the same time illustrates that the hyena's image had not improved since Buckland's time: "the presence of broken hyaena bones in the cave, would seem to indicate that the ancient hunters were reduced in times of scarcity to feeding on the flesh of this disgusting animal."[15]

Sollas was disappointed that no drawings on ivory or bone similar to those from the Continent were found at Paviland. Personal adornments had consisted "merely" of necklaces from wolves' teeth and seashells and bracelets of ivory. Sollas thought it likely that their garments had been made out of fur. However, in view of the fact that the ancient hunters of Paviland had known certain magic rituals and buried their dead ceremonially, Sollas consoled himself by allowing for the possibility that we simply cannot know all they had been capable of. Giving free rein for a moment to the fiction of worthy Paleolithic British, he thus fantasized: "in fairness to the vanished Aurignacian hunters we must admit that the information we obtain by rummaging their kitchen midden does not exhaust their story, the best part remains untold: could we know the whole, we might have occasion to admire their ingenuity, to applaud their courage and to sympathize with their aspirations towards the ideal."[16]

Despite this lack of evidence, Sollas did not find it too speculative to summarize about the Cro-Magnons and their division of labor:

> The hunters who found shelter in the cave were men of large stature, members of that tall Cro-Magnon race which occupied the greater part of habitable Europe during the Aurignacian age. They were men of capacious brains, and had made great progress in such simple mechanical arts as are essential to society in its most primitive stage, when subsistence depends wholly on the natural products of the earth, chiefly on roots and fruits which it is the allotted labour of the women to collect, and next on the flesh of animals killed in the chase and contributed by the men to enliven existence by an occasional feast.[17]

In agreement with the title of his main work, *Ancient Hunters* (1911), and the centrality of hunting as a pastime for the manly Anglo-Saxon, the scenes at Paviland were reconstructed through the lens of a hunter.[18] It was only in a speculative space, not in the artifacts of the cave, that women reentered Sollas's picture of the ancient hunters. Their role as gatherers would not have left traces for the archeologist, and the fashioning of stone and bone tools and weapons and even ivory and other ornaments was inferred to be a male sphere. All in all, Sollas resurrected the Cro-Magnons, who had inhabited "Welsh" soil in Upper Paleolithic times, as a race to be proud of, even if this demanded some magic on his own part and an allowance for invisible evidence.[19] As we will see, Sollas's ancestor glorification was modest compared with later reconstructions of the Red Lady's artist-warrior race for nationalist purposes. Nonetheless, the brutality and savagery inferred by Broca and Pruner-Bey are no longer part of the picture.

Sollas did not interpret the Mousterian and Solutrean artifacts as due to occupation of the cave by another race than the Aurignacian, but rather took them to show a Solutrean influence. This raises the question of how the Red Lady as Cro-Magnon fit into Sollas's system of human evolution at large. Neither within Sollas's frameworks nor in the anthropological literature in general was the category stabilized. Even before the publication of the subsequent editions of *Ancient Hunters*, Sollas's statements about the Paleolithic races were confusing. In his presidential address of 1910, he had speculated that the Cro-Magnons, whom he associated with the Magdalenian industry, had died out. The following year, in *Ancient Hunters*, he assumed that the Cro-Magnons, to whom he still attributed the Magdalenian, might today be represented by the modern Alonquins. Here, the Aurignacian was associated with the Negroid Grimaldi type and a second race, whose bones had not yet been found, who were the ancestors of modern Europeans. To confuse matters further, the Cro-Magnons showed certain Negroid features, but had been described as Mongoloid by Pruner-Bey. Although Sollas still somehow associated the Cro-Magnon race with the recent Native Americans in his 1913 article on the Red Lady (and by inference with one of the Magdalenian races), since he compared the Red Lady's remains to such a sample, it is due to his work on the Red Lady and its paraphernalia that he identified the Cro-Magnons less ambiguously also with one of the Aurignacian races.

～ Like many anthropologists, Sollas thus came to accept the Cro-Magnons as possible ancestors of (at least some) modern Europeans. They had possessed an Aurignacian and Magdalenian culture in succession and were thus the creators of stunning Paleolithic figurines, engravings, and wall paintings, among other things.[20] In the meantime, the Cro-Magnons had achieved iconic status in the work of Boule, who relied on the race's physically and culturally advanced state to highlight the primitiveness of the Neanderthals. In the aftermath, the two races mutually reinforced their respective images of caveman brute and Paleolithic superrace in the work of other anthropologists. Although Sollas envisioned the Red Lady as of a race of noble hunters, Boule's expulsion of the Neanderthal from human ancestry discussed in the following chapter wreaked havoc in his picture of hominid evolution. He would have to reconsider the place of the Neanderthals therein and their relation to the Cro-Magnons.

~ 10

Human Evolution as a Trunkless Tree

IN THE YEAR Sollas published his study on the Red Lady, he still adhered to a linear model of hominid evolution similar to Schwalbe's, according to which *Pithecanthropus erectus* had evolved into Neanderthals and eventually modern humans. As we have seen, de Mortillet had been the most influential nineteenth-century proponent of this progressive view of evolution in France. Also in Britain, paleoanthropologists went through a stage of acceptance of the linear model. However, the first decades of the twentieth century witnessed major changes in theories of human cultural and biological evolution. Among other things, Boule's work on the La Chapelle-aux-Saints Neanderthal was crucial for the fact that the anthropological community gave up the linear model and increasingly came to accept a "tree" in which all known hominid fossils were on side branches.[1]

The nearly complete Neanderthal skeleton had been discovered as part of a burial in a cave in the village of La Chapelle-aux-Saints (department of Corrèze, in southwestern France) in 1908. Boule received the specimen at the Muséum d'Histoire Naturelle in Paris, among other reasons because it was discovered by clerical prehistorians, the Catholic priests Jean and Amédée Bouyssonie. They consulted yet another clerical archeologist about where to send it. Breuil recommended Boule, his old friend and fellow student under Cartailhac. Part of the reason for this was the radical politics, materialism, and anticlericalism of the other possible destination, the École d'Anthropologie under de Mortillet's successors. De Mortillet himself had sat in the Chamber of Deputies on the extreme left, and his politics had been strongly interwoven with his view of human evolution. On the basis of the universal law of morphological and cultural progress, paleoanthropology and Paleolithic archeology were political weapons for radical socialist aims, with human history as an integral part and logical consequence of human prehistory. Striving for progress in a humanist sense, de Mortillet had reasoned that the political Left would eventually prevail by necessity. He had predicted an inevitable succession from the reign of the nobility to the reign of the bourgeoisie and finally to the reign of the socialists.[2]

On 14 December 1908, Boule's interpretation of the skull was presented to the Académie des Sciences. In the paper and succeeding monographs on the find that would prove formative of the scientific and popular views of the Neanderthals for many decades to come, Boule emphasized the simian traits of the skeleton, even postulating that Neanderthal Man had a rather stooping, not entirely upright posture.[3] He thus agreed with Schwalbe that Neanderthal was a separate species and not simply a pathological form of modern humans or a fossil human race, since the Neanderthals had differed significantly from modern humans, morphologically as well as culturally. However, he contradicted Schwalbe, Sollas, and de Mortillet by rejecting the Neanderthals as the ancestor of modern humans. Boule dated the La Chapelle-aux-Saints Neanderthal to the Middle Pleistocene, which, according to his interpretation of the Grimaldi finds, put it in close temporal proximity to the modern human type, both of which thus appeared to have inhabited Europe at the same time. The expulsion of Neanderthal from the line leading to modern humans also put into question the ancestral status of *Pithecanthropus*, which seemed to show Neanderthal specializations even more markedly. Hominid evolution must therefore have had more than one line of descent, and the ancestors of modern humans were again unknown.

To confer weight on his conclusions, Boule used imagery and rhetoric that distanced the Neanderthals from modern humans, on the one hand, and from Paleolithic *Homo sapiens*, on the other hand. Appealing to national pride, he verbally denigrated the image of the Neanderthals through juxtaposition with the noble contemporary Cro-Magnons:

It has to be remarked that this human group of the Middle Pleistocene, so primitive with regard to physical characters, must also, judging from the standards of Paleolithic archeology, be very primitive from an intellectual point of view. When, during the Upper Pleistocene, we find ourselves, in our country, in the presence of industrial manifestations of a higher order and of true art, the human skulls (race of Cro-Magnon) have acquired the principal characteristics of true *Homo sapiens*, which means beautiful foreheads, large brains, and faces with little prognathism.[4]

In the 1913 article, Boule visually juxtaposed the Neanderthal reconstruction with a modern Australian Aborigine skeleton with the expectation, it seems, that the viewer would immediately notice the obvious differences (Figures 10.1a and 10.1b). Even the primitives at the peripheries of the earth, the logic went, were considerably more advanced than this brute, which was closer to the apes than to any human race.

Even though the La Chapelle-aux-Saints specimen was well preserved, the parts that are dotted in the illustration had to be inferred. In 1909, Denis Peyrony (1869–1954) and Louis Capitan (1854–1929) discovered another Neanderthal burial in a rock shelter in the Dordogne, at La Ferrassie (La Ferrassie 1). In 1910, Peyrony unearthed La Ferrassie 2, a female. These skeletons, which were also brought to Boule's laboratory, were used to complement the missing parts of the La Chapelle-aux-Saints specimen.

Michael Hammond situates Boule's expulsion of the Neanderthals from human ancestry in the context of France and its anthropology of the time. In 1902, Boule had succeeded Gaudry as professor of paleontology at the Muséum d'Histoire Naturelle in Paris. Gaudry, having had to deal with the Cuvierian tradition and suspicion toward evolutionism at the museum, had strictly limited his evolutionary ventures to the animal kingdom. Boule, finding himself in a more relaxed atmosphere, meant to apply Gaudry's model of mammalian evolution as being branched to hominid phylogeny. The La Chapelle-aux-Saints discovery was Boule's chance to do so. Hammond argues that Boule was a great conciliator who aimed to disengage paleontology from politics. He worked with political radicals, as well as conservatives, and believed in cooperation with the more enlightened clergy, such as Breuil. All of this

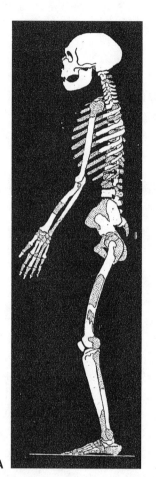 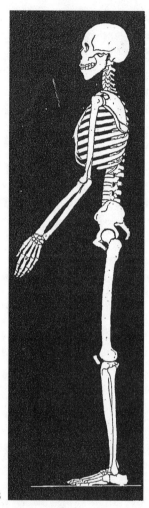

A **B**

Figure 10.1a and 10.1b. "Reconstruction of the skeleton of the Man from La Chapelle-aux-Saints, seen in profile" and "Skeleton of an Australian seen in profile"; from Marcellin Boule, 1913, figs. 99 and 100, pp. 232 and 233. Even though the La Chapelle specimen was well preserved, the parts that are dotted in the illustration had to be inferred. In 1909, Denis Peyrony (1869–1954) and Louis Capitan (1854–1929) discovered another Neanderthal burial in a rock shelter in the Dordogne, at La Ferrassie (La Ferrassie 1). In 1910, Peyrony unearthed La Ferrassie 2, a female. These skeletons, which were also brought to Boule's laboratory, were used to complement the missing parts of the La Chapelle specimen (Trinkaus and Shipman, 1993, pp. 188–189).

put him in clear opposition to de Mortillet's school of anthropology. Boule may thus have intended his interpretation of Neanderthal as a separate and unsuccessful branch not only as a demonstration of the validity of Gaudry's theory of evolution, but also as a blow to the politically explicit, purely materialist, linear view of hominid evolution associated with de Mortillet and the École.[5]

Boule's reconstruction of the La Chapelle-aux-Saints Neanderthal clearly had a great impact, and his work was widely publicized in newspapers. Although Boule distanced the Neanderthals from everything French, the prestige the discovery implied for French science and the French nation was used to further sell it to the press and the public. Nationalism was easily kindled at the time. The Third French Republic and the German Empire had arisen out of the hostility between France and Germany in the aftermath of the Franco-Prussian War (1870–1871), which had cost France not only men and money, but also large parts of Alsace and Lorraine. More and more French voices clamored to regain the territories lost to Germany. On 19 May 1909, the paper *Le Radical* seized the opportunity to slander Otto von Bismarck (1815–1898), who had strategically provoked the war between the North German Confederation and France to add the southern states to a united German Empire. Bismarck's skull had turned out to be the only one surpassing the La Chapelle-aux-Saints Neanderthal's in size. Because of the Neanderthal's poor image, this could serve to abuse Bismarck rather than signify high intelligence: "if Bismarck had a skull similar to the La Chapelle-aux-Saints Man's, it is because the iron chancellor displayed an atavism, by which the mentality of Paleolithic times was revived, when, already, force ruled."[6]

The trench between the French and German anthropological communities had been deepened by the archeologist Otto Hauser (1874–1932), who had sold the French Neanderthal discovery from Le Moustier to a German museum (Museum für Volkskunde, Berlin), and who, despite being Swiss, came to represent the Prussian threat. On 19 December 1908, *Le Temps* published Boule's demand for legislation on excavation rights at French prehistoric sites to avoid further theft.[7] However, the examples of La Chapelle-aux-Saints and La Quina, where further Neanderthal remains were discovered, strengthened the hope that this would never happen again:

This sensational discovery made by a French scientist will remain—needless to say—the property of France and of French science. Even under the regime of the liberty of excavation, which is the only one suiting the free and fertile activity of our prehistorians, there was no fear that the La Quina skeleton would end up—like the Les Eyzies skeleton, bought for 125,000 francs by Mr. Hauser—enriching the laboratories beyond the Rhine.[8]

In 1914, not unlike *Le Radical* in the example given earlier, Boule became more explicit in his use of paleoanthropology, in particular his Neanderthal interpretation and the associated glorification of the Cro-Magnons, as a weapon in nationalist discourse. It seems as though the emphasis on conflict and extinction that went with a dendritical human phylogeny appealed to Boule also for nonscientific reasons. He published an article in *L'Anthropologie* with the title "La guerre" in which he dehumanized the Germans in a manner reminiscent of his dehumanization of the Neanderthals of prehistoric times. In his own words, the mentality of the German elite represented "a kind of monstrous anachronism in the moral evolution of humankind,"[9] just as he believed that the Neanderthal was a primitive human state contemporary with more advanced forms. Boule regarded the Neanderthals as morphologically distinct from the modern human type, and he defined the Germans as being of a physiognomy clearly distinguishable from that of their neighbors, even though he denied them the status of a race. Boule claimed that the tendency to wage war was part of German nature and the result of a special course of evolution that favored militarism and imperialism over other intellectual faculties, which in turn had led to a regression to primeval savagery. The achievement of the German people, who were devoid of moral and higher spiritual capacities, lay solely in their materialistic and scientific advances, which Boule considered insufficient for inclusion within the rank of *Homo sapiens:* "The Germans are unworthy of the beautiful name *Homo sapiens;* they have degraded themselves to the order of primitive *Homo ferus.*"[10]

Boule opined that the German brain had lost its capacity for reason, and those cerebral areas had been taken over by instinct. The result was an untempered lust for all kinds of expansion, which again linked them to prehistoric species that had become extinct: "One may even say that paleontology is made up of examples of the rapid disappearance of beings that became the victims of their own gigantism or their overspecialization in a single direction."[11] In stark contrast, Boule argued, France had been the land of progress since Paleolithic times. The sculptures and murals of the Cro-Magnons that decorated the French caves gave clear testimony to an advanced aesthetic sense. These morphologically modern humans had been the pioneers of true culture. And just as the artistic Cro-Magnons had once triumphed over the brutish sub-*sapiens* Neanderthals, themselves victims of nonadaptive overspecialization, so would the French of today triumph over the not fully human German enemy in World War I.[12]

From the very beginning, Boule had thus placed his Neanderthal work in the service of current nationalism by juxtaposing the specimen from La Chapelle-aux-Saints with the Cro-Magnons, who represented everything worthy in humankind, and whom he associated with France. The press, too, found ways of appropriating the very primitive picture of the Neanderthal it drew from Boule's interpretation to France's national pride, as well as to the hatred for all things German. It is likely that this helped inscribe the brutish picture of the Neanderthal into our culture that would prove resistant to amelioration. Be that as it may, Boule achieved great fame through his study of the Old Man, whom he made the type of a separate species, *Homo neanderthalensis*, a name that had been coined by William King (1809–1886) many years previously. In 1910, Boule's dominance over the École was cemented by his appointment as head of the Institut de Paléontologie Humaine, endowed by Prince Albert I of Monaco. Boule tightened his connections, if not his school, by attaching his allies Breuil and Hugo Obermaier (1877–1946), another clerical expert in prehistoric art, to the institute.[13]

Boule's brutish Neanderthals were soon no longer welcome in the human pedigree, and they disappeared in most diagrams of hominid evolution from the line leading to modern humans. Thus originated the so-called Praesapiens theory, according to which the ancestor of modern humans had entered Europe from outside and exterminated the Neanderthals. As all then-known fossil hominids came to be seen as representing side branches of the line leading to modern humans, there had to be an as-yet-unknown ancestor deep in the human evolutionary past, the Dawn Man or Praesapiens.[14] Ironically, the theory implicitly brutalized the Red Lady's noble stock. On the other hand, as the discussion of Sollas's *Ancient Hunters* has shown, British imperialism had cost as much. Though "deplorable," it was the price paid for the progress of humanity. Although this notion was not in contradiction to his race-succession scenario, the more diverse and branched hominid phylogeny was diametrically opposed to his smooth linear development through the known fossil hominids, on the numericalization and visualization of which he had spent much time and energy. What was his reaction to the changes taking place in his community?

Luckily Sollas, whose preparation of the second edition of *Ancient Hunters* (1915) had been interrupted by the work on Paviland Cave, had his daughter Igerna and his wife Amabel to do the proofreading and construct the index while he was traveling to Italy, Adelaide, Melbourne, Sydney, and Brisbourne. When it appeared in print, the zoologist and geographer Herbert J. Fleure (1877–1969), successively lecturer, head of department, and professor in zoology, geology, geography, and anthropology at the University of Zurich (1904–1930), was so impressed that he chose a very similar title for an article in which he, too, discussed the modern races as descendants of Paleolithic precedents. In fact, however, the second edition, while providing an update on relevant finds of fossil hominids, contained no striking novelties.[15]

The most important find that had been made after the publication of the first edition was no doubt *Eoanthropus dawsoni*, remains of which had allegedly been discovered by the local solicitor and amateur geologist Charles Dawson (1846–1919), Smith Woodward, and others in the Weald of East Sussex from 1911 onward. Nine cranial fragments and the right half of a mandible, which seemed to date from around the Pliocene-Pleistocene barrier, had reportedly been unearthed in a gravel pit at Piltdown. Although the mandible was apelike, the braincase was modern looking. In fact, this hominid, informally called Piltdown Man after the quarry in which it had been planted, was made up of a modern human skull and an orangutan jaw. This chimera, which to the eye of the present-day scientist would seem most improbable, had a great impact on ideas on hominid evolution at the time and was only proved beyond doubt to be a forgery in 1953. Although some suspicion was voiced about the fragments representing one individual, the majority of anthropologists came to embrace the fossils that seemed to confirm the notion that brain expansion had been an early development on the way to modern humans. At the same time, Piltdown Man put the prevalent linear sequence of evolution from *Pithecanthropus* via Neanderthals to modern humans into question. Instead, it seemed to support a more diversified picture of hominid evolution as suggested by Boule, because with a European hominid around the Pliocene-Pleistocene boundary that had a nearly modern-sized cranium, the small-brained *Pithecanthropus* could hardly be a direct ancestor.[16]

Despite these developments, Sollas presented no striking new ideas on the genealogical relationship between the known fossil hominids and modern humans in the second edition of *Ancient Hunters*. To the contrary, in the main he simply added Piltdown to his discussions of *Pithecanthropus* and *Homo heidelbergensis*. He literally inserted Piltdown into the concluding remarks of the previous edition on the now-three genera, where he explained his indirect linear model, in which space may substitute for time, and which I have discussed for the first edition. Nonetheless, Sollas was acutely aware of the changes that were taking place in the views of his colleagues. In fact, he had to relegate a paper on *Pithecanthropus* by Elliot Smith, professor of anatomy successively at the Government School of Medicine in Cairo (1900–1909), Manchester University (1909–1919), and University College London (1919–1937), to a footnote. In the paper, Elliot Smith had argued for the expulsion of *Pithecanthropus* from the human line on the basis of the relatively modern-looking brain of *Eoanthropus dawsoni*. Most important, Sollas still regarded the Neanderthals as a race of *Homo sapiens* and not as a different species. He also abided by his interpretation of the Australians as the descendants of some lower Neanderthal tribes. That Sollas was unsettled by the trend of rejecting the known hominids as direct ancestors is nonetheless further highlighted by his inclusion of Boule's images reproduced as Figures 10.1a and 10.1b earlier, which with its poignant argument for the huge gulf between the two hominids stands squarely in the pages of the second edition alongside a discussion of the Neanderthals as of a race with the Australians.[17]

However, Sollas's incoherence with regard to the Neanderthal question did not damage his relationship with the French savant, who also assisted in the production of the last edition of *Ancient Hunters* (1924), so that Macmillan had to send the proofs to Boule, director of the laboratory of palaeontology, Muséum d'Histoire Naturelle, Jardin des Plantes, Paris. To consult with leading prehistorians, Sollas went as far south as Lyons and Monaco to make sure that he had all the relevant information. When the third edition finally appeared in 1924, it, like its predecessors, was carefully sent to important people and locations, from Basel, Switzerland, to the museum in Santa Cruz on Teneriffe. Possibly reflecting Boule's influence, in *Ancient Hunters* of 1924, Sollas unambiguously if reluctantly replaced his unilinear model of hominid evolution with a dendritical one and turned the Neanderthal race into a Neanderthal species that had died out in the Pleistocene. The change in opinion, once it appeared, was indeed abrupt: "The Australians are not only a different race [than the Neanderthal], they are a different species, and, notwithstanding the many characters which they share in common with the Neandertalians, they can no longer be regarded as directly descended from them."[18]

Yet even at this point Sollas revealed his unease about banning the Neanderthals to a dead-ending branch by an attempt to cover the move in terminological confusion. Undertaking a travel backward in time,

> He [*Homo sapiens*] disappears as we approach the beginning of the upper Monastirian age [geological time of the last glacial epoch], and we encounter in his place another and very different species, *Homo Neandertalensis* . . . But how are we to render the name of this older and extinct species in English? It would seem we have no alternative. We must say "Neandertal man."[19]

Obviously, the designation *Neanderthal Man* said nothing about taxonomic status and had been used by Sollas in the previous editions of *Ancient Hunters*. However, the discovery at Piltdown presented even more of a problem to Sollas's linear model because Piltdown Man and *Homo heidelbergensis* were not only likely to date from the early Pleistocene but were also both from Europe. There was therefore not enough distance between the two in horizontal space to substitute for the lack of distance in vertical space, that is, to signify the time necessary to allow an ancestral relationship. Sollas nonetheless tried to avoid contemporaneity: "This period [Chellean], however, was a very long one, so that it does not follow immediately that *Homo Heidelbergensis* and Eoanthropus were in existence at precisely the same time."[20]

Despite the attempts at smoothing the breaks, in the third edition of *Ancient Hunters* there was no linear sequence of hominid fossils anymore, and neither *Homo heidelbergensis* nor Piltdown Man was a direct ancestor of modern humans. Sollas even felt pressured to give up *Pithecanthropus* as a human ancestor, since the American paleontologist William King Gregory (1876–1970) had added his expertise on primate dentition to the already weighty arguments against it: "The reduction of the hypocone and metacone are a result of degenerative processes in the dentition of the most advanced and presumably later races of man, and their presence at this relatively early period in *Pithecanthropus* tends to remove that genus from the line of ascent leading to later human races."[21] Thus Sollas conceded grudgingly: "Its [the molar's] precocious appearance here [great reduction in the posterior moiety] is supposed to exclude Pithecanthropus from the direct line of human descent."[22] There was no fossil left to fill the cipher of human ancestor.

Speculating about the reasons for the change in opinion, one might offer Sollas's orientation toward the French prehistoric community, especially toward Boule and Breuil, even though he disagreed with Boule on certain points, including *Pithecanthropus*, which Boule interpreted as a huge gibbon. Generally speaking, Boule's work on the La Chapelle-aux-Saints Neanderthal had been well received in Britain, and with the advent of World War I the French and British communities seem to have moved closer together because their countries were allies, while the German community, whose country's relations to France, as we have seen, had been troubled for a long time, became further isolated. I have already discussed how Sollas at times felt isolated in his own country, and he also sported outspoken animosities against Smith Woodward, who had reconstructed Piltdown. Although Sollas accepted the Piltdown find as genuine, he never agreed to a Pliocene age for the fossil.[23] It is therefore difficult to make out any clear and strong influence, although his eventual adoption of his friend Boule's model in an atmosphere of general assent may not astonish. To the contrary, one may wonder about the delay; could it be that Huxley's influence was partly responsible for it?

SOLLAS DID NOT include a diagram of his hominid phylogeny in *Ancient Hunters* (1924) that might have helped visualize the Red Lady's new genealogy at one glance. Clearly, Paviland Man's identity had been significantly altered through the expulsion of the known hominids from his ancestry. Other anthropologists were not as self-restraining in their production of family trees, and these tended to place the Cro-Magnons close to the existing white races. The following chapters will explore how the distancing of anthropoid apes, fossil hominids, and living savages from the Caucasian or Nordic group moved the noble Paleolithic race and the highest existing races closer together. In combination with Boule's Cro-Magnon image set in relief through juxtaposition with the Neanderthals, the Cro-Magnons became emblematic of the apex of human achievement in a way reminiscent of de Mortillet's idea that after the Cro-Magnons, humanity had degenerated because of the debilitating influence of religion. This development is illustrated by Elliot Smith's work that tied in with Sollas's race-succession model.

~ 11

The "Evolutionary" versus the "Historical" Model

From the considerations discussed in Chapter 10 it appears that where human phylogeny was concerned, Sollas had been an early proponent of the linear model, to which he adhered longer than many of his colleagues. His main impact on the changes that took place in the initial decades of the twentieth century was rather through his migratory model of human racial evolution. In 1911, in the preface to the first edition of *Ancient Hunters*, Sollas observed that the view of human evolution promoted in that book—that migration rather than indigenous evolution was the engine of change—was generally regarded as heresy, although a similar approach was taking hold in cultural anthropology: "In reviewing the successive Palaeolithic industries as they occur in Europe, I find little evidence of indigenous evolution, but much that suggests the influence of migrating races; if this is a heresy it is at least respectable and is now rapidly gaining adherents."[1] In the preface to the second edition of 1915, Sollas claimed that this view had already become orthodoxy:

> Some views which were admittedly heretical, when originally brought forward [in 1911], have since ceased to be so, and have acquired indeed a dangerously orthodox complexion: at the last meeting of the International Congress of Anthropologists in Geneva much might have been heard of the effects of migration, but little of indigenous evolution.[2]

Sollas here referred to Breuil, who argued that the archeological sequence in Europe should be interpreted in terms of migrations and interactions of peoples rather than as being the result of local evolution, as de Mortillet had proposed.[3] He also had the work of Elliot Smith in mind, who began publishing on a new model of human cultural evolution as early as Sollas. Elliot Smith's thoughts had been strongly influenced by his experiences as professor of anatomy at the Government School of Medicine in Cairo (1900–1909), which informed his notion of ancient Egypt as an important center of cultural innovation and dissemination, including European Neolithic culture. His theory of cultural diffusion, which also entered his reconstruction of the European Paleolithic, found its culmination in *Human History* (1929), where he synthesized human prehistory and history to the extent of doing away with the demarcation altogether. Elliot Smith argued that each cultural innovation had been made only once and subsequently spread to different peoples and regions, in which process it would have been modified.[4] In an article originally published in 1916 and reprinted in the essay collection *The Evolution of Man* (1924), Elliot Smith repeatedly quoted from Sollas's *Ancient Hunters*:

> The issue raised in these quotations [from Sollas] has of late years intruded itself into almost every branch of humanistic study, ethnology and archeology, sociology and politics, psychology and educational theory. The divergence of opinion between the so-called "historical" [meaning cultural diffusion] and the misnamed "evolutionary" school [meaning parallel, local, and linear cultural evolution] is fundamental.[5]

Elliot Smith's ideas were in conscious opposition to the traditional model of Tylor, who regarded the same innovations as having been made several times in different locations because of the unity of the human psyche. Elliot Smith criticized that some ethnologists had influenced archeologists with their belief in a necessary series of cultural steps, with the result that some of the latter proposed an obligatory sequence of flint industries from the ruder to the more refined and polished as found in the archeological record. Thus the fact that the same cultures, for example, the Chellean, were found as far apart as France, South Africa,

India, and America, had been ascribed to parallel evolution. The unity of the human mind was supposed to have led humans in the same direction everywhere.[6]

In contrast, in Elliot Smith's view, it was more likely that the same people had fashioned the tools and/or that they had diffused their knowledge to other peoples and regions. Like Sollas, he envisioned a kind of epicenter of culture, from which ever-new waves of humans in possession of the newest technological inventions radiated outward. The first to reach the peripheries would have retained a culture that was long outdated at the center, so that if one had traveled out from the center to the peripheries, one would have met people of Magdalenian, Solutrean, Aurignacian, Mousterian, Acheulean, and Chellean culture in succession. Since some people of an older culture might have lingered and traveled with a more advanced people, it was possible in the end to find elements of different cultures in strata of the same time period, as was the case, for example, in the New World.[7]

In *Human History* (1929), Elliot Smith still praised Sollas for being essentially the only one who had discussed the entire prehistory and history of the human family in terms of migrations. However, in contrast to Sollas, Elliot Smith did not identify culture with biology. In his mind, a cultural innovation might spread across racial boundaries. Nonetheless, Elliot Smith characterized the modern Negro, for example, as being less able to communicate his feelings through facial expression and as deficient in enterprise and initiative, with some Negro races being nearly devoid of culture. In his phylogenetic "trees" of 1924 and 1929, he put the Nordic race at the apex and drew the parallel lines of the Negro, Australian, and common stem of less pigmented races further back in time than the occupation of Europe by the Neanderthals (Figure 11.1). The Australian appeared as the most primitive race and therefore as closest to the primitive type of the species. Next the Negro race was shown to separate from the main stem; it was considered more specialized but shared the black skin, common also to chimp and gorilla. Then the Mongol and Alpine races diverged. The remaining stem subsequently split into the Mediterranean and the Nordic races, in which the loss of pigmentation was observed to be most marked.[8]

The racial hierarchy visualized in the "tree" was thus justified on the basis of pigmentation:

I should make some explanation of my reasons for putting the Nordic Race at the apex of the main stem. In doing so I am not subscribing to those extravagant claims, so popular at the present moment, in virtue of which blondness is regarded as a character which marks this race as supermen. All that is intended in this scheme is to suggest that those bleaching tendencies, of which several distinct phases are found within the Human Family, are carried furthest in the Nordic Race, which also represents a number of primitive traits that other of the human races have lost. On the other hand, the Mediterranean Race has preserved a number of other primitive characteristics, and especially primitive features of brain, which differentiate it from the Nordic Race. But I have separated it from the main stem mainly on the ground of pigmentation.[9]

However, even though Elliot Smith opposed such Nordic superiority claims as those made in current expressions of the Aryan theory, the leveling notion that every race might have retained certain primitive characteristics is put in perspective by the fact that it was with respect not only to pigmentation but also to brain evolution that the Nordic race was set apart and above. This is particularly true in the case of Elliot Smith, who regarded brain increase as the driving force in hominid evolution.[10]

One of the most striking factors made visible in the "tree" of 1929 is that with the exception of the Mediterranean and Nordic races, Elliot Smith classified the modern human races as more primitive than the Cro-Magnons, and the modern Australians as even more ancient than the Grimaldi type. In other words, he regarded the earliest known specimens of *Homo sapiens*, the Paleolithic races of Grimaldi and Cro-Magnon, as of a higher type than the living Australian and Negro. They represented a relatively late phase in the history of the species (thus the justification of having the *Homo sapiens* stem going further down than the Neanderthal finds). Indeed, the Cro-Magnons were seen as showing affinities to "the Brown Race" and "the Blond Race" of contemporary Europe. Thus the Cro-Magnons again appear as a kind of Paleolithic superrace, budding off only shortly before the apex of the "tree," which is occupied by the Nordic race. Not surprisingly, then, when the Cro-Magnons had entered Europe, the Neanderthals had

The "Evolutionary" versus the "Historical" Model

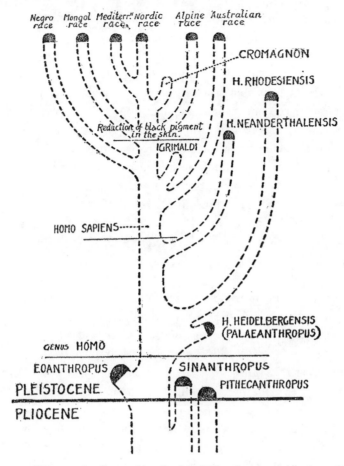

Figure 11.1. "Human family tree" by Grafton Elliot Smith, which shows long independent parallel evolution of the human races, and on which there is no fossil left to represent a direct ancestor of modern humans; from Elliot Smith, 1929, fig. 16, p. 54.

died out. As early as 1916, the advent of the Cro-Magnons in Europe was perceived as "on the cultural side the most momentous event in its [humankind's] history" and as "this great step forward in the history of mankind, when there are revealed for the first time men of essentially the same type as ourselves, endowed with the same intellectual qualities and artistic aptitude."[11]

Drawing on Boule's work, Elliot Smith characterized Neanderthal Man very unfavorably indeed, and clearly as a species distinct from *Homo sapiens*:

> Neanderthal Man is now revealed as an uncouth creature with an enormous flattened head, very prominent eyebrow ridges and a coarse face. The trunk is short and thick, the robust limbs are short and thick-set: the broad and stooping shoulders lead by a curve to the forwardly projected head set on an abnormally thick neck. The hands are large and coarse and lack the delicate play between thumb and fingers which is found in *Homo sapiens*. The large brain is singularly defective in the frontal region. It is clear that Neanderthal Man's limbs and brain were incapable of performing those delicately skilled movements that are the special prerogative of *Homo sapiens* and the chief means whereby he has learned by experiment to understand the world around him, and to acquire the high powers of discrimination that enabled him to compete successfully with the brutal strength of the Neanderthal species.[12]

Once again the primitiveness of the Neanderthals is accentuated by means of verbal juxtaposition with the tall, large-brained, and artistic Cro-Magnons, with an aesthetic sense and reason typical of (some) members of our species. Elliot Smith here also made use of the dominant solution to the problem of the Neanderthals' large brain: Although their brains were big, the most important brain regions were underdeveloped.

Where the non-*sapiens* fossils by then on the hominid "tree" were concerned, Elliot Smith considered Rhodesian Man (Broken Hill Mine, Rhodesia/Zambia, 1922) more primitive than the Neanderthals found in Europe. However, the skull of *Homo rhodesiensis* was likely to be of a later date than the Neanderthal finds, so that Rhodesian Man must have survived longer on the African continent than elsewhere, as many mammals had done, and stagnated in its evolution. Elliot Smith employed the same concept we have encountered in Sollas's scenario of hominid evolution, and which we will find also in Keith's work, when he explained *Homo rhodesiensis*, which he placed with the much older *H. heidelbergensis* in a new genus *Palaeanthropus*, as a remnant of an extremely ancient type that had survived unchanged in Africa. Similarly, although the Aboriginal Australians were not the descendants of the Neanderthals, they had retained primitive traits.[13] Since both *Eoanthropus*, which was close to the human stem, and *Pithecanthropus* were of the beginning of the Pleistocene, Elliot Smith reasoned that there must already have been considerably advanced hominids in the Pliocene or possibly earlier. Therefore, this tree might better be described as a cactus, with many arms rising in parallel to the main stem. As will be discussed in more detail in connection with human antiquity in the following chapter, the lineages of the known (fossil) human races, as well as species and genera, became increasingly viewed as of great antiquity, and their morphological similarities as due to parallelism rather than recent common ancestors.[14]

Elliot Smith's ideas highlight that Sollas's reluctance to give up the known hominid fossils as piers that once continuously connected the human to the animal world should not obscure the fact that the new view of cultural evolution was nonetheless related to the way morphological evolution was conceptualized. For Elliot Smith, a nonlinear model of cultural progress went hand in hand with a nonlinear view of morphological evolution. The "evolutionary" hypothesis therefore also did not apply for the hominid fossil record. Rather, Piltdown Man, Heidelberg Man, the Neanderthals, and the Cro-Magnons had not evolved from each other, but had all migrated into Europe, carrying with them new cultures. When Elliot Smith relegated the known hominid fossils to side branches of the phylogenetic cacti of 1924 and 1929, he thus not only thought of human morphological evolution as having taken place at a center (or centers), from which ever-higher forms had spread out, but also replaced parallel local evolutions of culture with the idea of the spread of culture from a center outward. In this process, he regarded modifications in the brain as the initiating steps from lower to higher forms, each increase in brain structure propelling a new wave of dispersal from the center of evolution. This model he applied to the evolution of the primates as an order and humans as a family of different genera and species, as well as to the racial variants of the human species.[15]

In general, the so-called central Asia hypothesis enjoyed popularity in the 1920s and 1930s. It postulated that humans had originated on the great plateau of Asia, where selective pressures had increased because of deforestation and a decrease in temperature caused by the uplift of the Himalayas. Those apes that remained on the now comparatively dry plains must have perished or adapted to ground living and meat eating; they must have become hominids. As a consequence of the progressively worsening conditions, successive waves of increasingly higher forms of hominids might have been forced to radiate outward. Thus "the successive phases of dawning humanity were passed through somewhere in the east, probably in central Asia. In that case, the periodical westward migration of peoples which is so familiar as a feature of historic times must have begun in remote prehistoric antiquity."[16] According to this view, the regions farthest away and most isolated from the center of evolution and dispersal were prone to harbor the most retrograde forms, such as the Australian Aborigines, Tasmanians, Veddahs, Negroes, Bushmen, Fuegians, and Eskimos today: "The remote lands of the southern hemisphere have always been the refuges in which old types of life have survived long after they became out of date and displaced in the more progressive northern hemisphere."[17]

During his scientific career, Elliot Smith proposed different geographical locations for the center of hominid evolution. He initially had put great weight on Darwin's cautious conjecture of an African origin because of the presence of the gorilla and the chimpanzee, but later reckoned that the center could be somewhere between the Himalayas and Africa. The displacement toward Asia was due to new discoveries. Asia had gained support as the cradle of humanity by Dubois's *Pithecanthropus* from Java and more recently by the Canadian Davidson Black's (1884–1934) *Sinanthropus pekinensis* (today *Homo erectus*) from near Beijing in China (1927–1929), or the Far East. On the other hand, at roughly the same time *Eoanthropus* had roamed England, or the Far West, and soon afterward *Palaeanthropus* had occupied Germany *(Homo heidelbergensis)* and South Africa *(Homo rhodesiensis)*, the southern periphery. Furthermore, in 1924 a very important fossil was brought to light from the limestone works near Taung (northern Cape Province) in South Africa, although its real significance was not recognized until decades later. Nonetheless, Elliot Smith estimated Raymond Dart's (1893–1988) Taung child, or *Australopithecus*, to be the fossil ape that resembled humans most closely, even though he did not regard it as a direct human ancestor because of its small brain. Within his biogeographical framework, Elliot Smith thus inferred a point of origin somewhere around the middle of these places of discovery, preferably southwestern Asia.[18]

∾ Although Sollas's model of human racial evolution and distribution thus shared many aspects with Elliot Smith's, there were also significant differences. For Sollas, the modern human races had their separate origins in Paleolithic races, and up to the 1920s the non-*sapiens* fossil hominids were arranged in a line of descent. He argued for more phylogenetic continuity and did not combine his cultural "catastrophism" with a similarly "nonevolutionary" model of morphological change until he felt forced to. Also, Sollas's cultural model did not go as far as that of Elliot Smith, since within the several lines of descent from Paleolithic to modern human races, "evolutionary" progress, rather than progress through replacement or cultural exchange, would have been possible. On the other hand, Sollas's vision of the human Paleolithic was darker than that of Elliot Smith, who believed in a golden age of peace and innocence before the invention of agriculture and irrigation in the Nile and Euphrates valleys. In the view of Elliot Smith, who attended shell-shock hospitals during the war, such a human aberration as large-scale warfare had been the result of the corrupting effects of civilization.[19] For Sollas, to the contrary, the racial violence of prehistory could explain, if not render necessary, the imperialistic behaviors and nationalistic feelings of the present.

Catalyzed by Boule's brutalization of the Neanderthals, the Red Lady's race came to represent "the Nordics of the Paleolithic."[20] As the antidote to the Neanderthals, the Cro-Magnons were the pride and hope of Western civilizations. However, the Cro-Magnons also served as a warning to civilizations engaged in a horrifying war and perceived as threatened by degeneration. The Victorian evolutionists' belief in progress had been inspired and proved true by revolutionary technological and scientific innovations, the expansion and stabilization of the British, French, German, and Russian empires, and growing economies. At home, there had also been advances in democracy and education. Under the leadership of Great Britain, the world had been increasingly integrated through global exploitation, cultural imperialism, and colonialism. Foreshadowing the trauma of World War I, the optimism associated with these trends had begun to falter toward the end of the century. Although Victorian Britain, the end of which was greeted by the Boer War, had been a time of relative peace and prosperity in Europe, Sollas's, Elliot Smith's, and, as we will see, Keith's model of human evolution originated in post-Victorian times. A fear of degeneration was born, among other things, from the negative effects the social changes forged by the Industrial Revolution had on parts of society. Anthropologists at times linked concerns about the quality of the British race, which became exacerbated in the context of war, with ideas on the struggle for survival of races in the course of evolution, again integrating the past with the present.[21]

This point will be illustrated in the following chapter, where Keith's opposition to Sollas in several respects will be examined. Keith worked with the concept of local evolution of the human races, which he combined with a pivotal role of war as evolutionary mechanism. He drove the notion of long independent evolutionary lines of the modern human races to an extreme. Particularly the British were put at the top of a long and noble pedigree, in which the Red Lady was granted a place. In general, with the modern human races thus reaching far back in time, modern human anatomy had to be considerably older than so far conceived, and none of the primitive hominid fossils yet discovered could serve as an ancestor to any of these. This gave the question of human antiquity great potential for controversy.

~ 12

The Moral Authority of Nature

As Elliot Smith's cactus-shaped phylogeny makes clear, the question of the correct structure of the hominid family tree was closely linked to that of "human antiquity." There was no consensus on how far back in time the independent hominid lineage and humans of relatively modern anatomy could be traced, nor on where Tertiary hominid remains were likely to be found. In this respect, the third edition of Sollas's *Ancient Hunters* (1924) introduced another surprising novelty. In the previous editions Sollas had vehemently rejected the so-called eoliths, supposedly human-made tools from Europe of Tertiary age, but he now accepted some of the Eolithic types from England as genuine tools rather than as mere plays of nature and compared them with the recent material culture of the Tasmanians. Also, Sollas's acceptance of some English eoliths as genuine was the result of long debates, and the arguments around their status as evidence for toolmaking hominids in Tertiary Europe involved the larger issue of the correct hominid family tree.

The controversy around the eoliths had originated in France in the second half of the nineteenth century. De Mortillet had been the most prominent supporter of the dawn tools and the creator of the term *Eolithique*, the Dawn Stone Age, which for him referred to the Tertiary period. In England, the matter was again brought to the surface toward the close of the century. In the early twentieth century, important scientific personalities became pro-eolith advocates, such as Smith Woodward and the former Oxford zoologist Lankester. They lent credibility to the eoliths to the extent of their inclusion in the Stone Age exhibit at the British Museum of Natural History in 1902, at which Smith Woodward had been keeper of paleontology since 1901, and which Lankester directed from 1898 to 1907. The Ipswich amateur archeologist James Reid Moir (1879–1944) and Lankester unearthed and described eoliths from supposedly Pliocene deposits in East Anglia. Lankester referred to the largest group of the boldly flaked flints, which took center stage in his arguments, as rostro-carinate because of their boat- or beaklike shape. Via the rostro-carinate type, Lankester and Reid Moir linked the various East Anglian sites and connected the English eolith industries to those on the Continent. Toward the other end of the timescale, the rostro-carinate was claimed to have evolved into Paleolithic implements. According to Reid Moir, this development was manifest not only in succeeding geological layers but also in the process of manufacture within one culture, so that a partly finished Acheulean hand ax was of a rostro-carinate form.[1]

From the beginning, Sollas had been among the outspoken English opponents. He found flints he considered to conform to Lankester's rostro-carinate type on present seashores that had been produced by the pounding of waves. For him, the type was associated with a wide geological range, from Upper Miocene to Pleistocene and present times, and "belongs to the same class of so-called implements as the eolithic scrapers which will not scrape, the borers which will not bore, and the planes which will not plane."[2] In correspondence, Lankester asked Sollas for a copy of his paper on the matter and to defer his judgment on the Suffolk eoliths until the French authorities Boule and Breuil had examined them. Unfortunately for Lankester, in September 1912, the French experts estimated that the eoliths were mere plays of nature. Lankester had also expressed his hope that Sollas would include a discussion and unambiguous verdict concerning the subject in the second edition of *Ancient Hunters* (1915). This wish was granted when Sollas devoted a long chapter to it in which he argued against the human workmanship of the "tools." Sollas took clear sides in the matter on which "anthropologists [were] divided into two opposing, almost hostile, camps."[3]

Unperturbed by the sharp criticism, Reid Moir continued his researches and discovered at Foxhall Hall, near Ipswich, two floors that contained eoliths. In these cases, the likelihood of flaking due to pressure or chance blows was small. Furthermore, this time the eoliths seemed to be part of a human settlement, and some even showed the action of fire. The combination of these facts inclined Breuil (but not Boule) to accept the intelligent workmanship of these tools, with the exception of the rostro-carinates, which did not seem to form a homogeneous group. However, Bonney, who had been between the fronts of Sollas and Lankester, still had his doubts. He wrote to Sollas in one of the last letters in which they discussed eoliths and "geological Mr. Harris" that "I shall be hard to convince as to the existence of a toolmaking Upper Miocene creature. Reid Moir made such blunders in his work near Ipswich that I cannot accept his evidence, and as to the rudest pre Chellean forms, I think, so far as I have seen, that natural and artificial fractures sometimes are indistinguishable."[4]

Nonetheless, because of Breuil's confirmation of Reid Moir's conclusions, Sollas wrote to the publisher Macmillan concerning the publication of the third edition of *Ancient Hunters* (1924) that "our acceptation of some Pliocene flints as human implements has destroyed the perception of the last edition."[5] In 1922, a commission appointed by the Institut International d'Anthropologie and composed of renowned authorities from France and America visited Ipswich and verified the human-made nature of the flints found there. The same year, Sollas also inspected the collection of Upper Miocene tools from France and accepted their signs of human agency. Consequently, he speculated on Tertiary Man in England in the third and last edition of *Ancient Hunters* (1924):

> The foregoing conclusions, should they be sustained by future inquiry, make us acquainted with the existence in Late Miocene and Middle Pliocene times of an Hominid possessing considerable intelligence, whose hands, preserved or liberated from special functions, were perfected as a universal instrument, while the lower limbs and the foot which had originally possessed the functions of a hand, had become strictly specialised as locomotor organs. Possibly such a being was not far removed from Pithecanthropus.[6]

Thus, once a staunch opponent of eoliths, Sollas came to imagine that *Homo erectus*–like hominids had roamed Tertiary England, where they fashioned tools similar to those of the modern Tasmanians.[7] The eoliths, now broadly accepted, engendered wild speculations about their makers.

From the very beginning, Reid Moir had been well aware that he needed the bones of Tertiary Man to decide once and for all on the human origin of his tools. In October 1911, he thought that his wish had come true when a partial human skeleton was found beneath the Chalky Boulder Clay near Ipswich. He sent the bones of the possible shaper of the Suffolk eoliths to no lesser person than Keith, conservator of the Museum of the Royal College of Surgeons. Although Keith found that the bones were of modern human anatomy, could not be said to be fossilized (mineralized), and still contained a comparatively high percentage of organic matter, he concurred with Reid Moir's ascription of the skeleton to the Pliocene. To lend credibility to his claim for an anatomically modern English skeleton of Pliocene age, Keith brought in other European human bones as collateral evidence, albeit ones that had been rejected by the majority of the scientific community because of their doubtful origin and age. Among these were the controversial Foxhall jaw, discovered in the Red Crag near Ipswich in 1855, the Moulin-Quignon mandible, discovered in a gravel pit at Abbeville in the Somme Valley in 1863 for Boucher de Perthes, and the Galley Hill skeleton, a human skull and limb bones found close by the Thames in 1888.[8]

The conceptual framework with which Keith approached the Ipswich remains was outlined in *Ancient Types of Man* (1911). Here he argued for the great age of modern human anatomy and established a long genealogy for certain Europeans. He contradicted Sollas's theory of blood ties between the Paleolithic Grimaldi type and the recent Bushmen and between the Paleolithic Chancelade specimen and the recent Eskimos. He considered both Grimaldi, which provided a link to the Galley Hill find, and the Chancelade type as mere varieties of the Cro-Magnon race that was later absorbed by the modern "tall men of commanding mien" in Switzerland, France, Germany, and Britain. Keith clearly respected the Cro-Magnons as ancestors to be proud of since "in strength, in stature, in physique as well as in size of brain the Cro-Magnon race represents one of the most stalwart human races ever evolved," "one of the finest the world has ever seen."[9] Keith constructed a line of descent from the Neanderthals to Galley Hill Man, which he claimed to be 200,000 years old, to Cro-Magnon varieties and to what he perceived as the highest modern European type. He still concurred with Sollas's 1908 article that the Neanderthals were not a separate species but the parent stock of the modern human races.

Keith's hominid phylogeny was at that point still unilinear. *Pithecanthropus erectus* had been ancestral to *Homo sapiens neanderthalensis* (under which he subsumed Heidelberg Man). However, this unilinear model had to account for what might seem paradoxical claims, such as his estimate that the big-brained Neanderthals had existed before the Pleistocene, that at the Pliocene-Pleistocene boundary there had been an intelligent toolmaker in Europe (inferred from the eoliths of East Anglia), and that Dubois had assigned a similar date to *Pithecanthropus*. He solved the problem of contemporaneity of forms the same way Sollas did, by suggesting that *Pithecanthropus* (according to Keith *Homo javanensis*) was either older than was generally thought or a remnant of an old form among further-advanced cousins.[10]

Already the following year, however, in his Hunterian Lecture of 1912, Keith switched to a dendritical phylogeny, to which he gave ample formulation in *The Antiquity of Man* (1915). He now built his case for the great antiquity of the modern human type, besides the Galley Hill remains, mainly on the eoliths from the Pliocene Red Crag of East Anglia and on Reid Moir's morphologically modern human skeleton from Ipswich, to which he assigned an age of approximately 600,000 years. The Piltdown discoveries of 1911–1912 were the strongest pillar in Keith's construction of the early occupation of Britain by a relatively modern-brained hominid only insofar as that, in contrast to the Ipswich Man and the eoliths, they were widely accepted. To Keith, the eoliths postulated on the Continent made it plausible that human antiquity had to be expanded to the Miocene, to almost two million years, according to his timescale. Of these European toolmakers he expected an average brain size of at least one thousand cubic centimeters, a size reached by the younger *Eoanthropus dawsoni*, *H. neanderthalensis* (now a separate species), and *H. sapiens*, but not *Pithecanthropus erectus*. With a relatively modern human anatomy going that far back in time, a dendritical model could accommodate the known fossils more easily than a unilinear one. Keith expelled all non-*sapiens* remains so far discovered from the line leading to modern humans and postulated an as-yet-unknown human ancestor in the distant evolutionary past.[11]

However, although *Sinanthropus*-, *Pithecanthropus*-, and Piltdown-like hominids were proposed as the shapers of eoliths and Tertiary inhabitants of East Anglia, Ipswich Man, Reid Moir's and Keith's favorite, never gained general acceptance as being of significant antiquity. Both sides, those who supported the skeletal remains of European Tertiary Man and those who did not, accused each other of interpreting the evidence with priority given to theory. Boule, Sollas, and others perceived in Keith's advocacy for doubtful bones an attempt to create evidence for his theory of the great antiquity of morphologically modern humans. On the other hand, Keith and Reid Moir argued that Tertiary human bones were rejected because they went counter to the expectation that such early hominids had to be nonmodern in morphology.[12]

In addition, in the last paragraph of *The Antiquity of Man* (1915), Keith made use of a strategy common among eolith advocates. The initial skepticism the pioneers of human antiquity had encountered from the scientific community in the first half of the nineteenth century was brought forward to gain something of the martyr aura (see Part I). At the same time, the parallel suggested that the eoliths and skeletal remains of Tertiary Man, too, were genuine, and that it was only the conservatism of certain scientists that kept them from their due inclusion among the evidence for a great human antiquity. In the nineteenth century, opponents of prehistoric humans had made use of the supposed absence of genuinely fossil human bones from Quaternary deposits as negative evidence for their nonexistence. The admittance of absence of evidence as evidence of absence by the scientists of the past could now be represented as a non sequitur and enlisted against the eolith deniers, since they, too, based their rejection at least partly on negative evidence.[13]

Both the human workmanship of the paleoliths in the first half of the nineteenth century and the human workmanship of the eoliths in the first half of the twentieth met with the same counterarguments: the lack of design (human intentionality), the lack of other kinds of artifacts (for example, nonstone), the nonfunctionality, the lack of progress, the possibility of experimental reproduction by chance blows, and the lack of human remains. However, the comparison strategy was unfair in the sense that although human antiquity clearly was at issue in the first half of the nineteenth century, the dividing line in the eolith problem did not separate proponents and opponents of Tertiary Man. It separated those who theorized Tertiary Man in Europe from those who found Europe a most unlikely place to encounter the remains of Tertiary Man, since humans were seen as having arrived in Europe at a relatively late stage in the course of their evolution. Even the latter scholars were positive that Tertiary human bones would be found in Asia (which may be seen as another parallel to the nineteenth-century debates as discussed in Part I).[14]

In stark contrast, Reid Moir perceived a linear development in the English Stone Age from the earliest to the most advanced industries:

> So far as actual evidence of man's former presence goes, we have in East Anglia, as those who have read these pages will, I think, agree, a wonderfully complete record of nearly every stage in human progress from the earliest and most primitive flint implements, to the advanced types made at the close of the Stone Age. Thus, it is possible, that what is now England was the home of the earliest men, and there can be little doubt that if a tithe of the money spent upon researches in other parts of the world were expended upon archeological work in Eastern England, still further and more important discoveries, bearing upon the question of man's origin, would be made.[15]

It was especially this local progression that was strongly opposed by eolith opponents such as Boule. Because of his migratory model of human evolution, he had no problem accounting for the sudden appearance of a relatively sophisticated culture in Europe, the Chellean. He did not need to desperately search for any precedents, such as eoliths, and their derivatives in Europe. As early as 1905, Boule made this very clear in an article titled "L'Origine des éolithes": "With regard to this question . . . great consideration must be given to a phenomenon that has played an important role in the history and development of all groups of fossil beings, the phenomenon of migrations. Nothing proves that the evolution of the human species or the human genus, as you like, has taken place locally."[16]

To the contrary, for Keith, the search for anatomically modern human remains in late Pliocene or early Pleistocene British deposits was embedded in his idea that each human race had evolved a long time ago in the area where it was found today. Keith thought that the modern human races differed enough from each other to suggest their divergence from an Australoid type before the beginning of the Pleistocene, since for modern Australians "with brains of smaller size than 930 c.c.

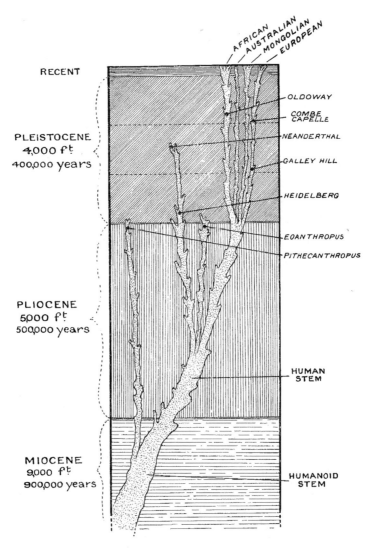

Figure 12.1. Phylogenetic "tree" from Arthur Keith, 1915a, fig. 187, p. 501.

we can scarcely expect a human intelligence. Of all the races of mankind now alive, the aboriginal race of Australia is the only one which, in my opinion, could serve as a common ancestor for all modern races."[17] In his genealogical diagram in the 1915 edition of *The Antiquity of Man* (Figure 12.1), Keith therefore not only relegated all non-*sapiens* fossils to parallel side branches of the main stem but also had the latter split into the parallel "twigs" of human races before the end of the Pliocene.

In the text, Keith refined his diagrammatic distinction between the African, Australian, Mongolian, and European types. Having evolved far beyond the original lowly Australoid stage, the British were provided with the following genealogy: the Eolithic Ipswich Man of modern anatomy some 600,000 years of age; the Chellean Galley Hill Man; the Acheulean Dartford Skull (Swanscombe); Mousterian Man (implements in Thames Valley, but no Neanderthal remains in England); the Aurignacian Red Lady; and Solutrean and finally Magdalenian Man (engraved horse, Cresswell Crags; Thames Valley). Keith's view of human racial evolution as local thus clashed with Sollas's migratory model, and Keith tore Sollas's theory of racial migration apart in a review of the third edition of *Ancient Hunters*.[18]

Although it is impossible to divide the eolith proponents and opponents into two well-defined camps where theoretical background is concerned, the eolith question was linked to the Praesapiens theory. This is indicated by, among other things, the negative responses to Ales Hrdlicka's (1869–1943) article "The Neanderthal Phase of Man" (1927). The Czech-born physical anthropologist at the National Museum of Natural History of the Smithsonian Institution in Washington, D.C., supported the view that the Neanderthals had been ancestral to modern humans. Hrdlicka claimed, among other things, that although some scientists traced the appearance of modern humans even further back in time than the Neanderthals, this was done without any trustworthy evidence.[19]

Reid Moir was upset by Hrdlicka's negation of Tertiary Man. He also objected that the Mousterian industry was clearly demarcated from the preceding Acheulean and clearly inferior to the succeeding Aurignacian. He seems to have functioned as Keith's mouthpiece when he expelled the Neanderthals from human ancestry primarily on the assumption of the great antiquity of humans of modern anatomy in Europe, particularly in Britain. Indeed, at one point they even argued that in Britain, the Mousterian culture had not been fashioned by the Neanderthals, but by a modern human type who had already mastered the art of pottery. One may speculate that they envisioned pre-Dawn Man to have eventually evolved into the anatomically modern "British," leaving behind the Suffolk eoliths and all succeeding industries along the way, bypassing *Homo neanderthalensis*, a form more primitive in morphology and culture that never made it into Britain.

Although this is an extreme scenario, until after World War II, the proponents of a Neanderthal ancestor were a lonely minority.[20]

The presence of early Tertiary tools in Europe could even support the so-called lemuroid or tarsioid theory, which might be seen as an outgrowth of the Praesapiens model. It postulated that the hominid line had bypassed the anthropoids completely, originating instead from a tarsier-like creature in the Eocene. Sollas was well aware of the credibility true eoliths gave to the idea of a European Tertiary Man of relatively modern anatomy, and thereby to the notion that human evolution never went through an anthropoid stage, but that humans had arisen through parallel evolution "uncontaminated by intercourse with the apes."[21] Because in the last edition of *Ancient Hunters* (1924) Sollas finally accepted some of the eoliths as evidence for human agency, he conceded: "if stone implements could be proved to occur low down in the Tertiary Series, in the Oligocene strata, for example, it [the lemur theory] would have a claim upon our serious consideration."[22]

Keith took more than a page in the review of *Ancient Hunters* (1924) mentioned earlier to celebrate his triumph in the eoliths question: Finally, even the former opponent of human Tertiary tools in Europe had to accept their genuine nature and to pay credit to those who had stubbornly continued to research the question. The review was written in such a tone of defeat for Sollas that the latter felt impelled to publish a reply. To Keith's mocking tone on Sollas's cutting of the duration of the Pleistocene period, Sollas responded irritably: "We now come to the chronology, and here again we encounter terms that seem more appropriate to party politics than the discussion of a scientific question: 'Progress and liberal'; 'niggardly and retrograde'! Does the Reviewer imagine that geological time is at my own disposal to give or withhold as I please?"[23]

On this point, however, Keith did not have the last laugh, and in the second edition of *The Antiquity of Man* (1925), he reduced the timescales for the geological epochs. This enhanced his dependence on parallel evolution. He gave up the idea that the early fossils represented anachronisms that had survived into later times in peripheral regions, but interpreted them as representing the evolutionary stages at which the human branches in different geographical regions had actually found themselves. Similarly, supposedly Pleistocene protoraces, remains of which had been discovered on several continents, were put on the parallel lines he traced from the modern races back to the Pliocene-Pleistocene boundary. From this mid-1920s outlook, Keith's evolutionary framework became increasingly marked by parallel lines through some of the hominid fossils known in different geographical regions. In the "tree" of 1948, the beautiful Cro-Magnons eventually came to represent a final stage in the independent ascent of the Caucasian race from an australopithecine precursor.[24]

Even though where the modern human races were concerned, Keith had always emphasized local developments, his scenario of their history was no less aggressive than the race-succession scenario. Keith's thoughts on human racial evolution had begun to take form while he was in contact with a foreign people after his studies of medicine at the University of Aberdeen (1884–1888). His three brothers had already emigrated to Australia to work as a Presbyterian minister and gold miners, respectively, when he spent several years (1889–1892) as a medical officer for an English company at gold mines in Siam (now Thailand). It was here that he realized that his own stock was of the competitive and aggressive kind in comparison with the docile Siamese, who had therefore little chance in the interracial struggle for supremacy.[25]

These ideas were substantiated during World War I, when Keith had long been back in England, where he found his world changing radically. The Museum of the Royal College of Surgeons, where Keith had been conservator since 1908, was crowded by officers who trained recruits for the Army Medical Service. When the Germans bombed Paris in 1914, the museum was closed, and a fireproof chamber was built underneath the college for the most precious specimens. Keith reinstalled his lectures on the basics of human anatomy, which were relevant for first-aid ambulance services, at the college and gave lectures on the medical treatment of men whose joints had been disabled in the war. Keith's sense of the needs of his country grew so strong during the war that he gained an understanding of the workings of patriotism: "By the latter date no inducement whatsoever could have tempted me to change my nationality; war hardened the sense of nationality in all of us."[26] Although war thus strengthened the feeling of unity, it had no doubt been a terrifying experience for Keith. He entered in his diary, "Rather ashamed of the way my muscles behave when the raid comes," and he recalled in his autobiography, "My teeth chattered, my limbs trembled, and a strange feeling went down my spine in spite of all my efforts at control."[27] It was this picture of war as terrifying, but as creating a sense of national identity, that entered Keith's evolutionary theory as an indispensable mechanism for human progress.

In a long series of lectures, articles, and books, Keith theorized on the central role of war and imperialism in human evolution as race-forming mechanisms, as catalysts of intratribal cohesion and intertribal competition. In an idiosyncratic interpretation of Darwinian and Spencerian argumentation, Keith reasoned that while new races arose through geographical isolation, isolation for humans ought not to be geographical, but naturally arose because of competition and inborn jealousy between neighboring groups.[28] When jealousy reached a certain level, or a group needed more territory, war would ensue. He concluded that even World War I was a race-forming war in the sense that although the nations involved were not pure races, they were in a process of isolating themselves from their surroundings, developing a strong sense of identity, and by inference constituting a breeding community. Keith not only talked of nations as incipient races, with nationalism being the equivalent of race prejudice, but went as far as treating race hatred, race riots, national jealousy, frontier disputes, patriotism, caste, class, and party formation as factors that lead to group isolation, and thus as essential to human evolution.[29]

For Keith, a struggle for survival between hominid genera, species, and races had been going on throughout hominid evolution. In this regard, history and the present were continuous with prehistory, and in an article on Darwin published in *Science*, he even seems to have implied that there could only be one survivor among the living races, as there had once been only one survivor among the ancient hominid genera and species.[30] In order to achieve this kind of continuity also with respect to imperialist practices, Keith eventually allowed for one Paleolithic migration as the exception to his general rejection of migration as an important factor in modern human origins. The Cro-Magnons had probably evolved in western Asia, from where they had spread to Africa north of the Sahara and where they had remained isolated from (and uncontaminated by) the sub-Saharan Negro race. They also invaded Europe, where they extinguished the Neanderthals:

> It is now clear that towards the end of the last ice age an evolutionary tragedy was enacted in Europe very similar in nature to

that which is now taking place under our eyes in the Continent of Australia. A vigorous people, the early Caucasians [i.e., the Cro-Magnons], colonised Europe and in the process exterminated its native Neanderthalian population.[31]

Obviously, in explaining the past through analogy with the present, the present appeared as the natural consequence of the past. In the prehistoric struggle for territory and survival, the Cro-Magnons had proved themselves the winners, the Paleolithic supermen, the Caucasians or even Nordics of the Paleolithic. However, Keith was no pure materialist in this respect and certainly no Darwinian in the present-day sense. Rather, for Keith, evolution was progressive and purposeful. Within a framework that increasingly worked with parallel branches that represented independent lines of development through some of the known fossils, anatomical similarities between these branches came to be accounted for by what Keith called "collateral evolution" or the "law of uniform evolution," which said that ancestors contain latent tendencies for development in certain directions that will only become apparent in their descendant lines. For example, the common hominid stock of the early Pliocene had a latent tendency to cerebral increase that exhibited its first structural manifestation in different branches of their descendants.[32]

This brings me back to Sollas, who, although he disagreed on nearly everything with Keith, had a similar understanding of the role of natural selection in hominid evolution and was not at ease with a purely stochastic view of the process. In fact, Sollas believed in some power or spirit behind the workings of the world that was mindful of the human lot, and although he kept science and religion mostly apart, he saw no contradiction between the two.[33] The passages in which Sollas explicitly indulged in thoughts about the actual mechanisms of evolution are few; one of them is the following:

> Here we are face to face with that mystery of mysteries, the problem of evolution, for which no ingenuity, however great, has yet furnished a solution. Natural selection, that idol of the Victorian era, may accomplish much, but it creates nothing. In matters of invention, discovery, the attainment of skill, we have some experience of the inner nature of the process; it involves the mind, with

its powers of observation, reflection, and imagination, and it is accompanied by a sense of effort. When the effort is light and the result appears disproportionately great, we speak of it as inspiration, and this is another mystery. If these experiences within ourselves correspond with a progressive modification of the substance of the brain, then it would seem possible that the fundamental cause in the whole process of evolution is in reality an affair of the mind.[34]

Sollas granted natural selection the power of destroying the unfit variants in the course of racial evolution, but it had not been the source of progress. Progress was seen to depend on an effort of the individual, or of the race, that would then feed back into the system in a "Lamarckian" or "Spencerian" way and survive into the next generation. Far from subscribing to the notion of randomness in Darwinism, he perceived humans as active shapers of their destiny. A race's stage in the hierarchy of progress was therefore at least partly a result of its own neglect or achievement. Once such a hierarchy had developed in the course of evolution, the supposed disparities in intellect between the races and in the propensity for cultural innovation meant that progress depended on warlike and imperialistic behavior.[35] Because these behaviors would ensure the prevailing of the fitter, they thereby achieved a legitimate qua inevitable status. For Sollas, these notions could account for the marginalization and extermination of those races that did not make the best use of their potential and of their land:

> What part is to be assigned to justice in the government of human affairs? So far as the facts are clear they teach in no equivocal terms that there is no right which is not founded on might. Justice belongs to the strong, and has been meted out to each race according to its strength; each has received as much justice as it deserved. What perhaps is most impressive in each of the cases we have discussed is this, that the dispossession by a new-comer of a race already in occupation of the soil has marked an upward step in the intellectual progress of mankind. It is not priority of occupation, but the power to utilise, which establishes a claim to the land. Hence it is a duty which every race owes to itself, and to the human family as well, to cultivate by every possible means its own strength: directly it falls behind in the regard it pays to this duty,

whether in art or science, in breeding or organisation for self-defence, it incurs a penalty which Natural Selection, the stern but beneficent tyrant of the organic world, will assuredly exact, and that speedily, to the full.[36]

Sollas reasoned that it was the responsibility of the race and its members to develop their potential to the fullest, whether through cultural inventiveness, militarism, or breeding. If societies did not live up to this, a more thriving race would take over, which in this case would mean only the fulfillment of natural selection.

⁓ As EXPLORED IN the essay collection *The Moral Authority of Nature* (2004), which has served as the inspiration for this chapter's title, throughout history, protean nature has appeared in many forms, bearing various kinds of authority.[37] Depending on the context in time and place, humans have used nature in their definitions of the worthy and the just. We have already encountered this phenomenon in Part I, Chapter 6, when Buckland claimed lawfulness for his social order through invocation of the natural order and vice versa. But there exists a significant difference between the moral authority of this nature and the ones discussed in this part. Buckland's nature ultimately derived its authority from God, mediated through the interpreter as savant and priest. Sollas's, Keith's, Boule's, and Elliot Smith's appeals, as discussed in the previous chapters, were to nature alone (even if animated by capitalized Natural Selection and intentional human minds), whose scientific expert interpreters they were. Nonetheless, in the rationalization of imperialism, the authority of nature in matters of organic progress was sometimes applied with nearly religious fervor.[38]

In theories of human evolution in the first decades of the twentieth century, nature's authority applied to all three of its main domains established by Lorraine Daston and Fernando Vidal: It naturalized values, defined necessities and freedoms, and arbitrated boundaries. It valorized the colonialist's European culture and biology higher than the philosophies and ways of life of the colonized. What was perceived as the human (or racial) potential, the "is," collapsed into the "ought" of the moral imperative to exploit (human and external) nature, so that the failure to do so turned into degrees of freedom wasted. Both of these aspects of nature's authority interdepended with the anthropologists' ability to naturalize the boundaries between different types, nations, and races and the hierarchies in which these were arranged. Among the boundaries that became more pronounced were those between Cro-Magnons and Neanderthals, hominids and anthropoids, *Homo sapiens* and other hominids, and higher and lower races or nations. The gaps between these categories and their occupants were deepened not least in the visual spaces of phylogenetic diagrams. In the hands and minds of Keith, Sollas, Boule, and Elliot Smith, the extant human races were ever more distanced from each other through the establishment of long independent lines of evolution. Simultaneously, the same was achieved for *Homo sapiens* with regard to fossil hominids, and for hominids in general with regard to living and fossil apes.

Historians of anthropology who have investigated these changes have pointed out that they originated in the contexts of imperialism and war.[39] Direct contact with and observation of nonwhite races subjugated under imperial rule, and the experiences of World War I, had strong impacts on paleoanthropologists' outlooks on human nature and evolution. Increasingly, racial conflict was viewed as having been or indeed as still representing a driving force in human evolution. It appeared to be in the nature of progress that some human races had to compete with and replace others less advanced in morphology and culture. The notions of race and of nation were thereby not always clearly differentiated.[40] Imperial expansion and national rivalry were rationalized within the Darwinian framework of struggle and the survival of the fittest, even though paleoanthropologists still allowed for alternative mechanisms for evolutionary change.[41] This credo could with one blow make prehistory, history, and the present continuous, naturalize imperialism and for some also war, and provide arguments for the stronger institutionalization and implementation of anthropology and its tools of race analysis and improvement. It is within this context that Sollas's work on the Red Lady and the conception of his evolutionary theory was situated, and it is this context that raised the meaning of the Red Lady's race to a link in the chain of arguments for British or French national superiority, or to a warning against what was perceived as degenerative tendencies in the most civilized societies.[42]

∽ Conclusion: Turbulent Times for an "Old Lady"

THE REANALYSIS OF the Red Lady and its associated artifacts within an evolutionary framework took place at a time of great changes within the young sciences of paleoanthropology and Paleolithic archeology. It seems that where the evolution of the human species was concerned, apart from Boule's work on the La-Chapelle-aux-Saints skeleton, other developments had furthered the general shift in the English community from a unilinear and indirect linear to a truly divergent model, or model of parallel evolution of apes and humans, different human genera and species, and/or the human races (diagrammatically visualized in Models A to D in Appendix B). New fossils had been discovered, the most important of which was Piltdown Man, which only many decades later was exposed as a forgery consisting of an orangutan jaw and a modern human skull. With a European hominid around the Pliocene-Pleistocene boundary that had a nearly modern-sized cranium, the small-brained *Pithecanthropus* could hardly be a direct ancestor.

Boule's diversified model of human evolution, as well as Piltdown Man, which also flattered English nationalism, fit the theories of the English paleoanthropologists, such as Keith's great antiquity of the modern human type and of the human races and Elliot Smith's priority of the brain as the Rubicon in human evolution.[1] The experiences of imperialism and eventually World War I had helped kindle evolutionary racism. Theories of human (racial) evolution that stressed conflict, such as the Praesapiens model, which could serve to legitimate imperialism to the extent of genocide and war, grew in a climate of general nationalism and white supremacist ideology.

I have come to regard Sollas as resisting rather than promoting a dendritical view of hominid phylogeny. However, drawing on earlier syntheses of Paleolithic archeology with ethnology and borrowing from the new trends in cultural anthropology, Sollas synthesized the interpretation of fossil human races, the analysis of Paleolithic industries, and studies of the biology and culture of recent races into an imperialist model of human racial evolution. He argued that cultural and biological progress, rather than being the result of linear local evolution, had been brought about by the migrations of new races into territories from which they drove the racially inferior inhabitants. He believed that the savage modern races and their cultures at the supposed peripheries of the earth were the descendants of the various Paleolithic races and industries that had successively been driven out of Europe during the Pleistocene. In cases in which fossil human bones had not been discovered to represent the shapers of a certain Paleolithic industry, Sollas inferred biological affinities between Paleolithic and living races on the basis of cultural similarities alone.

Thus Sollas interpreted the Australian Aborigines, for example, as closely related to the Neanderthal race. They represented relics of the primitive human type who had persisted into recent times at the antipodes so as to become contemporaneous with the most advanced European races. This concept is reminiscent of the *scala naturae* principle or the polygenist racial hierarchies established in nineteenth-century physical anthropology, rather than designating a model of evolution as truly dynamic and diversifying. It also played a crucial role in Sollas's ideas on the evolution of the human species. I have interpreted his notion that some pre-*sapiens* fossils represented relics of a stage of evolution preceding the one to which they were geologically/paleontologically dated as a last attempt at saving a more linear model of the evolution of the human species. Sollas's conception of space in both the horizontal and the vertical dimensions as a measure for time, and thus phylogenetic distance, allowed him to explain the contemporaneity of different hominid species or genera within the linear framework, since representatives of lower forms could survive into relatively recent times when they were marginalized in "forlorn" geographical regions.

To summarize Sollas's theoretical development, he was an early proponent of a model of the biological evolution of the human species structured along the lines of Appendix B, Model A, and changed to an indirect linear model when he was confronted with the contemporaneity of certain fossils (Model B). By the time of the third and last edition of *Ancient Hunters* (1924), he felt forced by the overwhelming support in the international community for a dendritical "tree" to adopt a diversifying model in which there were no fossils on the line leading to modern humans (Model C). However, Sollas continued to reject a Praesapiens model, in which the origin of modern human anatomy was pushed ever further back in time, in its most extreme expression as the tarsioid theory even sidestepping the anthropoid stem altogether (Model D). Nonetheless, Sollas's imperialistic theory of human cultural evolution, in which migrations and the replacement of lower with higher races were the reasons for the cultural progress observed in the archeological record of Europe, in fact corresponded nicely with a more diversified view of the evolution of humans as a species. As we have seen in the case of Elliot Smith, for whom all the known hominid fossils from Europe represented successive but not ancestral invaders into this region, the "nonevolutionary" or "historical" model could account for morphological as well as cultural progress beyond the human species.

Such views were generally associated with the belief in a center of evolution, most popularly in Asia, from which ever-higher hominids had spread across the globe so that the earliest and simplest forms would have reached the peripheries at a time when higher forms could already be found close to the center. The concept of certain fossils as evolutionary anachronisms was therefore also part and parcel of this framework; they would have stagnated in isolated regions, in which there seemed to be little struggle for survival. According to a widely held worldview in which the Western world constituted the intellectual center, the coexistence of higher human forms at this center with lower forms at the peripheries of the earth was as visible in the contemporaneity of the modern European and the modern Tasmanian or Australian as it was in the prehistoric record in the Javanese *Pithecanthropus* and its contemporaneity with the more advanced *Homo heidelbergensis* in Germany or Piltdown in England. Indeed, the waves of migration emanating from Europe to bring progress to other regions in present times appeared as a simple inversion of the prehistoric past. The difference between the concept of relics in this scheme of evolution and Sollas's early indirect model is that the relics no longer represented an earlier stage through which Europeans had actually ever passed. Rather, they were only marginally related cousins, and in the case of pre-*sapiens* relics they represented evolutionary dead ends.

The fact that Sollas did not adopt the parallelism inherent in the hominid evolution scenarios proposed by Elliot Smith, Keith, and others allowed him to be among the first to welcome the newly discovered australopithecines as significant missing links between apes and humans rather than as fossil apes. It is within this framework of interpretation that Sollas's exceptionally early acceptance of both *Pithecanthropus* and Dart's Taung australopithecine child as closer to hominids than to anthropoids is no longer surprising.[2] In 1924, when Dart received the Taung skull, Sollas was exceptional because he did not expect a large-brained Asiatic ancestor, but something closer to what Darwin had predicted and Dart's discovery represented, namely, a small-brained bipedal African ancestor. In the then-prevalent brain-first theory, supported by the Piltdown forgery, there was no place for the Taung child. In 1933, in another of his detailed craniological studies of recent and fossil hominids, Sollas pronounced Piltdown "a morphological anomaly and a palaeontological anachronism"[3]—it had never fit into his linear sequence of gradual brain development in hominids, even though it was a perfect illustration of his concept of mosaic evolution.

A close look at the works of Sollas, Elliot Smith, Keith, and to a lesser extent Boule has revealed differences, as well as similarities. In all cases, an attempt at synthesizing recent events and the personal experiences of war and empire with a picture of human prehistory and history did take place, albeit in markedly different ways. Although all four came to embrace a dendritical hominid phylogeny during the initial decades of the twentieth century, Keith envisioned the human races as having evolved in parallel in the areas in which they are found today, with racial prejudice and intertribal violence as engines of progress, while Sollas regarded the replacement of lower with higher races as the reason for the cultural progress observed in the archeological record of Europe. Elliot Smith, who was highly suspicious of such outgrowths of modern societies as slavery and war, nonetheless distanced non-European races because of their skin pigmentation and brain size and thus regarded the

Australian, for example, as further away from the European line than the Paleolithic European races of Grimaldi and Cro-Magnon.

Even before Sollas's reanalysis of the Red Lady, it had been caught up in arguments about the amount of diversity in the European fossil record and about whether this record would best be accounted for by a framework of successive invasions into Europe, according to the Aryan theory, or as constituting a linear and local development. Building on the insights gained into Paleolithic archeology in southern France and on the definition of the Red Lady as Cro-Magnon in *Crania ethnica*, Sollas's and Breuil's reexcavation of Paviland Cave authoritatively turned the Ancient British Red Lady into an Aurignacian Cro-Magnon Man. This work not only cemented Sollas's place among anthropologists but also "corrected" his system of ancient hunters and their modern representatives as presented in the first edition of *Ancient Hunters* (1911). The race of Cro-Magnon, once a chimera in morphology and behavior, capable of works of art and driven by instinctive violence, assumed the role of an ancestor of modern Europeans marked by a beautiful physique, high intelligence, and a refined mind. Within Sollas's new version of human evolution in the third edition of *Ancient Hunters* (1924), the great warrior race was no longer simply responsible for driving the Mousterian race of Neanderthals out of Europe to the peripheries of the earth, but had been the cause of extinction of the Neanderthal species, by now seen as different and rather brutish.

With their elaborate industry, including ivory, bone, and antler as materials, and their fascinating cave frescoes, statuettes, and engravings, many anthropologists perceived in the Cro-Magnons the beginning of true humanity. Within scenarios marked by parallel evolution, hyperdiversity, and competition, the Red Lady and her race symbolized the good in human nature, expressed through a sophisticated culture, with an art that some regarded as surpassing that of ancient Greece.[4] Because they had been the winners in the interracial struggle for survival, they gave hope for the prevalence of these qualities over the dark and non-progressive side of human nature, epitomized in the mindless force and primitive state of the Neanderthals and their supposed present-day analogues. The juxtaposition of the noble Cro-Magnons with the savage Neanderthals thus reverberated with earlier attempts to make evidence of Paleolithic humans fit the Aryan theory, as well as with contemporary ideologies and practices. With the expansion of modern human antiquity, the expulsion of all known fossils to evolutionary sidelines of the main line leading to modern humans, and the extension of the independent development of existing races, the Cro-Magnons became even more tightly linked to the Caucasian group, So much so that the nonwhite living races appeared to have left the stem leading to the existing white races before the advent of the Cro-Magnons. This notion was driven to extremes in Keith's racial phylogenies, in which the Red Lady represented an important stage in the long line of ascent of what he conceived of as the incipient British race, emphasizing the independent origins of the existing races. Increasingly, he reinserted the known fossil hominids on the parallel lines leading to the modern types, tightening the bond between "the Nordics of the Paleolithic" and the Nordics of today.

I thus leave this second turning point in the Red Lady's afterlife, which was a turning point in the young science of paleoanthropology at large, with the observation that as during Buckland's time, during Sollas's time the personal and educational background of the interpreter, his institutional and parainstitutional affiliations, his outlook on human evolution and practical approach to fossil bones, and the wider scientific and cultural context codetermined how the bones were read. As an individual representative of the Cro-Magnons, the Red Lady was instrumentalized, not only by Sollas, for a certain politically inspired vision of human origins. At the same time, the skeleton influenced Sollas's conception of ancient hunters and their affiliations with modern representatives.

~ *III*

An Interdisciplinary Team, an Early Upper Paleolithic Shaman, and a Definitive Report

The preceding parts have been structured around the history of interpretation of the Red Lady of Paviland, situated in what might be called chapters in the prehistory of mainly British paleoanthropology. The notion of contextualization has been broad indeed when applied to the careers of the scientists involved with the Red Lady's fate and their cultural backgrounds and scientific contexts, as well as some institutional history of Oxford University. This was considered worthwhile in order to gain an understanding of how the successive definitions were reached and kept alive, but also of what it meant to be postdiluvian or a Cro-Magnon within the larger meaning systems of a time. Therefore, the Red Lady has been caught up in many fundamental questions troubling those engaged with human prehistory that revolved around the binaries "human recency/antiquity," "taxonomic diversity/unity," "migration/local continuity," and "similarity/difference" between contemporary "primitive" and Paleolithic peoples. The stances toward these binaries of those involved with human origins have turned out to be complex and unstable, pointing to the fact that they are gross simplifications. Since William Sollas's times, no agreement has been reached on any of these issues, and as is the subject of this last part, the Red Lady continues to be implicated.

Indeed, the anthropological sciences have acquired an entirely unmanageable size and complexity through advancement, increase, and diversification with regard to theories, technologies, and the hominid fossil and archeological record, as well as with regard to practitioners, (sub)disciplines, and institutions. Thus, more than Parts I and II, this part depends on the Red Lady and the people, technologies, practices, and ideas revolving around it to stand in for what doing Paleolithic anthropology/archeology might mean today. The project under the archeologist Stephen Aldhouse-Green, then at the University of Wales, Newport, is well suited for such a case study. Culminating in a monograph in 2000, the project aimed at a reexcavation of Paviland Cave and at an encompassing reanalysis and recontextualization of its material. Since the Brixham Cave project (Part I, Chapter 6), the number of (sub)disciplines from which expertise may be brought to bear on a site has dramatically increased. Furthermore, new laboratory technologies have come to characterize the geological and anthropological sciences, most importantly in the realm of dating, but also in the determination of phylogenetic relationships. These have been applied to Paviland Cave as a Paleolithic site. In particular, the Red Lady has been an early beneficiary of such technological novelties as radiocarbon dating and DNA analysis. As we will see, these at times exacerbate rather than resolve controversy. The results of their application to the Red Lady's material set the stage for the latest Paviland project, and their development will therefore be sketched in the following in association with their implications for the larger issues of hominid evolution and modern European origins.

After World War II, Kenneth Oakley (1911–1981) of the British Museum of Natural History in South Kensington, London, pioneered the development of relative and absolute dating techniques for fossil human remains. Oakley and his collaborators removed from the Paleolithic record many of the fossils on which Arthur Keith in particular had built his case for a great age of modern human antiquity and the long independent evolutionary lines of the modern human races. The Galley Hill skeleton and the Moulin-Quignon mandible were among those shown to be intrusive (post-Pleistocene), whereas others were confirmed as originating from the strata in which they had been found. The infamous Piltdown cranium and mandible, too, were subjected to these procedures and were finally radiocarbon dated to a Holocene age. Impressed by this success, the Wenner-Gren Foundation granted Oakley financial support for applications of the combination of fluorine, uranium, and nitrogen dating procedures to approximately one thousand skeletal parts from Europe, Africa, the Americas, Asia, Australia, Indonesia, Malaysia, and the Pacific Islands. This was done in cooperation with museum curators, other scientists, and laboratories. A slip-index databank containing information on the dating and the name of the responsible analyst and laboratory was compiled at the British Museum.[1]

As this indicates, the fossil record was diversifying geographically, as well as chronologically and morphologically. Most spectacular were the apelike australopithecine remains from South Africa and eventually East Africa that began to be discovered from the mid-1920s. Their recognition as of importance in hominid evolution took some time because, among other things, as discussed in Part II, Asia was preferred over Africa as the cradle of humankind, and brain expansion was considered to have been the hominid specialization that led away from the pongid line, a belief substantiated by the Piltdown forgery. The small-brained but bipedal fossils from Africa thus did not fit the expected hominid pattern, and their morphological similarities to hominids were explained by parallel evolution rather than close phylogenetic relationship. Conversely, combined with the removal of Piltdown from phylogenetic trees, the acceptance of australopithecines as hominids turned the tide in favor of Africa as the place of origin and bipedalism as an early adaptation of the hominid branch. Their acceptance as hominids with predominantly human but also anthropoid affinities was also associated with a curtailing of the use of parallelism as an explanation for morphological similarity. Australopithecine anatomy showed resemblances to *Pithecanthropus/Sinanthropus*, renamed *Homo erectus*, and opened the possibility for their reinclusion in the line leading to modern humans.[2]

Simultaneously, during the 1950s and 1960s, the impact of the Darwinian synthesis was felt in paleoanthropology, and such mechanisms of change as use-inheritance and orthogenesis increasingly gave way to explanations by natural selection, adaptation, and genetic drift. The new synthesis also meant a shift from a typological to a populational approach. In contrast to a typological thinking that encouraged the creation of a new species, if not genus, for new fossils, the biological concept of species emphasized variation (polymorphic species as reproductive, ecological, and genetic units). With an understanding of species as polymorphic populations, multiregional unilinear evolution became possible within the single-species concept, which assumed that at any one time in the course of hominid evolution, there had been only one species. On the basis of potential or actual gene flow between geographical populations throughout evolution, the systematist Ernst Mayr (1904–2005), who also included the australopithecines within the range of variation of *Homo*, thus regarded all as-yet-known hominids as representing a single line of descent (*Homo transvalensis-erectus-sapiens*).[3] This hypothesis therefore differed markedly from the multilinear "trees" according to which the branches of modern human races (or species) have arisen in parallel and in isolation from each other. Such models, as discussed for Keith in Part II, Chapter 12, too, persisted into the years after World War II.[4]

However, these trends in turn were criticized, among other reasons, because of the worldwide rapidly growing fossil record that seemed to demand more complex phylogenies. Niles Eldredge and Ian Tattersall of the American Museum of Natural History applied the punctuated-equilibrium hypothesis as developed by Eldredge and Stephen Jay Gould (1941–2002) of Harvard University to the human data. The model postulates that in the absence of environmental pressure a population can remain static over long periods of time, while it may respond to environmental change by a comparatively fast process of speciation. Speciation does therefore not have to be gradual, but may be episodic, which could also account for the gaps in the fossil record. Rather than assuming smooth evolution within one or a few hominid species, Eldredge and Tattersall now suspected that human evolution had been marked by speciation and extinction, and that the clusters and gaps in the fossil record indicated real taxonomic proximity and distance. In combination with cladistics as the method to arrive at evolutionary relationships and patterns in the hominid fossil record (developed by the German entomologist Willi Henning in the 1950s), this approach resulted in a view of human phylogeny as of great taxonomic diversity, with numerous dead-ending side branches.[5]

These debates were accompanied by the application of technologies from molecular biology to anthropology. Comparisons of proteins and DNA (nucleic and mitochondrial) of living peoples were applied to the question of human phylogeny. The amino-acid molecular clock had been introduced in the 1960s. It was based on the idea that proteins, like organisms, evolve, and that two groups of organisms that at one point shared a common ancestor would at that point also have shared a molecular makeup. After separation, an ancestor molecule would have evolved independently in the two lines of descent while maintaining structural homologies. The concept and technology of the molecular clock further assumed that rates of amino-acid change were constant in different lineages and in the course of time. The assumption was justified by the great flexibility of protein structure in relation to function; mutations would therefore mostly be selectively neutral. When Vincent Sarich and Allan Wilson (1934–1991) applied the method to primates on the premise of a mathematical relationship between "the immunological distance" and the time of divergence of any two species, the results indicated that humans had separated from the apes as recently as five million years ago, a number that was subsequently further reduced.[6]

Further stunning news came out of molecular biology laboratories that lent support to a model associated with human recency and greater diversity in the hominid record. In particular, one influential study, during which mitochondrial DNA of approximately 150 people from African, Asian, Australian, Caucasian, and New Guinean populations was sequenced, confirmed the view that modern humans are the descendants of a population that left Africa relatively recently.[7] Because mtDNA is inherited exclusively from the mother, it was postulated that all human mtDNA could be referred back to a female who had lived in Africa some 200,000 years ago. This hypothesis was therefore called the Eve theory or the recent-African-evolution model and suggests that the transformation of archaic to modern *Homo sapiens* took place in Africa some 100,000 to 140,000 years ago. When modern humans spread across the globe, they replaced archaic forms of *Homo* in Asia and Europe.[8]

While this scenario therefore emphasizes the importance of migrations, the rival multiregional model stresses the significance of local continuity. In this view, local *Homo erectus* populations have given rise to the modern human geographical varieties. Rather than assuming a relatively recent last common ancestor and thus a human "racial" divergence so recent that it coincides with *Homo sapiens*, the process is regarded as reaching further back, to when the supposedly first migrations out of Africa were undertaken by *Homo erectus*. Hence the multiregional hypothesis continues to be associated with less taxonomic diversity in the hominid record and with a more linear phylogeny, including local Neanderthal populations as ancestral to modern humans. One of its advocates, Milford Wolpoff of the University of Michigan, has repeatedly pointed out the difficulties of the out-of-Africa hypothesis, as the recent-African-evolution scenario is also called. He criticized Rebecca Cann and her colleagues for basing their molecular reconstruction of modern human origins on the assumption of a too-high mtDNA divergence rate. If the same mutation rate is used to calculate the point of anthropoid-hominid divergence, the human and chimpanzee lines would have split as recently as about two million years ago, when australopithecines had long since evolved.[9]

Adherents to either scenario consider their model supported by paleoanthropological and archeological evidence. Additional strategies in the argument are to solicit support from other disciplines and anthropological traditions.[10] Another rather unfortunate one is the more or less outspoken charge of racism:

> Finally, both Cann *et al.* (1987) and Gould (1987, 1988) have indicated that the Garden of Eden hypothesis [Eve theory] should be regarded as an important justification for accepting the reality of "the underlying unity of all humans." This is because the recent common-ancestry interpretation is said to be a means of showing that all human beings are closely related. But the contrary is not true, and the opposing hypothesis about multiregional human evolution does not support a different precept. In fact quite the opposite. The interpretation of human evolution as a persistent shifting pattern of population contacts and shared ideas may provide an even stronger biological basis for accepting the unity of all humanity. The spread of humankind and its differentiation into distinct geographic groups that persisted through long periods of time, with evidence of long-lasting contact and cooperation, in many ways is a more satisfying interpretation of human prehistory than a scientific rendering of the story of Cain, based on one population quickly, and completely, and most likely violently, replacing all others. This rendering of modern population dispersals is a story of "making war and not love," and if true its implications are not pleasant.[11]

These observations, besides resonating with the insight that for many westerners evolutionism has become the new religion, are reminiscent of the opposing evolutionary theories of around the turn of the twentieth century discussed in Part II, even though under closer scrutiny the analogy does not hold. Furthermore, as the analysis of the latest work on the Red Lady will show, now, as then, neither the continuity model nor the replacement model is a priori more or less racist or Eurocentrist, and both general frameworks exist in various idiosyncratic expressions.[12]

While, as we will see, the analysis of the Red Lady's mtDNA was situated in the debates just outlined, its radiocarbon dating created other uncertainties. Even before the Paviland project discussed in this part, the Red Lady had been among the fossils that were subjected to direct radiocarbon dating at the Research Laboratory of the British Museum. For this purpose, twenty- to thirty-gram samples were drilled from the tibiae and femur, leaving "wounds" that were afterwards supposedly repaired to invisibility. In 1968, Oakley published the result of an approximate age of 18,500 BP (ca. 16,500 years BC). The Oxford Radiocarbon Accelerator Unit furthermore dated a sample of the left humerus found in Paviland Cave that Sollas had described as different in character from that of the Red Lady. This bone, by then putatively Upper Paleolithic Paviland 2, turned out to be Mesolithic. However, with the radiocarbon date of about 18,500 years for the Red Lady, a date Oakley reaffirmed in 1980, a problem presented itself. It suggested that the Red Lady and "his" contemporaries had lived at the height of the last glacial period, when the average temperature had been around zero, and there had been a tundra of permanent frost, with the glacier only six kilometers north of Paviland.[13]

This has led Theya Molleson, who has continued Oakley's work of dating fossil skeletal material at the British Museum of Natural History, to speculate that the burial had actually taken place after Paviland Cave had provided residual shelter to the Paleolithic people. The young man might have been buried there at a time of great cold when the cave had not been continually inhabited. This hypothesis seemed to be confirmed by her analysis of faunal material from the site (approximately 27,500 BP). The skeleton of the young male indeed appeared to be more recent than the cave fauna and the implements not in direct association with the grave. Had Buckland been right in this respect? John Campbell, now associate professor of archeology at the James Cook University (Queensland, Australia), reanalyzed the Paviland material in the course of his PhD research at Oxford University, for which he surveyed the British Upper Paleolithic sites. He, too, interpreted the lithics from Paviland Cave as of an Aurignacian type (approximately 40,000 to 28,000 BP) and on the basis of similarities to Continental ivory goods thought it possible that the burial was of the younger Gravettian.[14]

Roger Jacobi of the Natural History Museum (based in the British Museum, Prehistory and Europe) was similarly troubled by the fact that the burial had been dated to the glacial maximum, when any significant population was absent from Britain. Contrary to Molleson, however, he rejected Oakley's date for the burial and on the basis of the grave goods once again assumed contemporaneity between the Aurignacian artifacts from the cave and the burial event. Thus, despite the tightening technological and scientific network around the identity of the Red Lady, the traces of evidence from Paviland did not fall easily into place, and as recently as 1997 the skeleton was even ascribed to the early Holocene.[15] Jacobi's judgment related back to Wilfrid J. Jackson (1880–1978), who gave an account of the Red Lady and Paviland Cave in *British Caving* (1953). Jackson, who reproduced Sollas's results, described the site as the oldest ceremonial human burial found in Britain. Looking back to the Red Lady's beginnings as a scientific object, he observed that "at the time of its discovery there was religious prejudice against the immense antiquity of man and Buckland regarded the skeleton as a later intrusion among the bones of extinct animals."[16] However, the magic already felt by Buckland to be attached to the place was maintained, and Jackson added the conjecture that the ochre might have substituted for blood in the burial ritual.

Although Jackson attributed a spiritual life to the Red Lady's people, he continued the tradition discussed in Part II of putting it into sharp contrast with the other cavemen. In Britain, "the sub-human Neanderthal race" was associated with the Mousterian industry of Kent's Cavern, Creswell Caves, and the Wookey Hole hyena den. In Jackson's scenario, these caves had been inhabited shortly before and during the first phase of the last glaciation some 130,000 years ago, when cold fauna of the Late Pleistocene, such as the woolly rhinoceros, the mammoth, and the lemming, had replaced the warm fauna of the Early Pleistocene, such as the hippopotamus, ancient elephant, lion, and hyena. In the Upper Paleolithic some 70,000 to 20,000 years ago, after the first phase of the last glaciation, the Cro-Magnons had replaced the

Neanderthals. Also, the fully human type of Cro-Magnon, who had manufactured the Aurignacian to Creswellian industries, had lived in caves and hunted a cold fauna such as the horse, bison, and reindeer. Jackson assumed that about or soon after the second phase of the last glaciation, cave occupation in general had ceased in Britain—the noble race of the Red Lady had moved farther south.[17]

The most recent project discussed in this part had to address the questions left open by those who preceded it with their interpretations of Paviland Cave, and its place in the larger picture of the European Paleolithic: What was the chronology of human habitation in Paviland Cave, and how did it correspond with the climatic chronology? Had the Red Lady been buried during times of human occupation of Paviland, or did the burial represent a later event than the main human occupation period? What was the relation of these Cro-Magnon people to earlier and later inhabitants of Britain? Apart from the ongoing continuity/replacement controversy, and as Jackson's interpretation of Paviland suggests, the trope of magic was also continued. It remained linked to concerns about the appropriateness of ethnological analogy, which presented themselves with particular acuteness for a project that followed in Sollas's footsteps.

Furthermore, while already Buckland associated the Red Lady with Welsh pride, as well as with presumed Welsh backwardness, the "center-periphery" tensions have been exacerbated in the interval. As we have seen for the Cro-Magnons in general, the Red Lady is an object of contention in the discourse of "national" identity. Jackson may well call it the oldest British burial, but others regard it as specifically Welsh. Thus the related questions of ownership and access, which have been identified as typical of caving, are still an issue. Sollas's work at Paviland Cave had once again brought it to general attention. After the cave had undergone over one hundred years of digging for bones and artifacts by trained and amateur archeologists and paleontologists from the region, as well as from farther away, it was finally declared a protected ancient monument in the 1930s.[18]

Of most interest in the context of this book is yet another question: Has the Red Lady's story been brought to an end with this last rigorous analysis of the Paviland site? Despite the lingering uncertainties with regard to human evolution and the European Paleolithic in particular, practitioners may sometimes be motivated by a desire for completeness.

They try to arrive at a closure of a particular research subject by conclusively describing and defining it or by finally deciding what really happened at a site. Not entirely independent of these desires, and fittingly for the last part of my story, the beautiful and extensive monograph that presents the results of the most recent Paviland project carries the title *Paviland Cave and the "Red Lady": A Definitive Report* (2000). According to *The Oxford English Dictionary*, *definitive* means "having the function of defining, or of being definite," "having the function of finally deciding or settling, decisive, determinative, conclusive, final"; something definitive "makes or deals with definite statements." If used in conjunction with a textbook, *definitive* means "having the character of finality as a product, determinate, definite, fixed, and final," as well as "authoritative" and "the most complete to date."[19]

The monograph titled *Paviland Cave and the "Red Lady": A Definitive Report*, with its 314 glossy pages full of jargon, acronyms, tables, and curves, is certainly authoritative. It speaks with the voice of cutting-edge sciences, substantiated by technology-mediated numerical data. It is a beautifully made and richly illustrated book that indeed presents a convincing and powerful statement, based on an international and interdisciplinary approach, with each (sub)discipline bringing its particular methods and techniques to bear on the site. Held together by the black hardcover with golden letters and the equally golden outline of three bone spatulae (discovered at Paviland in the 1830s), the collection of individual contributions by different authors on different aspects of the site, supplemented by several appendixes, suggests a certain unity. In what follows, I attempt to dive below the smooth surface of the authoritative statement with which the Red Lady's story has seemingly been brought to a happy ending.

In order to gain a feeling for the monograph before a closer analysis of it, this part is structured like a virtual tour through the *Definitive Report*. In other words, in Chapter 13, I outline how the project came about and some of the difficulties it faced. At the same time, I will also provide a summary of the insights gained on the archeology and anthropology of the site by the different project participants. This will be done without further comment or analysis. Chapter 14 deals with the last two chapters of the Paviland monograph, chiefly with the overall reconstruction of the happenings at Paviland Cave by Aldhouse-Green. Again, the reader will be guided through the magic Paleolithic scene at Paviland as conjured up by Aldhouse-Green without me breaking into the story line.

In Chapter 15, visual reconstructions of the burial scene at Paviland Cave will be examined, and here the angle of observation will begin to broaden because the different reconstructions reproduced in the monograph stimulate thoughts about alternative interpretations on the basis of the evidence from the cave. These thoughts are encouraged by the monograph itself, where Aldhouse-Green points to the differences in the visions behind the reconstructions. Thus the critical analysis of the contents of the monograph and of the way in which they are presented begins within its pages. Once room for a step back is opened up, I will proceed to a discussion of "the making of" the most recent Paviland project and of the way in which it relates to its product. Chapter 16 will therefore ask on what grounds the monograph is presented as definitive. As we will see, this claim is largely based on the up-to-date approach to the site, the project's interdisciplinarity and internationality. In this context, it will also be of interest how far the monograph is conclusive in the sense of presenting an unambiguous and synthetic picture. Is it conceivable that the last chapter in the nearly two-hundred-year story has finally been written?

~ 13

The Paviland Project
and Its Results

THE LAST EPISODE to date in the Red Lady's biography was written by an international and interdisciplinary research team under the archeologist Stephen Aldhouse-Green, then at the University of Wales, Newport. He had been concerned with Paleolithic research in Wales since his appointment as keeper of archeology at the National Museum of Wales in Cardiff in 1976, and in 1978 he began a program called Palaeolithic Settlement of Wales Research Project. However, he did not work on Paviland then because he believed that there was nothing left in the site. Thus, even though he was the curator responsible for the Paviland artifacts, he did not think that it would be profitable to research them at that time. Not until about ten years later was his attention drawn to the site by an "amateur" archeologist with a particular interest in caves. He informed Aldhouse-Green that the storm tide had surged into Paviland Cave and had stripped out the modern beach pebbles so that there were undisturbed deposits on view. When Aldhouse-Green and a colleague from Cadw, which is the body concerned with the legal protection of prehistoric sites and monuments in Wales, went to the cave to verify this information, they indeed found visible deposits that had not yet been touched. Partly because of the difficulty of reaching and excavating the site because of the sea, the incident remained without consequences until several years and a letter later.[1]

The Paviland project was finally initiated in 1995 when the physical anthropologist Erik Trinkaus, then at the University of New Mexico, made the Red Lady the subject of a letter to Aldhouse-Green. Trinkaus had studied the Red Lady's bones in the 1980s, in part because many decades had passed since Sollas's work on the skeleton, so that in view of the changed ideas about its context, a new interpretation seemed desirable.[2] Looking for a possibility to publish his results, Trinkaus approached Aldhouse-Green, hoping that the Welsh would be interested in the analysis of the oldest human skeleton from their country. This hope was built on the fact that at least since Hellenic times, governments and peoples have been keen on the bones of their long-gone "ancestors."[3] From the very outset of the Paviland project, questions of ownership and access were thus close to the surface, and Aldhouse-Green considered it unlikely that the National Museum of Wales would fund the publication on a skeleton that, although originally from Wales, resided at Oxford. It was then that Aldhouse-Green had the idea of carrying out a "definitive study of the site."[4] The work on the skeleton by Trinkaus would be welcome as part of a publication that included reports on a reexcavation of the cave in southern Wales and a reanalysis of the Paviland collections at the museum in Cardiff. It was an advantageous moment to launch this project, since the time period of the Early Upper Paleolithic was experiencing a kind of renaissance of interest in archeology, and several analyses or reanalyses of major sites were being carried out.[5]

So it happened that in 1997 Aldhouse-Green began the fieldwork at Paviland Cave. Nothing had changed about the fact that excavation at Paviland was arduous because of the flooding of the cave at high tide, so that within the few hours during which excavation was possible, the fieldworkers sometimes had to wade or swim to the site. The work was funded by the University of Wales College and the National Museums and Galleries of Wales and aided by the Royal Commission on the Ancient and Historical Monuments of Wales, of which Paviland Cave is considered part.[6] Other contributors to the funding of the excavation were the Board of Celtic Studies and the Cambrian Archaeological Association. Permission had to be obtained from the landowner, Cadw, and the Countryside Council for Wales, the first of which donated the prospective archeological and paleontological material from Paviland Cave to the National Museums and Galleries of Wales.[7]

However, although Paviland Cave is one of the richest Paleolithic sites in Europe, it has been nearly emptied in the course of the repeated excavations since the early nineteenth century. Most devastatingly, it has been washed out by the sea. Therefore, the deposits still present in the cave turned out to be from far below the level of the Red Lady's burial. Indeed, they might refer back as far as 150,000 years. Thus Aldhouse-Green's initial apprehension with regard to the site proved true, and the excavation yielded no new archeological material. This meant that insights on the human lithics would have to be gained exclusively from existing collections. For this purpose, the circulation of artifacts through the networks of private and public collections needed to be at least partly reconstructed.

The Early Upper Paleolithic people who sporadically visited the Welsh peninsula, and at times Paviland Cave, had had their own networks of exchange of shells, ivory, ochre, and other rare and precious materials. According to Aldhouse-Green, when the climate worsened around 27,000 years ago, unrelated groups might have had to share food and develop even larger networks of exchange.[8] In order to retrace this earliest distribution system, the scientists applied the most recent technologies. They located the places of origin of the raw materials of artifacts and carried out petrographical and geochemical analyses to trace the origin of the ochre, iron, and manganese ore from the site.[9] After their excavation in the nineteenth and twentieth centuries, the artifacts tended to be widely dispersed. The gains from Sollas's excavation, for example, had been distributed among three museums, partly because he had been cosponsored by the National Museum of Wales, which consequently received part of the yield. Museums not only sponsored excavations but also bought existing collections from other museums and private collectors. Furthermore, the collectors' heirs time and again donated or sold some Paviland material to museums, such as the British Museum in London. Other private collections were probably lost. Since Buckland and Sollas had both been at Oxford University, its museum holds the largest Paviland collection.

The reconstruction of the history of excavations and the location of the collections at as many as seven museums and several private homes fell to Stephanie Swainston, who had an archeology degree from Cambridge with an emphasis on Paleolithic archeology, and Alison Brookes, then a research student at the University of Wales, Newport.

They consulted museum catalogs and archival and published material from excavators, as well as the collectors themselves or their family members, to retrace the history of explorations of and collections from Paviland Cave.[10] In general, even though it might not be done with this rigor, archeological projects need to reanalyze material previously gained from the site under examination by means of the latest technologies and background knowledge to integrate it into the new picture. Thus strictly archeological work tends to be preceded by a historical investigation into a site's excavation record.

Swainston also worked on the stone artifacts. For her MPhil, she had studied the Paviland lithics in the Cardiff Museum, and once the Paviland project started, she expanded this work, which was funded by the Friends of the National Museum of Wales from 1996 onward, to include other collections of Paviland material as a PhD thesis.[11] Overall, the Paviland project involved several laboratories, universities, museums, their staffs, and an impressive technological apparatus, such as that at the National Environmental Research Council (NERC) Oxford Radiocarbon Accelerator Unit. Twenty-one specialists from the many subfields of archeology, paleontology, anthropology, geology/geoscience, geography, environmental sciences, and genetics, associated with institutions in Europe, North America, and Australia, have been involved. Most researchers who cooperated in the project had been well known to Aldhouse-Green and had even done some previous work with him. Some of them brought in or recommended a colleague.[12]

The most precious part of the Paviland material is no doubt the Red Lady's skeleton.[13] Initially in the cave owners' collection, it eventually became part of the Oxford Museum. Within the Paviland project, a new description of the human bones, which had been carried out with financial aid from the Leakey Foundation, the BBC, and the University of Mexico, was presented in the final monograph. Trinkaus studied three sets of human remains from Paviland: Late Pleistocene Paviland 1 (the Red Lady; Middle Upper Paleolithic), early Holocene Paviland 2 and 3 (distal humerus and first metatarsal; Mesolithic), and Late Holocene Paviland 4 to 6 (molar and two manual phalanges).[14] Paviland 1 refers to the Red Lady as discovered by Buckland; Paviland 2 and 3 were discovered by Sollas and have been radiocarbon dated to the Holocene; Paviland 4 to 6 refer to human material recognized in a collection of animal remains by Elizabeth Walker, collections manager in the Department of Archeology and Numismatics of the National Museums and Galleries of Wales (now curator of Paleolithic and Mesolithic archeology).

The description of the Paviland 1 remains takes up by far the largest part of Trinkaus's contribution to the monograph. He confirms Sollas's judgment of a male skeleton in his later twenties, generally "healthy, if dead."[15] The main indication of masculine sex is provided by the Red Lady's pelvis. In fact, Trinkaus is surprised that it has taken so long "to correct Buckland's error" in determining the skeleton's sex, since at Buckland's time anatomists knew the differences between modern human male and female pelvises. What renders the sex determination even more robust in the eyes of contemporary practitioners is the fact that there exists a relatively large comparative sample of Middle Upper Paleolithic (hereafter MUP) pelvic remains that are distinctly female or distinctly male by modern human criteria, and that at the same time correspond with other skeletal markers of sex. Nonetheless, as with any population in modern humans, based on an isolated pelvis, the sex can only be determined with approximately 95 percent accuracy, so there must be a small percentage of MUP specimens that are on those criteria difficult to sex.[16] Overall, there seems to have remained a certain ambiguity with regard to the Red Lady's sex on the basis of measurements of the pelvis and other bones, and the general impression was that of a relatively gracile male body: "Despite the inconclusive nature of several of these individual features, the overall morphology of the Paviland 1 os coxae, combined with its femoral length, supports the designation of male."[17]

Possibly most spectacularly, mtDNA of the Red Lady's femur was sequenced by the Oxford geneticist Bryan Sykes, and it was found to correspond to the dominant contemporary European lineage.[18] The meaning of this finding relates back to Sollas's replacement model versus Keith's continuity scenario for European racial evolution discussed in Part II. It supports the view that rather than having been replaced by the invading agriculturalists who initiated the turn to the European Neolithic, local Paleolithic lineages have survived the agricultural "revolution" in sufficient numbers to have provided the main ancestral stock for recent Europeans. The discussion shares some of its vocabulary also with the contemporary out-of-Africa versus regional-continuity controversy with regard to the origin of *Homo sapiens:*

The interest in the recovery of DNA from the Red Lady lies in the rare opportunity it presents to distinguish two hypotheses about the colonisation of Europe by *Homo sapiens* (Sykes 1998). Much has been written about the relative contribution of the Palaeolithic and Neolithic intrusions from the Middle East to the modern European gene pool. The demic diffusion model of Ammerman and Cavalli-Sforza sees a slow but ultimately overwhelming advance of people, and their genes, from the Middle East fuelled by population growth arising from agricultural food production (Ammerman and Cavalli-Sforza 1984). More recently the pattern of variation in mitochondrial DNA found in modern Europeans has been interpreted as showing a persistence of Palaeolithic genes with only a minority component of Neolithic admixture due to farming . . . This could be called the continuity model and it would follow that, over time, farming was learned from a minority of agricultural pioneers and then adopted by the indigenous Mesolithic population who therefore survived the Neolithic transition genetically intact.[19]

Furthermore, the morphology of the Red Lady was analyzed and interpreted in relation to the out-of-Africa/multiregionalism debate. The theory that body proportions of recent humans vary with climate, and that those in cold areas weigh more (Bergmann's rule) and have shorter extremities (Allen's rule), was applied to the Red Lady's remains.[20] Since human body proportions are considered to have a strong genetic component and are relatively stable over long periods of time, they are taken as evidence for climatic adaptations, population movements, and/or gene flow in fossil hominids. As we have seen in Part II, René Verneau had argued already at the beginning of the twentieth century that early modern humans in Europe approached the modern "Negro" in their body proportions, and Sollas had confirmed this observation for the stature of the Red Lady. In more recent times, this has been understood as supporting the replacement model of modern human origins, with its claim for a recent African ancestry.

The anthropologist Trenton Holliday from Tulane University, New Orleans, compared the body proportions of the Red Lady with those of other MUP skeletons and contemporary Europeans and Africans.[21] He found that Paviland 1 has a slightly more warm-adapted body shape than recent Europeans and a slightly more cold-adapted one than recent sub-Saharan Africans. It is somewhat more cold adapted than the MUP Europeans in general, who were more tropically adapted than recent Europeans. Holliday argues that the MUP group as a whole suggests that extraneous gene flow and/or population spread from Africa into a cold region such as glacial Europe played a critical role in modern human emergence in Europe: "In fact, given that when the MUP sample is compared to a global sample of recent humans, they appear most similar in body shape to recent Africans, it has been argued that they represent descendants of a then-recent dispersal from Africa (Trinkaus 1981; Ruff 1994; Holliday 1997)."[22] The Red Lady is one piece in this pool of evidence: "Thus, the body shape of Paviland 1, while well within the range of recent Europeans, nonetheless lends itself to the interpretation of extraneous genetic influence in the origins of modern Europeans, as has been postulated elsewhere (Trinkaus 1981; Ruff 1994; Holliday 1997)."[23] Once in Paleolithic Europe, the African immigrants would have found themselves under selection pressures that slowly adapted their bodies to the colder climate.

More closely related to the events at Paviland Cave and among the most central results presented by the 2000 monograph is the establishment of a chronology of human cave occupation. In the absence of stratigraphy in the cave floor, artifact typology and radiocarbon dating, uranium series, thermoluminescence, and other technologies applied to bone, ivory, stalagmite, and sediment were needed to place the cave occupations in their wider Paleolithic European and more local cultural and climatic contexts.[24] Collagen was extracted from samples of human bone, artifacts of organic material, and charred animal bone. As part of the process of radiocarbon dating funded by the NERC Scientific Services, the stable isotope values of carbon and nitrogen in the collagen were measured to provide some information about the Red Lady's diet.[25] Unfortunately, none of the ivory artifacts from the grave could be dated because they had been treated with preservatives. In the case of the Red Lady, Swainston identified two pieces of human remains among the miscellaneous objects previously recovered from Paviland that almost certainly derived from the Red Lady, but had never been treated. They could therefore be used for dating.

The dates for the Red Lady posited earlier by the British Museum now appeared unreliable because the following three main human occupation periods emerged (in radiocarbon years): 29,000–21,000, around 7,000 (Holocene), and postmedieval. There was also an earlier human presence, around 30,000 years ago, indicated by Mousterian finds. The burial event fell within the human occupation period of 29,000 to 21,000 radiocarbon years, which seemed to contain five stages: I, 29,000 (charred bone); II, 26,000 (Red Lady); III, 24,000 (the pendant); IV, 23,000 (the bone spatulae); and V, 21,000 (ivory workings). Most likely, the Red Lady had not been associated with any of the bone and ivory artifacts from outside the burial (including the ivory pendant that has come to be known as Sollas's egg), which are of later dates. However, contradicting Buckland's hypothesis, the ivory tools had been made out of ivory that at that point was not fossilized.[26]

This general human occupation history was corroborated by the glacial history of Gower as revised by the expert in Quaternary geology of Cardiff University, David Quentin Bowen. His time estimates for interstadials (climate amelioration) correlated with the occupation dates for Paviland. The new radiocarbon date for the Red Lady (approximately 26,000 years) removed "him" from the last glacial maximum, an observation that remained true in the light of an earlier date also for that event. Although the burial of the Red Lady coincided with a cold event, Bowen observes that it is only sixty-one years too early for the following interstadial (interstadial 4), which seems negligible, considering the complexity of the problem at hand.[27]

Aldhouse-Green analyzed the perforated shells, ivory rods, and bracelet fragments found by Buckland in close association with the skeleton. He concluded that these formed part of the offering of a Gravettian ceremonial burial.[28] That the Red Lady's burial represented one instant in a long history of sporadic cave visitations as suggested by the dating program was further supported by Swainston's typology of the lithic artifacts, which established the following series of occupation: Mousterian, leaf point, late Aurignacian, the Fontirobertian facies of the Gravettian, Creswellian, Final Upper Paleolithic, and Mesolithic. While the Aurignacian is the most dominant, only one implement has been classified as Gravettian—that is, of the same culture as the burial—beyond doubt. Although William Sollas's expression of an Aurignacian station therefore seems to have some justification, the Red Lady had been a later addition to the cave sediments not associated with the lithic tools.[29] The various results converge to support the picture that when, around 29,000 radiocarbon years ago, anatomically modern humans had come to Britain for the first time, they had left traces at Paviland, in the form of Aurignacian artifacts and possibly burned bone. In southern Britain and at Paviland, they had been followed by the first Gravettians around 28,000 to 27,000 radiocarbon years ago (Fontirobertian), whose later presence was evidenced by the Red Lady's burial at 26,000 radiocarbon years ago.

~ 14

There Is Magic at Work

IN THE NEXT-TO-LAST chapter of the monograph, Aldhouse-Green draws on insights from the individual studies to provide a possible reconstruction of the one event that has repeatedly been of concern in this book. On the basis of the traces left by the event that have now been read by means of the latest technologies and methods and have been integrated into the current view of Paleolithic Europe, and by means of the knowledge gained about the climatic, geological, paleontological, and archeological history of Paviland Cave and Paleolithic Wales, he dares to go one step further and to reanimate the scene for the reader. What exactly happened in Paviland Cave some 26,000 years ago, and who had this man in his late twenties been who had once erroneously been dubbed the Red Lady?

To answer this question, Aldhouse-Green also draws on his interpretation of the perforated shells, ivory rods, and bracelet fragments from the grave as part of the offering of a Gravettian ceremonial burial.[1] According to current knowledge, burials were rare in the Aurignacian, while Gravettian ritual interments have been discovered across Europe. Aldhouse-Green summarizes their typical elements: repeated use of locale for burial, multiple burials, extended burial position, position of corpse along cave wall, megaherbivore remains, limestone slabs at head and feet, ochre, personal ornaments, perforated shells, male gender of the single burial, headless body.[2] Obviously, the majority of these elements correspond to the situation at Paviland as encountered by Buckland. Aldhouse-Green explains that the megaherbivore (the Paviland mammoth skull seen in Buckland's visualization) may have protected the grave or represented a totemic or shamanic animal. Another interpretation he deems reasonable is that it functioned as an instrument (osteophone), because bone flutes are known from the Aurignacian onward. With regard to the missing head of the Red Lady's skeleton, Aldhouse-Green conjectures that the face might have been of particular symbolic importance as the seat of identity and communication and therefore might have been given special treatment (separate burial and/or ritual use).

The interment at Paviland is an early example of this Gravettian burial culture, and Aldhouse-Green sees in it the beginnings of a new ritual tradition on the otherwise isolated British Isles; it may represent "an example of allopatric cultural speciation, of the naissance of a ritual tradition in the isolated tundra of southern Britain."[3] The question remains, however, why these Early Upper Paleolithic people would have ventured into that part of today's Europe despite the fact that the climate turned worse from 27,000 radiocarbon years ago onward, which would eventually again force humans out of the region. One attraction of northern lands might have been ivory, which was cherished by Gravettian hunters from such areas as southwestern and southern France/Liguria where mammoths were rare. Nonetheless, Aldhouse-Green feels that this cannot satisfactorily account for the fact that Paviland had repeatedly been visited and was even chosen for a ritual burial when humans were absent from the rest of the British Isles because of the harsh climate. Paviland must therefore have had a special significance:

> The selection of Paviland as a site for ceremonial burial and visits repeated through a period of increasing climatic harshness speaks of a meaning to the site that transcended everyday ideas of temporary shelter for hunters and task groups far from base. It speaks, indeed, of the concepts of *locus consecratus*, sacred revelation, perhaps even of pilgrimage. I have hinted that the presence of Aurignacian ivory-gatherers at Goat's Hole may have led to the site's acquiring a special ancestral status. I have suggested too that the use of the cave by hibernating bears could, itself, have been influential in a mystical perception of the site.[4]

Aldhouse-Green thus situates the Paviland event in a tradition of veneration of place, with Paviland Cave functioning as *locus consecratus* over 5,000 (from 26,000 to 21,000 radiocarbon years) or even over 8,000 years (from 29,000 to 21,000 radiocarbon years), with the Gravettian burial tradition itself extending over 5,000 years and stretching from Wales and Portugal in the west to Italy, the Czech Republic, and Russia ("European liturgy"). The interpretation of Paviland Cave as a sacred place over thousands of years is seen to be sustained by traces from the cave: the burned bone is taken to indicate that Paviland had already been a sacred site when the Aurignacians burned it there as part of a ritual; then there was the ritual burial; the ivory pendant had already been ascribed magical power by Sollas, and the bone spatulae are interpreted as anthropomorphic figurines such as have been found at other sites.

In connection with the notion of a sacred place to which spiritual and actual quests might have been undertaken for the sake of sacred revelation, Aldhouse-Green speculates that the burial ritual was of a shamanic nature.[5] As he understands the term, shamanism is not a religion and offers no integrated system of belief. Rather, it refers to ritual practices common in hunter-gatherers in which sorcerers, medicine men, or ritualists transgress the boundary between the human and the spirit worlds to negotiate with the spirits for the good of the community (for the curing of the sick or control of the weather or the movements of game). The shaman can travel to the world of the spirits through altered states of consciousness induced by dancing, drumming, drugs, meditation, or fasting. He or she is often assisted by ritual equipment, such as drums and staffs, and a symbolic animal, as well as by a special place for ritual such as a cave, where the three tiers of the shamanic cosmos meet—the upper world of the spirits, the middle world of the living people and animals, and the underworld of the dead.[6]

Aldhouse-Green also asks whether the Red Lady himself might not have been a shaman. This assumption has consequences for the system of status inferred. In archeology in general, rich burials that testify to ritual interment are seen as providing information about the person entombed. A special treatment of the body of a shaman would refer to status through merit rather than through inheritance, with burials honoring not only important dynasties but also leaders, founders, or heroes. This again would indicate a rather egalitarian social structure. However, the evidence is ambiguous and can be read in more than one way. Since in the so-called Gravettian burial tradition, children received as much veneration as adult males, it has been inferred that status was inherited rather than acquired. It has thus been suggested that in these Paleolithic societies, status was distributed according to complex social rules, giving evidence of the existence of intricate social systems before the economic system of agriculture.[7]

In this context, Aldhouse-Green's determination of the archeological gender of the Paviland burial is of interest. In agreement with the anthropological sex of the skeleton, Aldhouse-Green observes that besides the fact that single Gravettian graves are male, shell ornaments of the species *Littorina littorea* as placed with the Red Lady have only been found in male burials, although perforated shells in general appear with both sexes and all ages. On the assumption that the burial is that of a male, it seems, he warns that the term *bracelets* is used metaphorically, since it is unknown whether the ivory rings from the grave really had a function. In connection with a male body, they might rather have been shamanic rings, or they might have been used in musical performances. Correspondingly, Aldhouse-Green interprets the short ivory rods from the grave as pieces of former magical wands, reconnecting to Buckland's notion that the cave had been a place of magic, and the Red Lady a dealer in witchcraft: "On present evidence, we can conclude only that the rods were either unused blanks that were placed with the dead body in reflection of their perceived value and inherent status or that they were, as suggested elsewhere (chapter 11, Aldhouse-Green), fragments of magical wands."[8]

Although Aldhouse-Green's vision of the Red Lady and Paviland Cave strongly reminds one of Buckland's reconstruction of Paviland as a place of magic, he, like Sollas, allows for the possibility that there might once have been evidence of the hunters' artistry and ingenuity that did not survive. For instance, at Paviland the colors red, black, and white were present in the form of the ochre-stained bones, black manganese pigment, and calcined bone, as well as white ivory. The first is taken to universally signify blood and thus life, health, happiness, power, and resurrection; white means purity and goodness; and black, disease and death. To Aldhouse-Green this suggests a wider use of pigments, as in wall or body painting, even if there is no longer any direct evidence for this. Evidence of wall paintings, traditionally associated with shamanism, would have considerably strengthened his interpretation of Paviland as a sacred place to which the Early Upper Paleolithic Europeans undertook spiritual quests, and his vision of the burial as a shamanic ritual.[9] Conversely, the notion of the cave as a special place renders more likely the former presence of murals, since caves are seen as sites where humans typically seek connection to the underground world of shadows. Thus Aldhouse-Green thinks that at Paviland at least the specially endowed—the shamans—expected to have visions:

> Some such [adventurous journeys] were plausibly made by shamans who may have had visionary experiences in the underground world of darkness where the wall of the cave became not a rock barrier but a permeable boundary between layers of the tiered shamanic cosmos. Beyond the rock wall lay the world of the spirit animals who endowed the shaman with his shamanic power (Lewis-Williams 1997: 332).[10]

Aldhouse-Green seems to suggest that the experience of caves as special places of this kind is indeed universal, and because of its global distribution among hunter-gatherers, the associated shamanic ritual practices have been interpreted as a function of the modern human nervous system.[11] To analyze the sacred quest and the cave as place of revelation as universals, Aldhouse-Green introduces the promising concept of an archeology of adventure, which allows him to draw parallels between the importance of a spirit of adventure for the exploration of unknown areas in the twentieth century, the exploration of deep sanctuary caves in the Paleolithic, and the epic adventures with mythological significance undertaken by indigenous Australians. Furthermore, and again strongly evoking Buckland's time of early geological cave exploration, he situates his interpretation of Paviland Cave as *locus consecratus* in the traditions of classical Greek and medieval cave symbolism, oracle, and heroic journey and pilgrimage and of the Old Testament prophets (Isaiah, *mons sanctus*), from which he chooses the mottoes for his contribution to the Paviland monograph. The cave, it seems, functions as a kind of chronotope in which the discourses of quest, sacred place, and magic ritual from different times and spaces converge:

> Paviland Cave is such a site [of veneration] and its very size and situation—its lofty entrance below a high forbidding cliff—invoke

a sense of mystery in many visitors. Should we not consider the possibility that our Aurignacian forbears, who are known to have penetrated deep and dangerous cave systems for purposes of ritual, may also have conducted journeys overland to sites of special meaning to them? A pilgrimage to Paviland may have represented just such a *peregrinatio pro numinum amore*.[12]

The quote suggests the belief that in the cave, the different time-space levels converge when contemporary visitors relive the universal human sensations that the Aurignacian hunters expressed in spiritual quest and shamanic ritual. To further develop this notion, Aldhouse-Green asked for a contribution to the Paviland monograph by the Welsh-born Australian archeologist Rhys Jones (1941–2001). Jones's chapter on the social and ritual life of Australian hunter-gatherers is in fact the final chapter and therefore the seeming culmination of the *Definitive Report*. Aldhouse-Green's idea with regard to including this chapter—which might at first seem unrelated to the Paviland site in southern Wales—was that hunter-gatherer societies, in sharp contrast to our Western ones, are considered unlikely to show a clear division between the sacred and the profane. Rather, activities, objects, and places are infused with sanctity. Thus Jones's chapter is meant to provide some perspective on the preceding one by Aldhouse-Green on the shamanic interpretation of the Paviland site. Indeed, the final chapter ends by closing the magic circle when Jones as present investigator, like Buckland nearly two hundred years before him, experiences "a vision having his eyes open":

> At the moment of discovery of Kutikina [a large limestone cave, southwestern Tasmania], what hit me in some instinctive way, was the smell of peat underfoot, the damp mossy skins on the tree branches, and the slightly acrid overlay of crushed dark green leaves, which for some reason reminded me vividly of Gower . . . In my mind's eye, I felt that here at Kutikina at the deepest south, and also far to the north at Paviland, was the entire latitudinal span of human occupation of the Globe at that time [of the Paleolithic] (Jones 1996). These two sites were conceptually linked within the same narrative of the primary advance of modern *Homo sapiens* to the geographical limits of the inhabitable world (Aldhouse-Green 1998).[13]

In fact, Jones initially intended the chapter to construct a dance in time between the Red Lady and an Australian skeleton. The chapter as it stands, however, instead constructs a dance in space and time through the experience and meaning of caves as special places. The cave appears as a site with which the human belief in the supernatural and the desire to express it in ritual are closely interwoven. When explaining the assumption of a universality of shamanic ritual in hunter-gatherers, anthropologists have not only referred to 100,000 years of humans with the same cognitive apparatus, they sometimes also evoke notions from psychoanalysis; there might be archetypes of behavior, to which we all have access to a certain degree. Not without analogies to Sigmund Freud's (1856–1939) naturalization of the Oedipus complex through the claim of real prehistoric events as ultimate causes, the suggestion seems to be that the belief in the supernatural and the associated rituals, objects, and places are deeply ingrained in our phylogenetic memory.[14]

~ 15

Visualizing Paviland Cave

Whether we accept the notion of shamanic rituals linked to sacred spaces as human universals or rather conceive of the one of caves as special places as part of our own cultural heritage, it seems likely that the aura of the prehistoric cave, as already evoked by William Conybeare's cartoon of Buckland entering Kirkdale Cave (see Part I, Chapter 2, Figure 2.1), is attractive to a wider public. Correspondingly, there have been several visual reconstructions of the burial scene at Paviland Cave, and some of them have been reproduced in the *Definitive Report.* One reconstruction painting invites visitors to reconnect to their Paleolithic precedents in the National Museum of Wales in Cardiff. It has been repeatedly reproduced and is included in the Paviland monograph (Figure 15.1).[1]

When the archeological galleries at the National Museum of Wales were redone during Aldhouse-Green's time as assistant keeper in 1980, the idea had been to exhibit all the artifacts and the cases on one side and a series of reconstructions on the other side. The National Museum of Wales has a tradition of archeological diorama and reconstruction painting, a fact on which the idea built. The Paviland painting was commissioned from Gino D'Achille, one of a pair of Italian artists who were employed by the museum for such reconstructions.

D'Achille made additional reconstructions as part of the same commission. Two Neanderthal scenes were published in Aldhouse-Green

Figure 15.1. Reconstruction painting of the Paleolithic burial at Paviland Cave in acrylic on canvas by Gino D'Achille for the National Museum of Wales gallery from a brief by Stephen Aldhouse-Green, 1980; first published in Aldhouse-Green and Walker, 1991, fig. 29, p. 34; courtesy of The National Museum of Wales.

and Walker's *Ice Age Hunters* (1991): a Middle Paleolithic Neanderthal cave scene (approximately 50,000 years ago) and a Lower Paleolithic open site and hunting scene (approximately 300,000 years ago), with which Aldhouse-Green was very pleased.[2] In congruence with the conjectures about the social life of Neanderthals and modern humans given in the book, the illustrations show division of labor according to sex, with males hunting, producing tools, and lighting fires. In contrast to the males, who are indicated to roam far in search for prey, the females are shown sitting, bending over, or kneeling on all fours in front of the cave or tentlike structure. They are engaged in cooking, preparing hides, and picking berries. In the painting that shows the earlier life scene, their nakedness also serves to lend the picture a certain erotic touch. Further contrasting with the males, the females may be accompanied by children.[3] The underlying theory is that the Neanderthals, as well as the Paleolithic modern humans who followed them, showed a relatively marked sexual dimorphism that resulted in a clear division of labor according to sex. In analogy to chimpanzees, the dimorphism is interpreted as signifying male competition over females and thus polygyny. In this view, males were the main providers for females and the young.[4]

The second artist who occasionally worked for the National Museum of Wales, who was actually first approached for the Paviland reconstruction, is Giovanni Caselli. Beginning in 1975, he produced a series of paintings for *The Evolution of Early Man* (1976) that portray gender stereotypes similar to those of the images by D'Achille. This popular science introduction to human evolution secured Caselli's recognition as a specialist in the field of prehistoric reconstruction.[5] Despite Caselli's great gift and precision, Aldhouse-Green prefers D'Achille's work, which he considers particularly atmospheric. This might be due to the fact that D'Achille, apart from historical subjects, has considerable experience with cover images for science-fiction and adventure books, which also demand that the artist turn the imaginary into the real (and restore the dead to life) and are of the hyperrealistic or illusionary style typical of the prehistoric genre. For example, he has illustrated a children's version of Jules Verne's *20,000 Leagues under the Sea* and is the illustrator of the *Flashman* covers.[6]

Yet D'Achille stresses that the Paviland reconstruction, unlike the science-fiction illustrations he has produced, is based on an actual historical fact, for which he has also used the tighter style appropriate for scientific reconstruction. At the same time, he acknowledges that the sparse evidence at hand for prehistoric reconstructions needs to be turned into a complex scene through the use of imagination. This means that because of changing evidence and scientific ideas, such reconstructions are inherently transient, which D'Achille, however, considers the beauty rather than the problem of prehistoric reconstruction.[7] Also in the reconstruction of the Paviland burial, D'Achille, like any illustrator of Paleolithic archeology/paleoanthropology, was thus constrained by the data from the site and their interpretation by the commissioning archeologist. In general, a reconstruction is deemed trustworthy if it contains as much of the available evidence as possible and does not contradict any of it, a premise that still seems to leave great creative freedom in the reconstruction of a prehistoric life scene.[8]

D'Achille was instructed by Aldhouse-Green, who provided him with a photograph of the cave from the outside and showed him artifacts from the Natural History Museum collection in London. D'Achille, who had little scientific knowledge and had never been to the site, produced a sketch that was then discussed during Aldhouse-Green's visits to his London studio. Without much conversation or exchange of ideas, the painting got on quite smoothly. It was done in acrylic, which is easier to use for commercial art because it dries much faster than oil. The highly finished, photographic style was intended to give a sense of the textures of the materials represented, but it also confers objectivity and matter-of-factness on the scene. In reconstructions of prehistoric life scenes, condensation of time takes place, which in the case of the Paviland Cave painting means that it comes to stand for Early Upper Paleolithic humans in Britain. But there is also a condensation of action in that everyone is shown as engaged in a certain activity considered typical of the time period represented. To provide as much information as possible, there is little gratuitous detail, such as people who simply happen to be in the picture, and in the example of the Paviland reconstruction, all the figures are part of the burial ritual.[9]

At the time of the painting's production, the Red Lady's burial was thought to be Aurignacian, so the spear on the right is tipped with an Aurignacian spear point, while the one to the left has a leaf point, then sometimes suggested to be Aurignacian. Aldhouse-Green intended this cultural mélange to stimulate discussion with visiting parties to the museum and more widely. The fact that the image, based partly on outdated interpretations, was still included in the Paviland monograph might be due to the time-consuming and cost-intensive work invested in a reconstruction painting. As a consequence, they run the risk of becoming idées fixes, being reproduced again and again even after the data have changed or have been read differently. As in the case of the Paviland image, a reconstruction in the process might appear in widely differing contexts, from museum gallery and exhibition guide to popular textbook and scientific report. Depending on the social meaning of the space in which the reproduction appears and on the immediate context of caption, related text, and adjacent illustrations, the image will be understood differently.[10]

However, it seems that the survival of partly outdated images may also be linked to subliminal reasons and to some extent to aesthetic ones. After all, it is this reconstruction, which represents a vision of Paviland Cave as sacred place that Aldhouse-Green had before his research at the site, that still, after the international and interdisciplinary project has brought to light new evidence and insights, and after alternative reconstructions have been produced, comes closest to his view of what really happened: "My favourite is the National Museum of Wales one, and that is the one that engages my passion for the site." This passion is related to the relative freedom imagination has been given in the production of the image, and Aldhouse-Green remarks that "I guess that I find belief easy and so I saw this as a sacred occasion."[11] What kind of story does the image tell?

The Paleolithic people involved in what appears as a sacred occasion, just like the beholder of the reconstruction painting, may look out through the entrance of the lofty cave to see the open tundra of the Bristol Channel Plain where now there is the sea. Appropriate for the event and testifying to D'Achille's flair for atmosphere, torches throw the interior of the cave into a reddish light that provides the scene with warmth and magic (which is unfortunately not visible in the black-and-white reproduction, so that in my discussion of this as well as of the following images I appeal to the reader's imagination). Red ochre is ritually spread over the dead body. Two members of the group are holding magic wands (as we have seen, a rather free interpretation by Aldhouse-Green based on the much shorter ivory rod fragments from the grave that could be seen as blanks for beads). The cave is shown to be more spacious than it really is, probably to intensify the dramatic effect. Also for matters of dramaturgy, the burial is moved to the center of the cave to allow positioning the figures around it, although, as becomes clear from Buckland's sketch of the cave, it would have been up against the cave wall.[12] Overall, the painting attests to the burial ritual's symbolic significance. It visually conveys it as a sacred occasion, possibly even as the interment of a shaman. Paviland Cave appears as a worthy place for an Early Upper Paleolithic *locus consecratus* to which these and other people once migrated in a sort of pilgrimage. Inside the cave, where bears also hibernated, they could have expected to encounter the spirits of the dead.

One may wonder what Aldhouse-Green's vision would have looked like were the Red Lady still considered female. Would he have seen a female shaman? Would the bracelet(s) still have to be understood metaphorically? Would the people who interred the body still have taken upon themselves an arduous and possibly dangerous journey into a cold and inhospitable country to perform a ritual at a sacred place worth the merits of their female hero or leader? Although it seems unlikely that there will ever be such a vision for Paviland Cave, if we look again at the illustration as it stands, may we not recognize kneeling at the side of Paviland Man's open and shallow grave a witch performing the ritual that might be the true avatar of Buckland's Red Lady? At any rate, the image is atypical of visualizations of prehistoric scenes in that not only is there a woman among the few figures, but she also takes center stage in the space of the image and in the action conveyed. In the red color that dominates the image from the light, the ochre, and the female ritualist's red hair, the themes of magic and desire associated with the Red Lady since Buckland's times once again converge, this time connected to the locality of Wales in an empowering sense.

A second reconstruction painting, which was published for the first time in the Paviland monograph side by side with the one just discussed, was made by Angela Swainston from a brief by her sister Stephanie Swainston, who, as we have seen, was one of the project participants (Figure 15.2). In contrast to the first image, this cave is bleak and cold. The hurry of the group of Paleolithic humans, who are in haste to get the body interred, seems echoed in the rash style of the painting. It has become so cold in Paleolithic Wales that the group wants to move farther south. Perhaps the young man they are burying has died from the exertions of the journey. Correspondingly, any ritual importance of the mammoth skull that was discovered in close proximity to the burial is played down. Maybe in fear of animals, one of the men guards the entrance.

Thus, even though the artist clearly used the older reconstruction after Aldhouse-Green as a template (copying from existing images is general practice in prehistoric reconstruction, again, among other things, for lack of time and money), it tells a different, more secular story of the event in Paviland Cave some 26,000 years ago. Although one may interpret the second reconstruction as shamanic, the intent was to convey the insecure status such a representation must have on the basis of the rather sparse evidence. The beads on the clothes of the people are meant to show how ornaments would have been worn by everyone, without indication of special status. Instead of wands, the short ivory rods appear as found by Buckland. They are added to the grave close to the cave wall by the mourners, who are also filling it with ochre. However, apart from the general layout, it shares with Aldhouse-Green's

Figure 15.2. Reconstruction painting of the Paleolithic burial at Paviland Cave in watercolor and ink on paper by Angela Swainston from a brief by Stephanie Swainston, 1996; first published in Aldhouse-Green, 2000c, pl. 11.6; courtesy of Angela Swainston and The National Museum of Wales.

image the inclusion of women in the burial scene, an absence of sexual segregation that suggests that women would also have been part of other stereotypically male activities, such as hunting.[13]

The impression the image communicates does not deceive. Stephanie Swainston disagrees with the shamanic interpretation Figure 15.1 brings to life. She generally considers the evidence of the grave goods insufficient for the assumption of a special status of the person buried. She also regards it as statistically improbable that the dead person had been a shaman, since many different occupations could have been carried out, while shamans must have been rather uncommon. She attributes our knowledge of this burial to chance, while other burials that had taken place in the open or in sediment that has been washed away might have indicated an equal amount of care, but did not survive.[14] An idea of Swainston's more cautious approach to the Paviland material can be gained from her MPhil thesis, where she warns that "an understanding of the purpose of the ivory artefacts (rods and spatulae), shells and ochre associated with the burial is difficult to obtain. The function of these artefacts is an area of surmise rather than profitable enquiry."[15] Although the by-now-successful fantasy/science-fiction writer—some reviewers have crowned her queen of the new weird—obviously does not lack creative imagination, she believes in keeping the realms of science and fiction as far apart as possible.[16]

There also exists a three-dimensional reconstruction of the Paviland burial that was created in 1995 for the National Museum of Wales exhibition "Art, Ritual and Death" of 1996 (end of January to end of May) (Figure 15.3). It was photographed for inclusion in the Paviland monograph.[17]

The exhibition was produced shortly before Aldhouse-Green left for the University of Wales in March 1996, with the intention to engage people's interest in the role of ritual in the past.[18] Thus, even before his work at Paviland Cave, Aldhouse-Green had the idea of reconstructing the body of the Red Lady, or Ochred Man, as he sometimes calls him, as wearing skin clothing that had been stained by ochre, rather than with ochre having been strewn into the grave in the course of the ritual. Later he thought that the site evidence confirmed the interpretation that the red staining of bone and ivory had originated in this way. Notwithstanding this aspect, the three-dimensional reconstruction leaves more room to the imagination of the viewer than the two paintings discussed earlier because one is confronted by the grave without any social context of the burial act added, and the grave goods are shown as armlets, shell necklaces, and ivory rods next to the body.

In the "Art, Ritual and Death" exhibit, the visitor could not see the original paraphernalia or the real Red Lady skeleton. In fact, the Red Lady has never been exhibited in his "home nation." The Welsh museums would like to have the skeleton that was found close by and is regarded as part of the Welsh national heritage. It is said that to one of the requests for sending the Red Lady on a vacation west, the Oxford University Museum responded with "not until the Elgin Marbles go back to Greece."[19] In 1998, the Swansea Museum and Art Gallery finally asked the Gower Society for the gift of a cast of the desirable skeleton to celebrate the society's fiftieth anniversary, and in the 1996 exhibit in Cardiff, the three-dimensional model had to stand in for the real thing. For political and economic reasons, the debate has been drawn into the public sphere, where it is continued with no loss of pointed historical analogy. The Gower Society is behind a campaign initiated by Swansea

Visualizing Paviland Cave

Figure 15.3. Three-dimensional model of the Paviland grave produced for the National Museum of Wales exhibition "Art, Ritual and Death" of 1996; courtesy of The National Museum of Wales.

officials to get the Red Lady "back to Wales." An interpretive visitor center near Paviland Cave or at the Gower Heritage Centre is envisioned as the future home for the ancient Welsh skeleton and at the same time as a wet-weather and all-year-round tourist attraction.

Possibly influenced by these events, the Oxford University Museum has granted the genuine skeleton to the National Museum of Wales on loan for a year, beginning in December 2007, as part of its centenary celebrations. At the same time, the Red Lady is intended as the main attraction for the new archeology gallery called "Origins: In Search of Early Wales." The good news instigated BBC Wales to broadcast a short documentary on the history of the Red Lady, providing a platform for the demand for repatriation and celebrating its "return home," even if temporarily.[20]

One might add that curiosity will also not be satisfied at the Oxford University Museum, where the genuine Red Lady cannot be seen either, but resides in a drawer behind the scenes. As is common practice, the seemingly priceless skeleton exhibited behind the glass of a showcase is a cast. Nonetheless, in the redoing of the galleries of the Oxford Museum in 2005, the Red Lady was allocated a prominent place. The casts of artifacts and bones from Paviland Cave appear in the last showcase of a sequence from prosimians, monkeys, apes, and fossil hominids to recent varieties. The Red Lady could be read as a representative of the apex of this progress, and "his" showcase informs the visitor of the newest developments in research related to Paviland Cave.

The last visual argument to date with regard to what really happened during the Paleolithic burial scene has, however, been made by Stephen and Miranda Aldhouse-Green, and it is clearly in a shamanic language (Figure 15.4). Miranda Aldhouse-Green, former professor of archeology and head of research in the school of humanities at the University of Wales, Newport, and currently professor of archeology at Cardiff University, shares her husband's interest in the archeology of religion and ritual, Druidism, shamanism, and magic and is the author of several popular books on Celtic mythology.

The drawing was produced by the illustrator Anne Leaver, then at the University of Wales, Newport. It first appeared in *Gwent in Prehistory and Early History* (2004), edited by Miranda Aldhouse-Green and Raymond Howell, and was reproduced in *The Quest for the Shaman* (2005) by Miranda and Stephen Aldhouse-Green.[21] Leaver contributed considerably to the latter publication through the collation of photographs and other original images and the drawing of the maps, plans, and rock art. For this image, she followed instructions from Stephen and Miranda Aldhouse-Green, although the process of its production involved considerable dialogue with Stephen Aldhouse-Green and many suggestions on her part.

The reconstruction shows the funeral ritual under way partly outside the cave. For the public event with a fair number of participants, the corpse, which is that of a high-status person, if not a shaman, and has been dressed in ochre-stained clothes, has been placed at the entrance to the cave. It occupies the space between this world and the world of the

Visualizing Paviland Cave 257

Figure 15.4. Reconstruction painting of the burial at Paviland Cave in pen, ink, and watercolor by Anne Leaver for Stephen and Miranda Aldhouse-Green; first published in Aldhouse-Green, 2004a, fig. 1.8, p. 19; courtesy of Anne Leaver.

ancestors. Some of the Paleolithic people dance around the body to music; others play on flutes and drums. The grave goods—shells, wands and bracelets of ivory—appear as magical objects spread out on the cave floor. In the course of the ceremony, they will be broken and placed in the grave, a ritual act that might at this moment be initiated by the young man standing closest to the body, while the one behind him follows ready with ochre. The skull of the specially revered mammoth is awaiting its role in the burial layout. On the walls of the cave, one perceives paintings in the form of negative hand stencils, dots, and bisons that identify it as a sacred place. Again, it is worth pointing out that women as much as men are part of the picture and the action it portrays. They dance, make music, and maybe perform a part of the magical ritual, as in the case of the woman (the Red Lady?) who wears a skin dress and is richly ornamented. She is closest to the observer and slightly to the right, the circle of magical objects being on the central axis.[22]

The landscape visible outside is indicative of a relatively mild ice-age climate (interstadial 4), an interpretation made possible by Bowen, as discussed in Chapter 13. In the tundra and tree countryside of the Severn Valley, hide shelters are visible along the river, interspersed with peacefully grazing mammoths. Obviously, this is not the inhospitably cold and desolated Wales of the Swainstons' image. In contrast to Paviland Cave's present location in the forbidding coastline, it is here shown in its Early Upper Paleolithic context with the estuary at the far distance, when the sea would have been about one hundred kilometers to the west. The majestic position of the cave already pointed out by Sollas is here taken to add to its sacred aura:

> Indeed, the locale of the cave, overhung as it is by sheer and towering cliffs, frequently overawes the modern visitor who has made his or her precarious way to the wave-cut platform beneath Goat's Hole. In prehistory—in an age of belief—it must truly have seemed to be a place of oneiric (dream-like) power where terrestrial humans could approach and even enter the very *mons sanctus*, the holy hill, and in the Stone Age darkness engage with the world of spirits.[23]

The burial festivity is reconstructed as having taken place partly outside for lack of space inside in the case of the actual ceremony happening 26,000 years ago, as well as for the illustrator. Leaver was to include a ceremony with a considerable number of people attending. There was to be music from flutes and drums and dance besides the burial, the mammoth skull, and the cave paintings. The list of objects to be included of course also contained the actual grave goods of ivory rods, shells, and ochre. This could best be accomplished by moving some of the activity to the outside. As inspirations for the ceremony, Leaver studied photographs and films of this kind of ritual in African, Indian, and Inuit cultures. This was the incentive for including women, who could be shamans and perform shamanic rituals in the material studied. In contrast to the other reconstructions, this one suggests that children might have been among the dancing crowd, and Leaver is specific about the intent of leaving the identification of the dancers to the viewer's imagination. None of the figures is facing the viewer, who enjoys an untroubled and disconnected (and hence objectified) gaze at a scene from a time long past. In the same interesting inversion as in the other two two-dimensional reconstructions, the beholder is looking out from the cave back in time. Leaver in particular has chosen this approach, aware of the fact that D'Achille had done so before, for dramatic effect. The cave entrance's nearly triangular form is put into relief most effectively through this inside-out view.[24]

Figure 15.4 is made in a far less illusionary and seemingly less authoritative style than Figure 15.1 and thus seems to be more open about its speculative nature. The pen, ink, and watercolor technique gives it a certain provisional character, as though representing work in progress. At the same time, it owes more to imagination than to direct archeological/paleoanthropological data, since one cannot know whether the body is that of a shaman, if a shamanic ceremony was performed, and if so, what it looked like. Most speculative of all might be the cave paintings, which lack direct evidence and represent an older conjecture in support of the shamanic interpretation that has now been given visual expression. Possibly they are also a concession to Welsh pride and a certain jealousy of Continental Paleolithic cave art. Correspondingly, the reddish hair on the heads of the two largest figures can be understood to signify Welsh ancestry. Overall, even to those who prefer more data-sustained reconstructions, the lively social scene that engages both genders and different ages must come as a welcome change.[25]

In general and as supported by the Paviland Cave reconstructions, such visualizations are not mere ornaments of texts. They constitute theories and contain elaborate arguments, feeding back into the scientific debate.[26] The scientific expertise of the artist and his or her cooperation with the scientist(s), in combination with other extrinsic and intrinsic persuaders, provide visual reconstructions of prehistoric human life scenes with credibility with regard to "what really happened" or "how it really was." The context in which a prehistoric life scene may appear is that of the scientific (popular) book, the museum, or similarly authorized spaces, where it might be accompanied by captions, recorded tapes, and explanations and depictions of scientific methods. Within the picture or diorama, the situating of the scene at a real-world site, the inclusion of scientifically verified flora and fauna and human-made artifacts, and the familiar shapes of hominids engaging in everyday activities, gestures, and facial expressions can support the realism of a naturalistic style.

It has been suggested that the amount of distance between archeological, anthropological, geological, and other data and the visualization correlates with the "genre" of reconstruction, that is, skeletal/anatomical reconstruction, facial/full-body reconstruction, life-scene reconstruction and diorama.[27] However, this does not mean that at one extreme of the axis there is pure science and at the other pure fiction, but rather that different genres may be located at different points on the continuum. Reconstructions are always influenced by the purpose of the visualization, their immediate and wider context, contemporary conventions and technologies, and artistic and scientific idiosyncrasies. The location on the science-fiction axis will also depend on the richness and the age of the evidence (and availability of comparanda), the amount of research carried out by the reconstructor/artist, and the collaboration between experts, among many other factors. Location on a science-fiction axis can be further obfuscated because allocation to a certain genre may be ambiguous, as, for example, when different genres are combined, such as in facial reconstructions incorporated into life-size dioramas. Furthermore, there are many conceivable ways of dividing reconstructions into genres. Classification could also be based on, or be complemented by, such criteria as technique of production, including painting, drawing, anatomical reconstruction, replication, restoration, modeling, photography, or even reenactment, and increasingly photorealistic graphics, multimedia, and virtual-reality computer visualization. These might be broken down further according to genre-specific styles.[28]

In fact, where the Paviland reconstructions are concerned, the genre of reconstruction does not seem to correspond to a position closer to or further away from the science end on the science-fiction axis. For all the reconstructions of the Red Lady's burial, the data were similarly scant. Apart from the burial goods, there is only one skeleton, of which the head is missing. We will therefore never see one of the creepily realistic sculptural or computerized reconstructions of the Red Lady's face. With regard to "genre," technique, and style of reconstruction, the fact that there are three remarkably different two-dimensional-renderings of the same event suggests that there is considerable variation possible within the "genre" of life-scene-reconstruction painting or drawing. Furthermore, the three-dimensional reconstruction appears to be more rather than less "factual" than the life-scene paintings. The exhibition piece is comparable to a full-body reconstruction that in this case happens to be situated in a grave close to a cave wall, so that it is difficult to ascribe it to either full-body reconstruction or diorama.

Where purpose is concerned, the fact that one scientist was involved in three different visualizations of the Paleolithic burial scene hints at a use of the images besides conveying information to and/or persuading scientific and lay consumers. Aldhouse-Green's vision of the Paviland burial changed in the course of his engagement with the site, and this change is reflected in a series of reconstructions that functioned, beyond ways of communication with the scientific community and the public, also as creative tools in his own thinking about the archeology of Paviland Cave.[29] On the other hand, the reproduction of the early (1980) rendering of the Red Lady in the 2000 monograph (Figure 15.1) could be indicative of a case where an "old" view persists (and possibly limits reinterpretation) in the form of a reconstruction, a notion supported by the fact that even with the results of the latest Paviland project, Angela Swainston still used it for the layout of a new visual interpretation, and it has remained Aldhouse-Green's favorite.

Clearly, the images are powerful scientific arguments with their peculiar persuasiveness, and the life-scene reconstructions in particular leave little room to imagine the happenings at Paviland differently.[30] In contrast to text, where the author may express uncertainty with regard to particular aspects of her or his interpretation more easily, the Paviland visualizations are no exception in simulating a view into the past that appears as if little mediated, even though the intensity of their claims is nuanced by differences in style. Thus the style of both the Swainstons' and Leaver's drawings (Figures 15.2 and 15.4) are less realistic, and as a consequence they may appear less closed than the first one carried out by D'Achille for Aldhouse-Green (Figure 15.1). However, none of the reconstructions conveys a strong attempt at communicating uncertainty or subjectivity through experimentation with style. Many such attempts are conceivable, for example, the use of impressionism rather than naturalism, the blurring of the image, the inclusion of blank spots, or the introduction of alienation by showing an archeologist at work among the Paleolithic ritualists. All these tools could engage the addressee of the reconstruction self-reflexively in the process of telling one story to the exclusion of other legitimate stories on the basis of a restricted set of traces.

Nonetheless, something along these lines is achieved through juxtaposition. None of the images appears isolated, as the definitive and complete visual report. Rather, they are engaged in conversation. In the 2000 Paviland monograph, right next to Aldhouse-Green's 1980 reconstruction (Figure 15.1), we find the Swainstons' alternative reconstruction telling a different story (Figure 15.2). In *The Quest for the Shaman* (2005), where the latest image is reproduced (Figure 15.4), Stephen and Miranda Aldhouse-Green make it clear in the accompanying text that although certain things can be said with great likelihood about what really happened at Paviland, other things, such as whether the Red Lady was a shaman, are speculation.[31] The image thus appears as a personal vision accompanied by an invitation to share in it, to suspend disbelief, even if only for the moment of contemplation. If, in continuation of the 2000 monograph, all the now-existing Paviland reconstructions were exhibited simultaneously, the visitors would be invited to partake in a discussion about how the same data might be interpreted differently over time, or indeed at the same time by different scholars. The juxtaposition of different genres, techniques, and styles and the subjective expression or objectified statement they convey would further allow reflecting on the role of these aspects in persuasion and/or communication.

~ 16

The End of the Red Lady's Story: A Definitive Report?

> The present volume offers, if such a thing is possible, "a" definitive report that summarises the past 175 years of research on this site.[1]

AFTER A DISCUSSION OF the results of the Paviland project as presented in the 2000 monograph, I would like to return to its title and the question of definitiveness. It will be of interest in this chapter to what degree the monograph represents a synthetic statement on the site, which I take to be a prerequisite for the claim of its definitive nature. Furthermore, from the introductory texts of the *Definitive Report* it becomes clear that its authority is based on the interdisciplinary cooperation of scientists from many (sub)disciplines who have analyzed specific aspects of the site by the most up-to-date approaches. It is also stressed that this team of participants was drawn from the international scientific community. The interdisciplinary approach of the Paviland project does not come as a surprise in view of the fact that this approach lies in the nature of the investigation of archeological sites, and as we have seen, even the reconstruction of the history of excavations and of the associated collections is seen as an essential part of such projects. Nonetheless, the interdisciplinarity and internationality of the project are stressed throughout the preface, acknowledgments, and summary: "Quaternary excavations of this kind necessitate the formation of a specialist team of people. In the event, experts were drawn from three continents of the world. Expertise included archaeology, human remains, animal bones, dating, isotopic analysis of bones, geology and cave sediments."[2]

Of course, the meanings of interdisciplinarity and internationality transcend the necessity that arises from the complexity of the problems the field presents to the anthropological sciences. Science policy studies have shown how interdisciplinarity and internationality in general are tied up with notions of innovation, progress, and also legitimation. Interdisciplinarity is assumed to be a somehow more holistic approach than the supposedly reductionist and limited perspective from one discipline. It may even seem to promise a kind of knowledge that is more socially robust. Similar observations can be made with regard to internationality, since it has been suggested that the cooperation of scientists from not only different disciplinary but also cultural backgrounds allows a closer approximation to the "truth" because cultural diversity may be effective in the avoidance of preconceived ideas due to common sense. Like interdisciplinarity, internationality is an epistemological asset that seems to render possible more comprehensive pictures of the phenomena under consideration.[3] Furthermore, in the age of global science, internationality seems to be a guarantor of cutting-edge standards. To give just one example, the Volkswagen Foundation, the most important German science policy maker, has the advancement of interdisciplinarity and internationality as one of its goals. In the words of its general secretary, it demands of the scientists whose interdisciplinary endeavors it finances

> the readiness and ability to develop and apply a common terminology; the openness to free communication among all members of the team; the readiness to become engaged in a permanent process of mutual learning; the acceptance of a problem-oriented, instead of discipline- or person-oriented approach; and last, but not least, the readiness to give priority to the project's objectives instead of pursuing one's personal interests.[4]

Clearly, great expectations can be bound up with an interdisciplinary approach. However, the notion itself is notoriously fuzzy since too many ideas of what it really means exist. I will refrain from providing my own definition of a term that has proved as slippery as *interdisciplinarity*, let alone clearly demarcating it from the neighboring terrain of *multi-* and *transdisciplinarity*.[5] For my purpose here it will suffice to say that it may involve common interests and problems, shared facilities, instrumentation, and databases, and the circulation of such entities as tools, methods, results, concepts, and theories.⁶ With a main emphasis on the exchange of material and immaterial entities in the attempt to arrive at a certain synthesis attached to the notions of interdisciplinarity and internationality, I shall once again return to the Paviland project.

To begin with, there is a certain tension between the Paviland project's ostensible internationality and its considerable local involvement. As we have seen in Part II in particular, anthropology and the politics of national identity have a long history of interaction. Aware of the fact that proof of ancient ancestors strengthens claims to land, power, and status, Aldhouse-Green, who has always been impressed by the way in which French prehistorians have been able to call on their former presidents, would like Welsh politics to become associated with local archeology and in particular desired that kind of high-level association for the Paviland project. One cannot help thinking that this strategy might be especially profitable in a country that partly desires greater political and cultural autonomy. In general, contemporary anthropologists/archeologists are acquainted with the phenomenon of national pride being tickled by fossil ancestors—the older the better—a phenomenon that is limited neither to the present nor to Wales. We all share in one of its effects when in newspapers and magazines we read again and again of the oldest hominid or human being having been discovered, sometimes even contrary to the anthropologists' descriptions of their finds.⁷

It was lucky for Aldhouse-Green's purpose that Jones, who wrote the final chapter on tribes of Australian Aborigines, was the cousin of Rhodri Morgan, first minister of Wales. Jones agreed to approach Morgan on the subject of the Paviland project and the Red Lady. The chance presented itself when the first secretary of the National Assembly of Wales came to Australia on official business, a trip that nonetheless included a baseball match between an Australian and a Welsh team. After the game, the two met in an Italian restaurant in Sydney. The meeting ended by Jones sketching out the foreword Morgan would write to the monograph on Paviland Cave and the Red Lady.⁸ This foreword is announced in Aldhouse-Green's acknowledgments as "honouring Wales's earliest known modern human inhabitant."⁹ From the start, the Red Lady was thus implicated in a discourse on Welsh origins and identity, and the meaning of the Red Lady for the region of Wales is made explicit: "The 'Red Lady' represents the only Early Upper Palaeolithic ceremonial burial in the British Isles and the extensive European comparanda set out in this volume show how Wales has been an active participant in European developments for more than 25,000 years."[10]

Although this might sound like a statement about the role of Wales in the European Union, it in fact concerns the sparse occupation at the skirts of the glacial ice during the Early Upper Paleolithic in the region now known as Wales. Apart from Welsh assertion vis-à-vis England, there is therefore another undertone audible. The comment can probably be fully understood only in the context of French dominance in European Upper Paleolithic archeology. Particularly the work on the painted caves and archeological wealth of the Dordogne, where excavation techniques were perfected, has functioned as the baseline and benchmark for European Paleolithic archeology in general. Swainston observes that "the north-west European sequence was considered peripheral, and, presumably with a low population density, a poor relation of its classic French counterpart. It was thought to lack detailed stratigraphic sequences and artistic—hence cultural—complexity."[11] In view of the influence of French anthropology and archeology and such episodes as the Piltdown forgery (Part II), the ongoing attempt to prove the importance of the British Isles (or in this case Wales) for the prehistory of Europe does not seem entirely unexpected.[12]

In the foreword by the first secretary, the Upper Paleolithic Paviland Cave and the Red Lady become even more bound up with local Welsh scientific and cultural traditions:

> This monograph presents the result of this exciting new interdisciplinary work carried out partly by Welsh scientists. It establishes the site and its ancient occupation within the context of the climatic and environmental conditions of that era. After the passage of almost 30 millennia, this extends our understanding of the religious beliefs and customs of our forebears. I am delighted that the present and future generations of people in Wales will welcome this understanding of our distant past.[13]

Once again, the autonomous contribution of Wales to the (prehistoric) development of Europe and the supposed continuity between the Paleolithic site and Welsh history are simultaneously underlined. Wales's internationality and national identity are mutually constitutive, a fact that is highlighted when the monograph written in the common language of science, English, ends with a multilingual summary. Although the summary is given in Welsh to stress the local concern of the work, the monograph is at the same time meant as a contribution to the international field that shows the importance of the locale of Wales for the European Paleolithic, and the summary is thus also provided in English, French, German, and Spanish. This coexistence of the Welsh tongue and culture with the main European languages on equal terms is underlined by the fact that the foreword, too, ends with a paragraph in the native Welsh, and the monograph opens with local poetry.

In view of the significance of the Red Lady for the area of Wales, it does not come as a surprise that these are bones of contention. Because the skeleton was discovered in 1823 by the first reader of geology at Oxford University, it has become part of the Oxford Museum collection. As though to strengthen its claim to the ownership of the Red Lady's bones and the artifacts associated with the burial, Sollas as Buckland's successor at the chair of geology reexcavated and reevaluated Paviland Cave in 1912. Historically, then, the skeleton's ties to the old and influential university are stronger than its local bonds. However, the Paviland project intensified Welsh relations to the site, and the dispute about where the Red Lady belongs, in the Oxford University Museum or in a museum in Wales, is not yet over. From the Welsh point of view, the Red Lady is of Welsh origin. As the mtDNA analysis suggests, it is so not only by place of birth but also by genealogy.

Although in the case of the Red Lady and the associated artifacts circulation is impeded by intra-British competition, and although the Paviland project as a whole had to serve the superficially paradoxical desires of internationality and national identity, the interdisciplinarity of the project is put in perspective by the fact that there was minimal to no contact among the experts involved, many of whom had not been to the site. Indeed, one foreign participant recalls that his contribution consisted in "spen[ding] one day at Oxford studying the skeleton."[14] The participants who were asked had the impression that, predominantly, everybody worked on his or her topic and wrote a paper on the results without significant exchange. It seems that the participants had not in every case known of each other. This had the effect of making especially the younger ones feel isolated. The need was felt for a conference at the beginning, the middle, and the end of the project in order to get to know the participants and exchange insights. Only at the end, however, when the monograph was already on its way to publication, was a conference organized in Newport, at which mainly the participants from Wales and London presented their results on Paviland.

When asked about this aspect of the project, Aldhouse-Green recollected that he circulated a list of participants to all the specialists involved, so that in his mind anybody who wished to get into contact with somebody else was able do so. However, he conceded that for the most part the individuals or groups of individuals worked on their own themes, reports on which were then partly circulated among colleagues who had a direct interest. This rushing through was partly motivated by the deadline for the British Universities Research Assessment. In order to be taken into account for this assessment, the report had to be published by 31 December 2000. Aldhouse-Green recalled how in the end the monograph was produced under considerable pressure. In comparison, for the project that followed, there was plenty of time until the deadline, and he was also slightly older and no longer under the same amount of career pressure. He felt more at leisure to ensure the desired quality and had the time necessary to circulate the developing volume to everyone involved to comment on in order to obtain the level of feedback that he would have wished for in the Paviland project.[15]

Because of the lack of this kind of interaction in the Paviland project, there was no attempt at synthesis, which again puts in perspective the notion of a definitive report. In fact, some participants felt that the overall interpretation of Aldhouse-Green had been put at the end of the report as though naturally emerging from the individual partial analyses preceding it, while in fact it was perceived to be contradicted by some of the research pieces. However, there are also tensions between the individual contributions to the monograph. For example, while the mtDNA analysis is presented as uncontroversially placing the Red Lady within the most common extant lineage of Europeans, or indeed Welsh, at least one of the participants sees cause for distrust toward this technique, particularly because of the danger of contamination. The problem is seen in the fact that if the expectation exists that the DNA under study is generally within modern human ranges of variation, which in the case of the Red Lady, on the basis of the skeleton's age and anatomy, makes perfect sense, then it is methodologically impossible to determine whether the DNA sequenced is indeed ancient or modern. Instead of ancient DNA of the Red Lady, one might be looking at modern contamination from one of the many people who have handled the bones, some of whom, such as Buckland and Sollas, are already dead.[16]

Although this participant was joking when he added that for all he knew, the DNA analyzed might be his own, since he, too, had worked on the skeleton, the insinuations need to be taken seriously. This morphologist expressed his resentment over the intrusion of molecular biology into a territory traditionally belonging to physical anthropology. I use the term *territory* consciously here, since rather than having grown on an interdisciplinary common ground, the contributions to the Paviland monograph partly emanated from the isolated turfs of established disciplines. In the argument between the traditional morphological and the newer molecular way of determining phylogenetic relationships, the tables may sometimes be turned, as when it is the molecular biologist who is accused of being concerned with the genetic purity of living Europeans (as supposedly illustrated by the fact that the main extant line of Europeans can be traced back to the Red Lady in the Upper Paleolithic and has not been radically "contaminated" by later Middle Eastern invaders), and the physical anthropologist shows himself scandalized that "the stud book of living peoples" should still be of interest in this day and age. What is more, the molecular biologists even grind up precious fossils for the purpose.[17]

However, the public appeal of the young and sexy technologies of molecular anthropology cannot be denied. In 2001, Sykes, who, as we have seen, also carried out the mtDNA analysis of the Red Lady, founded the University of Oxford spinout called Oxford Ancestors. The company offers personal DNA analysis services and was immediately overwhelmed by requests. The service MatriLine™ informs customers about their own maternal ancestry and their place in the family tree of all humanity, whereas the Y-Clan™ and the Tribes of Britain™ services provide information about paternal ancestry. Obviously, Tribes of Britain™ works only for men of British ancestry and is supposed to determine Celtic, Anglo-Saxon/Danish Viking, or Norse Viking ancestry. Customers wishing an mtDNA analysis for £180 will receive, apart from a readout of four hundred base pairs of their mitochondrial chromosome, a certificate that indicates to which of the clans of "the seven daughters of Eve" they belong, that is, from which of seven postulated European female lines they have supposedly descended.

Sykes has given each clan the personal name he coined for the hypothetical last common female ancestor of its members, and the customers will even be provided with information on their clan mother's life and times.[18] An additional certificate will show how their clan is related to other world clans and to African Eve herself. Furthermore, those customers who wish to remain among their genetic peers may search Oxford Ancestors' databases for people with matching DNA sequences.

Throughout the services and the discourse of advertisement, attribution of a customer to a particular clan or tribe goes hand in hand with establishing his or her uniqueness. Similarly, the employees of Oxford Ancestors are identified on the website by one of the clan names of the seven daughters of Eve, but also by means of a photograph, name and job description, and interests. This coconstruction of group membership and individual uniqueness is epitomized by the information on the genetic positions at which mutations have occurred in their mtDNA control region. DNA sequences thus appear as the ultimate fulfillment of the anthropological dream of finding a means by which to identify a particular person and at the same time classify him or her as part of a "race." From this point of view, it makes particular sense to refer to individual DNA sequences as genetic fingerprints, since this is exactly what one of the pioneers of genetics, as well as the inventor of the term *eugenics*, Francis Galton (1822–1911), tried to achieve in vain with fingerprints.[19]

Contrary to the accusation mentioned earlier, however, the company is careful to prevent any arousal of the sleeping dogs of eugenics or association with the British obsession with pedigree. On an earlier version of its website, it claimed to be able to prove that we are all humans. Thus the advertising discourse is of the family-of-man genre and forestalls the question why the British—and which section of the British—should be interested in their ancient genetic ancestry in the first place; presumably not to find out that they are human. Nonetheless, on the same version of the website, the former chief executive of Oxford Ancestors deplored (if jokingly) with regard to his own DNA analysis that "although both father and son are blond-haired and blue-eyed they, disappointingly, do not carry a Y-chromosome of probable Viking descent."[20]

If we apply the model of the seven European clans to the Red Lady, he would have to be attributed to the so-called clan of Ursula, supposedly the oldest of the seven native European tribes that was founded some 45,000 years ago by the first modern humans establishing themselves in Europe. Again on the basis of mtDNA comparison, Sykes claims that these modern human pioneers did not interbreed with the local Neanderthals.[21] Eleven percent of recent Europeans are considered to be Ursula's direct maternal descendants. Many of them are found in western Britain and Scandinavia. However, the Red Lady's mtDNA seems difficult to fit into the seven-daughters-of-Eve scheme, since although agewise he must be of the oldest lineage, he is also seen to belong to the commonest extant European lineage, which would in fact be the Helena branch, said to be "only" 20,000 years old. Be that as it may, the Red Lady supports the seven-daughters scenario insofar as his genetic ties reach back beyond the agricultural invaders from the Middle East some 8,500 years ago, who according to Sykes are responsible for a mere 12 percent of the current European population, namely, the descendants of the Jasmine clan. The Middle Eastern invaders' clan is said to be the only one to have its origins outside Europe. Here the family-of-man discourse based on the notion that ultimately we are all Africans obviously breaks down, and the biblical notion of a single Eden gives way to seven different Edens with seven different Eves, only six of whom are native Europeans.[22]

With this background information on Sykes's business and ideas on modern European origins, one is able to fully appreciate the excerpt from his Paviland contribution quoted in chapter 13. There, Sykes uses the Red Lady for his arguments in the wider evolutionary debates of concern to him. His view of the origins of Europeans agrees with the genetic-replacement theory in that it assumes complete replacement of Neanderthals by the modern human invaders from Africa. At the same time, he challenges the "demic-diffusion" or "wave-of-advance" model's postulate of a successive massive replacement of the Cro-Magnons by invading agriculturalists from the Near East.[23]

The morphological analysis of the Red Lady as presented in the monograph also touches on these controversial subjects. The analysis and interpretation of the Red Lady's body proportions in the context of recent and MUP Europeans have implications for the Neanderthal question, even though these implications are not made explicit in the monograph. As we have seen, the discussion of body proportions takes place within the out-of-Africa theory, which postulates that modern humans first emerged in equatorial Africa between 100,000 and

150,000 years ago, and that after having remained for a considerable time in that ecological zone, they started spreading out across the rest of the Old World some 50,000 years ago. Although this is currently the model most generally accepted, Trinkaus belongs to those scientists who assume that there was admixture between these modern human newcomers and the local archaic populations such as the Neanderthals in Europe and the Near East about 40,000 or 30,000 years ago.[24] He has been working on the morphological evidence for this interpretation for some time and is of the opinion that also the modern human genetic data can only be explained by some degree of admixing.[25]

This is where the Red Lady's body proportions come in, because with their intermediacy between equatorial Africans and recent Europeans, they can be seen to support the idea that the heat-adapted population from Africa to some degree intermixed with the local cold-adapted Neanderthal population. Trinkaus's collaborator in the morphological study of the Red Lady believes that although to him the genetic data seem to indicate replacement of the Neanderthals by modern humans without admixture, some Neanderthals might still have been absorbed into recent human bands, but any genetic trace of this event would have been swamped. It has been mathematically shown that such swamping could result from the sheer superior number of the expanding African population. The absence of genetic evidence of intermixture therefore does not necessarily mean inability to interbreed or a competitive disadvantage or lack of intelligence on the part of the Neanderthals. As a result, the Red Lady (and the Cro-Magnons in general), rather than being put into sharp contrast with the Neanderthals, as he predominantly used to be (see Part II), appears as a descendant of African immigrants and to a more modest degree of local Neanderthals. Obviously, this is a nuanced scenario that shows elements of both the extreme version of the out-of-Africa theory that does not allow for any admixture and the local-continuity model, which regards local Neanderthal populations as ancestral to existing human geographical varieties.[26]

The continuity-replacement debates with regard to European prehistory and the associated Neanderthal question have thus remained controversial to date and have informed the most recent approaches to the Red Lady skeleton.[27] Even though the lines do not run strictly along disciplinary boundaries, the debates have been heated by the fact that, to borrow the imagery, the molecular biologists have invaded the territory of traditional anthropology in many cases without due respect to the locals. The self-confidence the new technologies inspire is remarkable, and Sykes is no exception:

> Without new evidence from a completely different and independent source, genetics, the debate about whether native Europeans were descended from Neanderthals or from the apparently distinct later arrivals, the Cro-Magnons, would have rumbled on unresolved. In all fields of human endeavour where there is a shortage of objective evidence, opinions and people inevitably become polarized into rival camps. Once entrenched, the occupants will not be dislodged; they would rather die than change their minds. Such was the situation when we set out to apply our powerful genetic tools to the conundrum; so we knew the path ahead would likely lead us into a minefield.[28]

The *veni vidi vici* attitude and the war metaphors make it difficult to discern whether the intention really is to put out the fire, but clearly gasoline is involved. The mental image inspired by the passage may well be one of a stagnating scientific branch that is caught in a stupefying deadlock (through internal conflict) being rescued by progressive invaders from another field. This is not unlike the way some early twentieth-century anthropologists thought of human evolutionary progress and their situation as imperial power (see Part II). Today we would probably speak of the attempt of a hostile takeover.

Just as controversial, but more tightly bound up with the Red Lady, is the issue of the main paleoecological and technoetic (a combination of the Greek words for art and consciousness) interpretation of Paviland Cave as a holy place associated with shamanism. A voice differing from it has already been heard in the discussion of visual reconstructions. Was the Red Lady a shaman? An important aspect pointed out by Trinkaus is that the vast majority of the human bones of the time of the Gravettian burial tradition in Europe were not ritually buried with this kind of elaborateness. There are sites across Europe from this time period where both very elaborate ritual burials of one or more individuals and bits and pieces of humans are found side by side. The remains of the latter were probably simply left on the surface or placed in bare, shallow graves and were subsequently consumed by scavengers. Although some individuals were therefore being accorded higher status than others, Trinkaus thinks that no obvious pattern of who was ritually buried and who was not can be recognized. For example, although there seem to be a few more male than female burials, there is evidence of both having been ritually interred. With regard to the Red Lady, he can therefore think of many reasons why this individual could have been of high status without invoking shamanism.[29]

However, although Trinkaus sees no pattern of status distribution with regard to sex in what is known of the Gravettian tradition, the preponderance of male graves has been read as difference in status, and one of the patterns described is that Gravettian solitary graves are male.[30] This interpretation presupposes that the sexing is done accurately, and that the sample we have discovered shows the same sex ratio as was present in the burials that actually took place. Both anthropological and archeological sexing of prehistoric bodies and graves have received some critical treatment, and both have been found to be circular to a certain degree because they set out from the gender assumptions of the day.[31] Put simply, since the first half of the nineteenth century it has been assumed that males were the weapon users, and therefore weapons define a grave as male, while ornaments tend to define a grave as female. In ambiguous cases, the male marker, that is, weapons, is determining. While a grave with weapons is thus male even if it contains ornaments, the gender is more ambiguous in the absence of weapons and still leaves the possibility of female or male, a fact that might in part explain the larger male sample (which again somewhat circularly is interpreted as the consequence of the higher status of males).[32] In this context, it is worth remembering that Buckland, who discovered the ivory rods, fragments of bracelets, and perforated shells with the burial, probably defined the Red Lady as female because he concentrated on the archeological rather than the anthropological sex of the burial.

In the most recent evaluation of the Paviland burial, despite the fact that it contained no weapons but only ornaments, the species of shells that has so far been found only in graves considered male, in combination with the fact that there was only one body in the grave, which is also taken to indicate male sex, was decisive. However, with regard to cases such as the Red Lady, of which the anatomical sex had been known since Sollas's time, another problem arises for the determination of artifact function. As we have seen, the anthropological sexing of the Red Lady is presented as unproblematic in the monograph. For the Paviland burial, the relative certainty of the anthropological sex determination could mean that what in a female grave would likely have been interpreted as bracelets might become seen as ivory rings used for ritual breaking during the funeral ceremony. In other words, once the grave was seen as male, it seems, a shamanic interpretation presented itself for the "ornaments"; they were now read as indicative of a particular kind of high status rather than femininity, with ritual rather than purely ornamental function.

A similar double standard has in fact been identified for our expectations of Neanderthal/archaic versus modern human finds. With respect to the Paviland material, it has thus been pointed out that "poorly documented modern human remains can become the relics of gravettian pilgrimages to a *mons sacra*, while repetition in a Middle Palaeolithic context is interpreted as habitual, and animal-like in nature."[33] Aldhouse-Green himself makes the reader of the Paviland monograph aware of the speculative nature of at least the vision of the Red Lady as shaman:

> We cannot say whether the "Red Lady" may have been a shaman. What we can say is that the corpse was that of a person of high status, a status possibly gained through membership of an important lineage, social or familial group. Moreover, the last rites—in that they must have constituted a minor theatrical event—would have resembled the liturgy of a shamanic performance, and may plausibly have been presided over by a shaman.[34]

While this quote gives the impression that the burial ritual is of more interest than the precise identity of the Red Lady, one may venture one step further and ask whether the star of the report and his burial, which quite naturally have attracted the most attention from researchers (including myself), have biased the way in which work has been carried out at the site. Of course, one may call it a happy coincidence that the skeleton was found at all, considering the fact that it had been covered by only a few inches of sediment when Buckland unearthed it, which means that it would in all likelihood soon have been washed away by the sea tide. It is nonetheless a provoking thought how the interpretation of the cave might have looked without the evidence from the burial, with excavators focusing more strongly on other material and searching in areas of the cave farther away from where the burial once has been. It is possible that an excavation of the posterior of the cave would have yielded some implements after all.[35] Thus Swainston reasons that "the Red Lady burial as a single brief activity cannot provide much information about the nature of the human behaviour in the Upper Perigordian [Gravettian], whereas the lithic implements have the potential to provide information about a much wider range of behaviour—for example distances involved in raw material transportation can be extrapolated to tell us about the size of group ranges."[36] However, the analysis of the existing artifact collections has revealed a certain bias, too, because early excavators tended to select artifacts for size, material, and retouch, because activities in the cave disturbed sediments, and finally because the collections have partly been lost, contaminated, or scattered.[37]

One difficulty with the interpretation of Paviland as having involved shamanism and pilgrimage, of which Aldhouse-Green is well aware, is that with so little datable organic material from the site, it was only possible to obtain a small number of quite widely spaced dates, which do not necessarily show the kind of continuity of visitation that would support the idea of pilgrimage. One reviewer of the monograph was particularly suspicious of the suggestion of a wide communality of social discourse, a common identity and belief system, at this point of climatic downturn in Paleolithic Europe. With regard to Aldhouse-Green's idea that the spatulae from Paviland had been shaped after the female body and in analogy to the eastern European figurines had had an encoded meaning, he retorts: "That Upper Palaeolithic societies had extensive social networks, with trade links extending over 700 km or more, is well established, but the recent push to turn this into evidence for a Eurasian communications network perhaps wishes the coca-cola age of globalisation onto the Palaeolithic."[38] However, we have seen that in search of discursive communalities, Aldhouse-Green reaches across even longer distances in time and space. To argue for a universally human experience of caves as special places associated with shamanic ritual and spiritual quest, he also draws analogies to ancient Greek and medieval traditions, as well as to recent hunter-gatherers and finally contemporary Western cave visitors.[39]

Aldhouse-Green's drawing on collateral evidence, such as the use of pigments for parietal art at Paviland, is meant to further support the notion of the cave as a kind of chronotope.[40] In this line of inferences, the universal human mind in altered states of consciousness produces a sequence of equally universal images. These images have been identified in Paleolithic cave art, which is considered indicative of trance states typical of shamanism and easily induced in caves. In contrast, although Paleolithic cave art has historically been associated with totemism, sympathetic hunting magic, initiation rites, and fertility cults, some contemporary archeologists have become wary of any transcendent and universalizing explanation or interpretation and demand a more thorough inclusion of individual contexts (such as the particular time and place of the archeologists, as well as the Paleolithic Europeans) in its analysis.[41]

Similar reservations are expressed toward the use of ethnographic analogy in the interpretation of the Paviland material, and not only where wall paintings are concerned: "Ethnographic accounts cannot be used to match and highlight a specific archaeological patterning, since many patterns of behaviour are possible with the present evidence."[42] In fact, Aldhouse-Green is careful not to be understood as saying that "they did this in Australia so it happened at Paviland . . . But rather saying that this is the kind of world of which Paviland was a part."[43] Jones, too, was aware of the legacy of a tradition of such analogies in anthropology and archeology, particularly with respect to "the Australian Aborigine," and his chapter begins by establishing distance between that past and his present essay:

> I wish to make it clear at this point what I am *not* saying. I am not saying that Australian Aboriginal people, some of whom I know well, or others who had lived during the last century are in any sense surviving cultural relics from some Palaeolithic age, neither am I saying that the western Europeans who lived prior to the last glacial maximum had beliefs identical to those of modern hunting peoples, whether they be from Australia, North America, southern Africa or the northern steppes of Siberia. What I *am* saying however is that from a study of comparative ethnography . . . there are many profound communalities in human responses to the great issues or problems of life. These especially relate to the mysteries of birth, the passage of an individual though [*sic*] the various stages of life and finally death. There are also the relationships between the human and both the natural and the supernatural worlds.[44]

Obviously, Jones wanted to emphasize that Sollas's analogizing of "Australian Aborigines," "American Indians," "Bushmen," and "Eskimos" with Neanderthals and different anatomically modern races of Paleolithic Europe to the point of identification was an altogether different matter. Nonetheless, the reviewer of the Paviland project monograph quoted previously has been prompted to comment: "It is an interesting and useful chapter, but its inclusion as the final chapter of this book might raise a few eyebrows."[45] This suggests that although it is interesting to learn about ritual practices, sacred sites, and color symbolism, as well as about spiritual quests among "Australian Aborigines," it remains uncertain how the information relates to the interpretation of the Paleolithic site in Wales.

Although the raising of eyebrows in amazement might be one of the intentions of the magician qua ancient shaman or contemporary scientist, it cannot obfuscate the fact that the drawing of inferences from recent indigenous peoples to prehistoric humans demands elucidations beyond disclaimers. The recourse to "profound communalities in human responses," supposedly due to essentially the same cognitive apparatus and a universal deep structure of linguistic and visual languages, as well as of myths and shared archetypes of behavior, is vague.[46] It leaves open the awkward question, if there is something like human psychic unity after all, why should we use an Australian Aboriginal people, of all humans, to help understand behaviors of Paleolithic Europeans? Indeed, there is no agreement on the right kind of ethnographic comparison with regard to Paviland, and some think that "a new model is needed, one which incorporates ethnographic detail only where it can be fairly applied. Unfortunately, such a model cannot at present move beyond a general picture of small, highly mobile and migratory groups of hunters in a steppe or tundra environment."[47]

Conclusion: An Unfinished Life

THIS APPEAL FOR a new model to interpret the events at Paviland Cave and the Red Lady seems the perfect vantage point from which to embark upon a conclusion. Although at first glance the monograph and the visual reconstructions might appear stable, consistent, and comprehensive, a close reading of the *Definitive Report* has shown that below the surface of unity and finality, it is internally contradictory. Rather than being "finally deciding" or "settling, conclusive, and final," it reaches out to unresolved and ongoing discussions about central issues of human evolution. In fact, the Red Lady is still read within some of the frameworks of debate Sollas and his contemporaries were concerned with, such as questions of continuity or replacement, the antiquity of evolutionary lines, the role of indigenous peoples in the interpretation of the past, or the diversity or unity of human populations in past and present. The international and interdisciplinary approach, one of the pillars on which the *Definitive Report*'s authority and suggested completeness are built, has furthermore been implemented only in the sense of more than one nationality and discipline contributing insights to the collection of results. Overall, the latest report does not "make definite statements," but can be definitive only in the sense of being "the most complete to date," which, as we have seen in the course of the Red Lady's history, is a transient notion indeed.

Thus participants in the Paviland project have pointed out that the report on the Paviland site cannot be definitive because additional material may still be found in the future, and because as scientific methods advance, additional and different information will be gained from existing collections. Aldhouse-Green, too, is perfectly aware that the report cannot be definitive; he was merely aiming at rendering it as definitive as possible. He wanted to produce a report that presents a thorough analysis of the body of surviving evidence, one that incorporates a field assessment, a dating program, stabilized information on the faunal and human remains, and some overarching idea of what it might all mean. In this he no doubt succeeded. He feels that certain parts of the report could be more elaborate, and on certain aspects of the project more work could have been done. Quite contrary to how the title might be understood, Aldhouse-Green was hoping that the published report would function as a stimulus to others to become involved with the Paviland site, to take investigations even further, and to eventually render the *Definitive Report* unsatisfactory.[1]

However, the choice of title remains, as well as the preference of some participants for a title that would have acknowledged that in science anything can only ever be a progress report. The discrepancy between tacit knowledge and articulated claim suggests that this could be another instance of humor at work in the Red Lady's story. The title might be a jocular play with our desires to bring stories to an end. At the same time, Aldhouse-Green invokes humor in a way reminiscent of Buckland to distance himself in retrospect from the claim of definitiveness. As with Buckland's use of humor, the claim does not lose its power when spoken only half-seriously; it still serves the desire for completeness that might be one motivation in the search for definitive answers and solutions. In this way, bold statements of finality, such as "it has taken archeologists nearly two centuries to unravel the mysteries of this remarkable site, with the definitive report published only last year," exert their magic in matters of funding and publicity.[2]

These intuitions seem supported by the fact that the implementation of interdisciplinarity, which, as we have seen, is associated with great expectations and therefore seems to be the right approach to reach a lofty aim, has been curtailed by an evaluation system that creates time pressure. In other words, partly the same considerations that are likely to have motivated claims to interdisciplinarity and internationality in the first place seem to have hindered their full development. Overall, such factors as the necessity of thinking of one's own career, disciplinary rivalries, and political involvement have been set against a more synthetic outcome. As Peter Weingart observes:

> The thesis is that, on the *socio-psychological level*, the values of universalism and organized skepticism suggest openness to all relevant knowledge, whether supportive or contradictory, as crucial to innovativeness (i.e., the highest value in scientific activity). This translates into "interdisciplinarity" as a programmatic value tantamount to innovation . . . the actual activity of knowledge production contradicts this squarely. The prevailing strategy is to look for niches in uncharted territory, to avoid contradicting knowledge by insisting on disciplinary competence and its boundaries, to denounce knowledge that does not fall into this realm as "undisciplined."[3]

The Paviland project agrees more with Weingart's observations on the actual realities of interdisciplinarity than with the Volkswagen Foundation's idealistic demands. Consequently, its results as presented in the monograph are comprehensive only in an encyclopedic and not in a synthetic sense. All contributions were produced by experts working within their (sub) disciplines, mostly in isolation from the rest, and the contributions to the monograph stand rather disconnected. As with the visual reconstructions included in the monograph that bring to life different visions of the burial scene, this lack of agreement and synthesis—for the overarching last two chapters do not represent a common effort—puts a greater responsibility on the reader and viewer. If the different chapters and reconstructions are understood as in conversation with each other, it is the reader's task to retrace the various arguments and, if desired, to achieve his or her personal synthetic vision. Read this way, the *Definitive Report* is definitive in the sense of presenting some of the most up-to-date, if not always consistent, data and opinions with regard to one site. It is part of larger ongoing discussions, not an end to the Red Lady's story. In fact, at this moment its history is being rewritten as a result of renewed radiocarbon dating of its bones and that of some Paviland animal remains carried out as part of a large program at the Oxford Radiocarbon Accelerator Unit.

~ *Postscript*

A SURVEY CARRIED out by the anthropologist Monique Scott as a doctoral candidate at Yale University on visitors of the British Museum of Natural History and the Horniman Museum in London has brought to light remarkable attitudes toward the idea of African ancestry and Africa as the cradle of humankind. Many respondents revealed their conception of "Africans" as living emblems of early human evolutionary history. Even among the colored respondents, some identified themselves as closer to the African ape-man ancestors than to people of non-African descent. Apart from this testimony to a survival of nineteenth-century evolutionist ideas, many visitors also expressed a belief in long independent lines of "racial" development, with the more progressive stages having taken place in Europe. Where the place of human origin is concerned, Africa was accepted only grudgingly (for example, as the place of biological but not cultural origins) or denied outright (as when perceived as a political act "to grant the Africans something"). Among those who accepted Africa as the place of origin, the reasons given were again "because we evolved from a black race" or due to a romantic perception of an African Garden of Eden.

Scott explains that her findings cannot be attributed to these two particular museums having old-fashioned exhibits on display. The human-origins section at the British Museum of Natural History does not support a linear, teleological narrative. Similarly, the Horniman Museum features an "African Worlds" exhibition aimed at subverting stereotypes of "Africa" as primitive and prehistoric.[1] Rather, the example serves to illustrate how much the stereotypes that have been cultivated throughout the history of anthropology have become deeply ingrained in our cultural heritage and memory. As we have seen, in the history of anthropology, Asia was generally preferred over Africa as the continent where the first hominids evolved, an idea that survived into the 1930s, and nonwhite "races" have traditionally been described as deficient and backward. Through innumerable reiterations, the presumed progressive line from primitive (associated with nonwhite/non-Western) to advanced (associated with white/Western) in biology and culture and the conception of a huge gap between "them" and "us" have been inscribed into our system, and it will need widespread conscious effort to get them out.

We are all affected by the history of views on human origin and diversity as discussed in the chapters of this book. At the same time, a note of caution against overemphasis of historical constants is appropriate. Among the histories of anthropological thought, there are attempts at extrapolating recurrent ideas and eternal schemata, such as the heroic-quest structure of origin narratives.[2] Similarly, the models of degeneration or progress as two opposite types of origin stories can be traced back to Hesiod (ca. 700 BC) and Lucretius (ca. 98–ca. 55 BC), are encountered in the works of the philosophers Jean-Jacques Rousseau (1712–1778) and Thomas Hobbes (1588–1679), and, as we have seen, have been part of nineteenth- as well as twentieth-century anthropology. To see this as an oscillation between a timeless couple of binary ideas or as two straight lines of development might be an obvious case of decontextualization on the loose. Nonetheless, in the course of this book, I too have identified the recurrence of opinions that emphasize one side of such dichotomies as human/hominid antiquity and recency, continuity and replacement in hominid evolution and distribution, linearity and diversity in hominid phylogeny, and similarity and difference between, for example, modern humans and the Neanderthals.

However, although already the preevolutionary counterparts of monogenism and polygenism might be seen as incorporating specific sides of these opposites, successive theories of hominid evolution have not amalgamated the same set of ingredients. For example, monogenism combined an emphasis on migration with the idea of the recency and unity of humankind, whereas later migrationist views as expressed by William Sollas and his contemporaries tended to stress current human diversity and its antiquity. On the other hand, today's out-of-Africa scenarios rather combine relative diversity of the hominid record with the recency of the hominid branch and its present representatives. A similar argument could be made with respect to those accounts of human (pre)history that rely more on local continuity. The polygenist idea of different creations of humans for the various geographical regions resonates with the linear progressionism of the post-Darwinian nineteenth century discussed for Edward Tylor and John Lubbock, which, however, was positioned against polygenism.[3] Then again, the cactus model of parallel evolution did show polygenist tendencies through the claim of a relatively great antiquity of the modern human type and the existing "races" (Praesapiens theory)—it was envisioned in the form of both the monocentric and migrationist models, such as advanced by Grafton Elliot Smith (and epitomized in the central Asia hypothesis), and even more distinctly the multilinear local-continuity model of the older Arthur Keith. The postsynthesis unilinear views anticipated by the German anatomist and Gustav Schwalbe student Franz Weidenreich again, in stark contrast, worked with little diversity in the hominid record through gene flow between locally continuous populations. So does the current multiregional hypothesis.

Theoretical genealogies can therefore be arbitrary, as when they are established on the basis of such notions as migration and diversity in evolutionary scenarios and hominid taxonomy. As discussed in Part III, they may serve to denounce the opposite view as racist. In reality, it is impossible to tie one specific model of hominid evolution to a particular set of ideological beliefs, even if general tendencies might become recognizable, as in the rationalization of the practices and consequences of imperialism in theories of human origin that drew direct comparisons between expansion into new territories associated with the (violent) replacement of indigenous peoples in past and present. A closer look at structurally similar scenarios of human origins thus also reveals their situatedness in specific historical contexts. For instance, the successive-displacement models of hominid evolution certainly worked with different theoretical and practical tools, as well as with a changed set of evidence, at Sollas's time than they do today.

We have also seen how even at one time level, paleoanthropologists who agree on the general model of hominid evolution will show idiosyncratic variations on the theme. They might also apply a different model to hominid evolution in general than to the evolution of the modern human "races." Moreover, most scenarios allowed for a range of mechanisms of development, including local continuity, diffusion, migration, intermixture, and replacement, with Sollas and Elliot Smith at one end of the spectrum with their overemphasis on migration and diffusion. In the case of Sollas, for whose work both his training under Thomas Henry Huxley and his orientation toward Marcellin Boule and Henri Breuil were relevant, personal and professional background and networks facilitated the embracing of new ideas in one area, but hindered the following of the general trend in another. William Buckland, too, played an important role in the formation of a historical geology, while his institutional and wider scientific and cultural contexts were among the factors that biased him against evidence for antediluvian humans in Europe.[4]

This discussion about the right way of looking at the history of ideas in the anthropological sciences risks perpetuating the supposed continuity/replacement binary. As a middle way, while rejecting the appeal to archetypal ideas, deep structures, and universal human concerns, and while emphasizing the context specificity of views on hominid evolution, it seems paramount to allow for the accumulation of a cultural heritage. This way, we do not run the risk of fatally denying the possibility even of radical change, nor are we forced to restrict the meaning and impact of each event exclusively to its immediate context. Some historical continuities, it seems, such as the influence of a teacher on a student or of one friend on another, can be documented relatively reliably. Furthermore, it has been convincingly shown that some stereotypes are remarkably resistant to change in the scientific as well as in other cultural realms, such as the markers of "primitiveness" and "progressiveness," as exemplified by Scott's results mentioned earlier. In fact, the conceptual backwaters of humanity have acquired such a connotation of stagnation that the historian of science Robert Proctor has dubbed the out-of-Africa theory "out of Africa: thank God!"[5] Again, the implicit notion transported with the theory might be that progress took off once early modern humans left Africa.

One of the advantages of the approach I have taken in this book, that is, the use of the Red Lady as a red thread through chapters in the history of scientific engagements with fossil human bones, might turn out to be a certain protection against a too-directed search for either historical breaks and constants or continuities. The microstories of the skeleton's successive interpretations, situated in scientific theories and practices, institutional settings, and scientific networks, have functioned as a warning against a strong mainstreaming and simplification as general trends. Nonetheless, to arrive at a coherent story, some such reduction of the complexity present at each moment is necessary. It is a great art to find the right balance between the general picture and the underlying diversity; the zooming in and out from the detailed interpretational history of the Red Lady and the larger context has hopefully been of help in this respect. It has enriched the picture in other ways. It has allowed glimpses at the use of fossils as national icons, at the role of magic in scientific practice and the subject of science, at questions surrounding the interdisciplinarity of human-origins research, and at the workings of visualizations of prehistoric scenes.

The trope of magic in the Red Lady's story is interwoven with the use of ethnological analogy. Although this method was not unknown at the time of Buckland, and he in fact made fledgling use of it, in the interpretation of the Red Lady he chose to draw his analogies from accounts of the Welsh in past and present, building on another center-periphery axis. As sketched at the beginning of Part II, the comparative method was an important tool for the early evolutionists who conceived of the indigenous peoples as being at Stone Age stages in the general progress of culture. One might argue that it reached its apogee in Sollas's comparison to the degree of identification. Ironically, the description of the Red Lady's Aurignacian tribe lacked a chapter on the living equivalent in his *Ancient Hunters* because it was perceived as ancestral to Europeans, whose cultures were, however, not used to elucidate these particular Paleolithic hunters (while the other Aurignacian "race" of the Grimaldi type was revived by a chapter on the "Bushmen"). No doubt, he presumed that the cultural distance accumulated between the two was too big. In contrast to the "Bushmen" and other "primitives," Europeans were not fossilized.

After ethnological comparison had been carried to excesses, which, as discussed in Part II, was already controversial at the time of the publication of the successive editions of *Ancient Hunters*, the comparative method continued to be debated in archeology. Its history has troubled its applications in the most recent Paviland project, where, among other things, disagreement on the right ethnological analogies or on the uses of such analogies in general has informed radically different reconstructions. Although for Stephen Aldhouse-Green the comparison of the European Upper Paleolithic to living Australian hunter-gatherers supported an interpretation along the lines of shamanism, Rhys Jones's work was distanced from the Sollasian precedent.[6]

Beyond reasoning by analogy, the successive interpreters of the Red Lady employed different strategies of persuasion. Buckland used his magic also in the creation of a self-image, mobilizing synergies with the romantic infatuation with caves as places of inspiration and imagination, prophetic or mysterious vision, and archives and annals of history. He drew on the revival of the romance genre and the chivalric quest in styling the practice of the geologist. These lofty visual and rhetorical strategies contrasted, however, with the emphasis on a down-to-earth approach to a subject that Georges Cuvier, as well as Buckland's colleagues in the Geological Society of London, wanted to base on a sound empirical foundation. Buckland's humor might be seen as mediating between the two discourses, as when he advanced unorthodox views in a quipping mode in order to allow his imagination to be tested against the evidence before transferring it to the realm of the serious. But humor could also serve to distract attention from data that might have presented a problem for Buckland's vision of antediluvian Britain and Europe.

Sollas, though himself an eccentric personality and the cause of many anecdotes, preferred to create credibility through instruments, measurements, numbers, and the visual abstraction of schematic drawings, even if for a wider public he did not refrain from a grand narrative of human progress in *Ancient Hunters*. Finally, the *Definitive Report* (2000) is by a specialist group, is addressed to specialists, and includes new technology-driven approaches, such as dating techniques and DNA analysis. But it also dares to tell verbal and pictorial stories that appeal to a larger audience. National pride, too, is among the points of contact. As we have seen, this is not a recent invention. Buckland had been aware of the tradition and romantic appeal of establishing long national and regional pedigrees and implicated the Red Lady in these debates. Cranial types figured in the Aryan theory that had nationalistic overtones even before such early twentieth-century anthropologists as Boule,

Keith, and Elliot Smith instrumentalized fossil human "races" for the foundation of national and/or "racial" identities.

The latest Paviland project indeed suggests that the curiosity about personal and "national" genealogies has by no means subsided. Rather to the contrary, apart from paleoanthropology and archeology, molecular anthropology and systematics and their popular implementation in genetic ancestry tracing respond to the desire for roots. A reviewer of *The Tribes of Britain* (2005), an account of the history of the British Isles by the Oxford archeologist David Miles that begins with the Red Lady, has observed slightly sarcastically that

> the contemporary British have discovered they don't know who they are, and so they keep asking the question. Why we should have succumbed to this identity crisis now is not easily explained. It probably arises from some malign combination of modernity, loss of empire, devolution, globalisation, America, Europe and eight years of peculiarly identity-free government. All of which have both prompted the question and denied us the possibility of an answer.[7]

Into this void enter books such as the one under review and those by Bryan Sykes discussed in Part III to allow for answers with scientific authority. Although the answer to why we are interested in personal, national, and phylogenetic histories is likely to be as complex as the answers to the questions of who we are or where we came from, devolution might also be one of the cues with regard to the *Definitive Report*. As is analyzed in Part III, the importance of the Welsh language and culture is emphasized, and continuity between the Welsh Paleolithic and present is suggested, among other things, by the leader of the Welsh Assembly Rhodri Morgan's foreword, but always within a pan-Europeanist discourse.

This presentation of the results of the Paviland project ties in with political and cultural meanings of the term *Celts/Celtic* that range from an ancient integrated European people as a kind of archetype of the European Union to a more restricted connectedness or specificity of regions such as Brittany, Ireland, Scotland, Wales, the Isle of Man, and Cornwall. Both meanings pay credit to the importance of roots, with the pan-Celtic movement working for the recognition of a unique Celtic identity (increasingly supported by genetic evidence) and the preservation and celebration of a linguistic and cultural heritage. In the popular realm, this heritage is associated with a spiritual counterculture and ancient wisdom, tropes celebrated in films, music, art, and books, in festivals, exhibits, and fairs.

However, the popular notion of contemporary Celts as the heirs of a prehistoric culture is in growing discrepancy with those scholarly opinions that see in the "ancient Celts" of the Iron Age a modern invention. Nonetheless, since the happenings at Paviland Cave 26,000 years ago are presented as part of Welsh tradition, the shamanic paleoethnological reconstruction of the Paviland site provided by Aldhouse-Green connects to the popular notion. However, although Celticity is linked to Irish and in a lesser degree to Scottish and Welsh nationalism, it was historically used by a group of ethnicities in resistance to English and French hegemony and thus may not lend itself to convey the individuality of regions such as Wales. Nonetheless, the notion of Celtic ancestry is politically and emotionally still powerful, and its deconstruction by archeologists at times meets with hostility, as one response by First Minister Morgan shows: "Celtic and proud of it! It's just English jealousy. We were civilized first. The earliest poem composed in a post-Classical language was Welsh. The Celts were here when Caesar landed. It's modern Anglo-Saxon propaganda."[8] This quote also reminds us that one must not lose sight of the fact that there is also a long tradition of Anglo-Saxonism, manifestations of which I have discussed in Part II.[9]

Overall, the Red Lady's biography as a scientific object, though the specificity of each individual interpretation has been highlighted, has been given a clear story line. It started out as a set of bones of contention in the midst of the controversy around human antiquity and was protected by Buckland's chivalry, magic, and humor from being slandered as antediluvian; even a prostitute seemed less scandalous. Its second awakening was just as turbulent, when theories charged with racism and white supremacist ideology propelled it onto the throne of Paleolithic humankind as an exemplar of the noble Cro-Magnon. The most recent Paviland project has closed the circle in many ways. The magic power of Buckland's witch has been returned to the Red Lady as shaman, and because of the results of mtDNA sequencing, it has in a certain way been proven to be "Ancient British or Welsh" after all. The Red Lady's story also shows some institutional continuity. It was discovered by an Oxford geologist and was reinterpreted by an Oxford geologist-anthropologist. Although it is still at the Oxford University Museum, there have been increasing demands with regard to rights of access and ownership. The last project on Paviland Cave and the Red Lady has shifted the weight to Welsh institutions, and the skeleton has become symbolic for Welsh roots.

Because of the Red Lady's age and place of discovery, once it had been incorporated into an evolutionary framework, it remained implicated in the ongoing debates about theories of modern human origins and diversity, as well as about the role of the Neanderthals therein. These changing debates involved the constant negotiation and drawing of boundaries between "self" and "other" and in their current expression involve mutual accusations of racism through the establishment of genealogies for one's own and others' theories, among other means. Even at Buckland's time, the antediluvians were alternatively interpreted as early descendants of the original pair (monogenism), as in his scenario of human dispersal from Asia, or as evidence for more than one divine creation of humans (polygenism), as in the case of Jacques Boucher de Perthes's early model. From the start, when Buckland had to come to terms with the possibility of a great human antiquity and its implications for religious belief, the Red Lady has been in the midst of the troubling question of what it means to be human. The moral dilemmas surrounding human-origins studies have absorbed researchers from Buckland to the present time, and religious overtones and biblical metaphors are still strong in current theories.

Although or even because the Red Lady is not as big a star as, for example, the australopithecine Lucy, it allows insights into the everyday business of doing human-origins research. The skeleton has seen different approaches applied to its bones, possibly from Buckland's tongue test via Sollas's strongly instrument- and number-driven physical anthropology to radiocarbon dating, present-day comparative anatomy, and mtDNA analysis. The preoccupation with measurements reaches back to physical anthropologists, such as Paul Broca, who had already developed techniques to classify bones of recent humans in a hierarchy of "races" before they were trusted with the analysis of fossil human bones. When fossil human bones were finally accepted as such, they became enmeshed in the ongoing debates about especially European "racial" origin(s) and diversification and were analyzed by the tools developed to classify extant "races," the central interest of the newly institutionalized discipline.

In view of the eventful biography of the Red Lady, the belief that we will ever arrive at a closure of interpretation of the Paviland site loses probability. As participants of the most recent Paviland project were quick to point out, with new evidence from the cave or from already-existing collections hidden in some private house or museum vault, with every new methodological approach and technological innovation, the questions that can be asked of the skeleton and other material will change, and so will the answers. In the transforming dialogue, the picture that emerges of what happened at Paviland Cave some 26,000 years ago (or the then-accepted date) will also change. Beyond this, personality, institutional setting, and national context, among other factors, have been found to influence what is seen. Without invoking the notion of an ever-closer approximation to an out-there truth that will someday be fully understood, which has been strongly contested from within paleoanthropology and especially Paleolithic archeology, one might say that the white and blurred spots in the polychrome picture of the Red Lady's burial will change number, shape, and location. The project, exemplary in many ways and carried out just in time for the millennium, produced in a certain sense a definitive report, but has not written the final chapter in the Red Lady's adventurous story—the happiest ending for anyone interested in the anthropological sciences.

Appendix A
Archeological and Geological Series

European Archeological Series Reconstructed after the Text of *Ancient Hunters* (1911)

Period	Culture; characteristic tools	Climate; characteristic fauna	Fossil hominids
Paleolithic or Old Stone Age (flaked stone tools; hunting and gathering way of life):			
Lower Paleolithic	Mesvinian; simple flakes as scrapers and rude knives		No human remains
	Strepyan; coarse scrapers (core tools)		No human remains
	Chellean; boucher (a core tool)	Warm climate; southern fauna; *Elephas antiquus* (elephant), *Rhinoceros Merckii* (soft-nosed rhinoceros), hippopotamus	*Homo heidelbergensis?*
	Acheulian (Upper, Lower); a boucher of finer workmanship	Colder climate with severe winters; mixed fauna; *E. antiquus*, *E. primigenius* (mammoth), *R. trichorhinus* (woolly rhinoceros)	*H. heidelbergensis?*

(continued)

(*continued*)

Period	Culture; characteristic tools	Climate; characteristic fauna	Fossil hominids
Middle Paleolithic	Mousterian (Upper and Lower); Levallois flake, Mousterian point, side and other scrapers, lancehead	Cold climate; cold fauna; *E. primigenius*, *R. trichorhinus*, reindeer, musk ox, Arctic fox, glutton, marmot, Arctic hare, ibex, lemming	Neanderthal race
Upper Paleolithic (Age of Reindeer)	Aurignacian (Upper, Middle, Lower); Chatelperron point, grattoir (carcinated scraper), Gravette point; bone, horn, antler, and ivory begin to be used; appearance of engraving, sculpture, painting, and drawing	Warmer climate; mixed fauna; horse, bison, cave lion, cave hyena, *E. primigenius*, reindeer	Ancestors of modern Europeans? and Grimaldi race of *H. sapiens*
	Solutrean (Upper, Lower); bone, horn, and ivory weapons; arrowhead, leaf point	Colder climate; mixed fauna; horse	Ancestors of modern Europeans? and Grimaldi race of *H. sapiens*
	Magdalenian (Upper, Lower); bone, horn, and ivory weapons; point, arrow- and spearhead, harpoon; engravings on bone and ivory, carvings, statuettes	Cold climate; northern fauna; reindeer, Irish deer, bison, horse, ass, musk ox, Saiga antelope, glutton, Arctic hare, lemming	Cro-Magnon and Chancelade races of *H. sapiens*
Neolithic or New Stone Age	Polished stone tools; domestication of plants and animals (agriculture)		
Historical Record			

The Paleolithic corresponded roughly with the Pleistocene (see below). The Magdalenian and so forth are different Stone Age cultures the succession of which was seen as becoming more elaborate toward the present time. However, they were often also used to designate time periods (e.g., the time of the Neanderthals might be called the Mousterian) and even for Palaeolithic races (e.g., the Mousterians for the Neanderthals). The time periods in years assigned to geological and archeological series differed widely. So did the dating of hominid fossils and even the estimate of the first appearance of human-made tools.

Geological Series Reconstructed after the Text of *Ancient Hunters* (1911)

Geological Period		Fossil Primates
Tertiary (Age of Mammals)	Eocene	Lemuroid primates
	Oligocene	Old World and New World monkeys and apes (*Propliopithecus*)
	Miocene	Primitive anthropoid apes (*Dryo-* and *Pliopithecus*)
	Pliocene	Anthropoid apes Hominids outside Europe? *Pithecanthropus erectus* (today *H. erectus*)?
Pleistocene or Quaternary, period of Great Ice Age divided into four glacial and four interglacial epochs	Lower Pleistocene	*H. heidelbergensis* *P. erectus*
	Middle Pleistocene	Neanderthals
	Upper Pleistocene	*H. sapiens*
Recent		

Appendix B
Schematized Views of Human Evolution

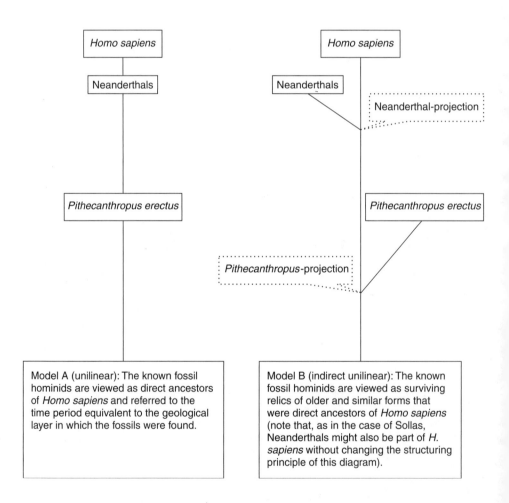

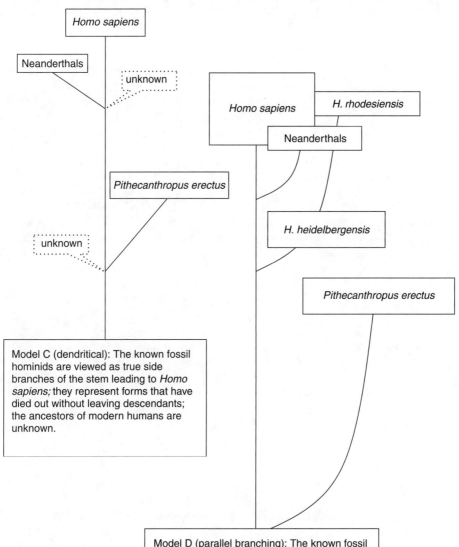

Notes

Introduction

1. Throughout this book, I use the term *race* and its adjectives in the way they were employed by the scientists under consideration. Neither their application to fossil and to recent human groups nor the value judgments associated with them represent my own views. The same applies to such terms as *savage* and *primitive*, all of which I retain to convey the large morphological and cultural differences anthropologists perceived between human groups—a central topic of this book. However, I have chosen to employ the adjective *human* and related forms instead of *man*, which was the predominant designation used for humanity by the earlier scientists under consideration. I adopt the original usage only in expressions such as *Neanderthal Man*.

2. As-yet-unpublished article by Aldhouse-Green for the *Western Mail* (completed 31 March 2006).

3. Daston, 1998; Daston and Galison, 1992, 2007; see also Daston, 2001a, on the scientific virtue of attention, and Daston, 2001b, on rationality; Foucault, 1969 (1961); on Foucault's *Histoire de la folie* as containing the seeds of his concepts of archeology and genealogy in the attempt to divest scientific objects of their actuality by tracing them back to the structural and historical conditions of their formation, see Sarasin, 2005, p. 22. Kindred reflections on the generation and nature of scientific knowledge and objects have a considerable tradition in the history and philosophy of science (on Ludwig Fleck, Gaston Bachelard, and Georges Canguilhem, see Rheinberger, 2006, sec. I).

4. Hagner, 1997. In another history of a generic object, Hagner, 2004, has worked what was originally meant as a chapter in the history of the change from organ of the soul to brain into a comprehensive history of the engagements with brains of "geniuses." As in the case of the search for an organ of the soul and an anatomical substrate for genius in the brain, Evelyn Fox Keller's *The Century of the*

Gene (2000) refers to the fact that the concept of the gene preceded clear ideas about its material substrate and was indeed realized by more than one substance before, as she argues, losing its overwhelming power as a research tool (both concept and material object). These examples point to a specificity of histories that follow the material object(s) rather than the concept, that is, focus on the concepts successively attached to the same object(s) rather than the other way around, which will surface again later.

5. Schiebinger, 2004; see, for example, Balick and Cox, 1996, pp. 132–141, on quinine and nutmeg; Coe and Coe, 1996; Pendergrast, 1999; Zuckerman, 1999; on the notion of agnotology, see Schiebinger and Proctor, forthcoming. As the histories of plants suggest, the emphasis on material objects is not restricted to the history of science. Rather, because of the price of exacerbating the notion's inchoateness, it represents an interest that increasingly brings together curators, historians of science, historians of art, artists, and writers (see, for example, Daston, 2004b; Heesen and Lutz, 2005).

6. James Secord, 2000; forty years after the publication of *Vestiges* (1844), the author was officially revealed to be the publisher, journalist, popular science writer, and natural historian Robert Chambers.

7. Darwin, 1859.

8. Rheinberger, 1997, p. 4; on epistemic things, see particularly pp. 28–31.

9. Rheinberger, 1997, distinguishes epistemic from technical things or objects. The technical objects are the experimental conditions that embed the epistemic objects. They simultaneously embody local and temporal research context and participate in the shaping of epistemic things. Conversely, once stabilized sufficiently, an epistemic object might take on the role of a technical condition. The epistemic units in which these transformations occur are what he refers to as *the experimental system* (see also Rheinberger, 1992). Although the main interest here is the experimental sciences, the descriptive and systematizing sciences are seen to bring into being their epistemic objects through processes of dislocation from their "original" contexts and reappropriation within the scientific theoretical and epistemic-practical contexts, as discussed later (Rheinberger, 2006, p. 336).

10. Daston, 2000; Latour, 1993, 1999.

11. Indeed, a fossil human skeleton might be viewed as *epistemologicum par excellence* (Rheinberger, 2006, p. 343), in that it has gained a certain stability in a material as well as symbolic sense (perhaps as type specimen) but retains the constant possibility for reinterpretation and reclassification as individual sample. The ambiguity with regard to subsuming skeletons under nature or culture of course also applies to living bodies, not to mention the problematic juxtaposition of these as independent entities in the first place (see, for example, Fausto-Sterling, 2000). For a recent focus on material objects in the history of ethnology, see Penny, 2002, whose analysis of the early history of German ethnological museums brings to light the worldwide trafficking of cultural objects along the axes of individual, institutional, economic, and political interests; on material cultural objects and the anthropological museum, see also Stocking, 2001, pp. 244–259.

12. On the notion of talking things, see Daston, 2004a. That human bones require technologies to be made to speak has been explored by Hagner, 2005. The project of inscribing anthropological meaning into bones that in this respect had previously been dead and speechless began only in the late eighteenth century. In

order for skulls to refer to individuals or types of people, to become objects of anthropology, systems of reference had to be established. The large collections accumulated in museums in the course of the nineteenth century were a sine qua non for the empiricist search for distinctive markers through comparison. In the process, anatomical details came to point to places in the taxonomic system. It is because of this traditional focus on detail, besides the reconstruction of the past from sparse traces unintelligible to the untrained eye, that paleoanthropologists like to see themselves as Sherlock Holmeses (or, as we will see, as magicians and diviners). Indeed, one of the practices a physical anthropologist might engage in is forensic anthropology (see, for example, Schwartz, 1993, pp. 27, 58–59, 85–86). Carlo Ginzburg, 1983 (1979), describes the reconstruction of the past through what he calls, with Thomas Henry Huxley, *retrospective prophecy* as characteristic of paleontology, which thus shares with human sciences the evidential paradigm (Huxley, 1915 [1880]; on the notions of trace, detail, and symptom, see particularly Ginzburg, 1983 [1979], pp. 83–84). Ginzburg's distinction between the experimental natural sciences as guided by laws and standards and the human sciences as oriented toward traces and signs has been refined by recent work in the history of science, which by inference renders Ginzburg's subsumption of paleontology under the human sciences less problematic (see Schaffer, 2005, pp. 345–349 in particular).

13. On the history of archeological epistemology, from positivist processualism to relativist postprocessualism, and intermediate frameworks as developed in feminist epistemology, see Wylie, 2002.

14. Latour, 1999, p. 158; similarly in Latour, 2000, p. 257.

15. Zihlman, 1981, p. 87.

16. Haraway, 1997, p. 68. Indeed, the paleoanthropologist Milford Wolpoff, 2007, is skeptical about the power of paleoanthropological evidence, not to mention the notion of data speaking for themselves. He regards the influence of the interpretational framework as so strong, and the data as so ephemeral and unrepresentative, that it is rather the theories that constitute the "facts" of those who work on human origins.

17. *The Oxford English Dictionary Online* (Oxford University Press, 2006) has an entry for *paleo-anthropology* of 1908 and one for *paleoanthropological* of 1909.

18. Chiarelli, 1992, p. 63.

19. *The Oxford English Dictionary Online* (Oxford University Press, 2006); see also Stocking, 1990, on the history of the discipline of anthropology in various traditions and its inclusiveness and exclusiveness, respectively; and Spencer, 1997b, on physical anthropology.

20. See, for example, Shipman and Storm, 2002, on Eugène Dubois as founder; Eiseley, 1957, on the significance of the early Neanderthal finds; Gundling, 2005, on the australopithecines; Delisle, 1995, on the impact of the synthesis; and Spencer, 1981, 1982a, for a focus on institutionalization.

21. For a historical introduction in the early canon, see, for example, Boule, 1923 (1921a), 1923 (1921b), chap. 1; on paleoanthropological preface history, see Corbey and Roebroeks, 2001.

22. The anthropologist George Stocking has published extensively on the history of the field (on biological anthropology, see, for example, the edited volume Stocking, 1988); other histories written by practitioners include Spencer, 1990a, 1990b; Tattersall, 1995; Trinkaus and Shipman, 1993; and Walker and Shipman,

1997. Among the most famous human-origins popular books are Lewin, 1997 (1987); Reader, 1981; and Willis, 1989, 1992; on the Piltdown forgery, see, for example, Walsh, 1996. Just in time for me to add a note, the paleoanthropologist and philosopher Richard G. Delisle, 2007, has published an encompassing history of ideas in paleoanthropology, particularly on primate phylogeny, in which he argues for a progressive development and relative unity of the field under an evolutionary approach since its "beginning" (1860). In this history, the differences in theories about mechanisms of evolution are seen to have played a minor role in the establishment of phylogenies. The history of paleoanthropology is said to have been preoccupied with the unilinear versus multilinear views of evolution, without adequate consideration given to parallel and polyphyletic trees. That these labels need to be complicated is part of the insight of this study. However, in contrast to this study, Deslisle is not interested in the historical, social, and cultural context of the knowledge on human phylogeny and evolution, which leads him to depreciate what he understands to be science studies. Interestingly, Delisle's strong stance is balanced by the preface of the book, contributed by the paleoanthropologist Milfrid Wolpoff, 2007, as in fact more generally by the self-reflexivity evidenced by paleoanthropologists, primatologists, and prehistoric archeologists and their occasional cooperation with social scientists (see, for example, Strum and Fedigan, 2000; Strum, Lindburg, and Hamburg, 1999; Tattersall and Eldredge, 1977).

23. The historian of science Peter Bowler has published extensively on the history of theories of human evolution in cultural context (see, for example, Bowler, 1986); the historian of science Brian Regal has made use of the genre of biography to provide a broadly situated understanding of one scientist and his work (Regal, 2002; see also Theunissen, 1989); the philosopher of science Raymond Corbey has been concerned with the conceptual boundary between apes and humans in the anthropological sciences for some time and has recently published *The Metaphysics of Apes* (Corbey, 2005); the sociologist Michael Hammond has been particularly interested in the social settings of events in the history of (paleo)anthropology (see, for example, Hammond, 1980). There are a number of historical analyses of developments in one particular national setting (see, for example, Spencer, 1982b; Zimmerman, 2001) and explorations of the relation of the anthropological sciences to specific historical moments (see, for example, Proctor, 2003, on the influence of the Holocaust on the development of ideas on human origins and diversity; Schmuhl, 2005, on the context of German national socialism).

24. See, for example, the historians Elazar Barkan, 1992, and Nancy Leys Stepan, 1982; and the biologist Stephen Jay Gould, 1981b.

25. For a comprehensive discussion of visualizations of human origins, see the archeologist Stephanie Moser, 1998, who has also collaborated on edited volumes on archeological visual reconstruction (see, for example, Smiles and Moser, 2005); on narrative structures in human-evolution scenarios, see the paleoanthropologist Misia Landau, 1991; for a case study of exhibits, see the cultural anthropologist Mary Bouquet, 1995.

26. See, for example, the historian of science Claudine Cohen, 2003; and the archeologist Lori Hager, 1997b. The role of gender in the institutional structures, approaches, and contents of archeology has been the focus of several studies (see, for example, Cros and Smith, 1993; Gero and Conkey, 1991; Wright, 1996; Wylie, 2001).

27. For a recent comprehensive history of geology of the late eighteenth and

early nineteenth centuries, see Rudwick, 2005. It would be even vainer to attempt to list all the relevant sources on the history of the fields influencing paleoanthropology than those on paleoanthropology itself. Both will therefore be referred to where they are drawn on in the course of this book. For historiographies of the human-origins sciences, see Goodrum, 2000, 2004; Proctor, 2002.

28. For the expression of the need for a historical engagement with the human-origins sciences, see, for example, Corbey and Roebroeks, 2001.

29. For histories of particular finds or taxa, see, for example, Gundling, 2005, on the australopithecines; Johanson and Edey, 1981, on Lucy in particular; on the Piltdown forgery, see Spencer, 1990a, 1990b; on the Neanderthal history and controversy, see Stringer and Gamble, 1993; Tattersall, 1999 (1995); Trinkaus and Shipman, 1993.

30. The ensuing summary is annotated only rudimentarily because all the relevant references can be found in the detailed accounts in the following chapters.

31. Latour, 1999, p. 158; similarly in Latour, 2000, p. 257.

32. On the concept of boundaries and boundary objects, see Bowker and Star, 1999, chap. 9 in particular; Star and Griesemer, 1989; on the monster and vampire figure, see Haraway, 1997, chap. 6.

33. Darwin, 1859, 1871.

34. Sollas, 1911, 1915 (1911), 1924 (1911).

35. Buckland, 1823.

36. On gender stereotypes in archeological reconstructions, see, for example, Gifford-Gonzalez, 1993; Moser, 1993.

37. Aldhouse-Green, 2000c.

38. On the cultural history of red hair, see Roach, 2005; on the history of the meanings of *scarlet* and *red*, see *The Oxford English Dictionary Online* (Oxford University Press, 2006).

I. William Buckland, an Ancient British Witch, and the Question of Human Antiquity

1. On the development of antiquarianism focused on the pre-Roman history of Britain from the late sixteenth to the early nineteenth century, see Piggott, 1989.

2. Blumenbach, 1790–1828, 2001 (1798).

3. See, for example, Shapiro, 1959, pp. 371–375; for a comprehensive history of ethnology, see Petermann, 2004.

4. For a concise history of modern attempts at racial classification and theories of racial origin(s) and diversification, and their connection to the early institutionalization of physical anthropology, see Sommer, 2005a; on the monogenism/polygenism debate as related to the question of human antiquity, see Grayson, 1983, chap. 7.

5. Buckland, 1840, p. 51.

6. Cuvier, 1812; de Luc, 1779. There are several histories of geology, which will be referred to throughout Part I; for an overview of theories of the earth from antiquity to the present, see Oldroyd, 1996; for a comprehensive history of the development of a historical geology in the late eighteenth and early nineteenth centuries, see Rudwick, 2005. Although the terms *catastrophism* and *uniformitarianism* were used only in the 1860s, I employ them for simplicity. The expressions *geolog-*

ical catastrophe, geological uniformity, uniformitarian, and *catastrophist* were already in use in the 1830s and 1840s.

7. Cuvier and Brongniart, 1808; on Smith, see Thomas, 1947, especially p. 337; see also Winchester, 2001; and *Memoirs of William Smith* by John Phillips (London: John Murray, 1844), republished with additional material by Hugh Torrens (Bath: Bath Royal Literary and Scientific Institution, 2003).

8. Briggs, 2000 (1959); Niedhart, 1987, pp. 15–39; on the contributions of mining engineers, mineral surveyors, and inventors to knowledge on stratigraphy in the context of the developing industrial economy, see Torrens, 2002.

9. On the degree of professionalization in geology, see the statistics in Morrell and Thackray, 1981, fig. 2, p. 15; on the influence of colonial career opportunities on the development of geology, see, for example, Stafford, 1989; on the process of professionalization in geology, see also Porter, 1978.

10. See Robert Jameson's (1774–1854) mineralogical notes in Cuvier, 1813. Jameson was regius professor of natural history at Edinburgh University from 1804, in which position he introduced the Wernerian mineralogical system and founded the Wernerian Natural History Society (1808). Although Cuvier's work is important to the topics of this part, this is not the place to provide a detailed discussion of his work, for which the reader is referred to Rudwick, 2005 (on *Recherches,* see pp. 499–512, with pp. 510–512, 596–598, on Kerr's translation in particular; on effects of the Napoleonic Wars on connections between Britain and France, see p. 377; although travel was obviously impeded, communication between savants was maintained).

11. Rupke, 1983, pp. 51–57, 233–240; on the BAAS, see Morrell and Thackray, 1981.

12. As a period, romanticism roughly covers 1770 to 1830, although the period between about 1740 and 1790 has often been considered separately as "late Augustan" or "preromantic" (e.g., Booth, 1989, pp. 225–250, on p. 234). *Romanticism* is a contested notion that covers a wide array of different phenomena. Its meaning in this context will be elaborated in the course of this part. On the development of regimes of time, see Nowotny, 1989; on natural philosophy during the age of romanticism, see Knight, 1990, pp. 13–24; Lawrence, 1990, pp. 213–227; Schweber, 1981, pp. 3–4, 18–19; on the flair for historicity, see Cunningham and Jardine, 1990, pp. 1–9; Lepenies, 1976, pp. 106–114; Rupke, 1990; Schweber, 1981, pp. 14–15; on the "romantic cave" in the "golden age" of geology, see Sommer, 2003b, 2005b. For reasons of consistency, I use the term *scientist* throughout to denote the philosopher of the late eighteenth and early nineteenth centuries, even though it was coined only as a reaction to Samuel Taylor Coleridge's abnegation of the honorable title *philosopher* to the members of what he considered the plebeian BAAS (Schaffer, 1990, p. 93; see also Wyatt, 1995, p. 6).

13. National Museum of Wales, Department of Geology (subsequently NMW, Bu P), Box: DLB 5, Buckland W, May 1816–May 1831 (subsequently Box 5), Folder: Buckland, William (4.) February 1823–January 1826, 84.2OG. D167-174 (subsequently Folder 4), p. 167: Letter from Buckland to Lady Mary Cole, 15 February 1823 (this letter is partially reproduced in North, 1942, pp. 107–110, 112, 116, and in its entirety in Edmonds and Douglas, 1976, pp. 149–152; the letters from Buckland held in the NMW are cataloged in Sharpe and McCartney, 1998, pp. 23–27). Contrary to Buckland's suspicion that the author was one of his under-

graduates, the poem was written by Philip Bury Duncan (1772–1863), fellow of New College (1792–1842) and keeper of the Ashmolean Museum at Oxford (1829–1855), and is reproduced in Daubeny, 1869, p. 122.

14. NMW, Bu P, Box 5, Folder 4, p. 167: Letter to Lady Mary Cole, 15 February 1823.

15. This footnote, which was included in Daubeny's reproduction of the poem, was Buckland's addition, since it first appeared in the letter to Lady Mary Cole.

16. Grayson, 1983; Van Riper, 1993. The ensuing discussion of Buckland and his interpretation of the Red Lady draws on Sommer, 2004a; on Buckland, see also Rudwick, 2005, pp. 600–638.

1. William Buckland

1. Gordon, 1894; Rupke, 1983, p. 71; on Buckland anecdotes, see also Gratzer, 2002, pp. 24–25; for poems inspired by Buckland, see Daubeny, 1869, for example, *The Last Hyena* on pp. 119–120; Buckland also reproduced several poems inspired by his person and work in his letters to Lady Mary Cole (NMW, Bu P, Boxes 5 and 6).

2. The term *amateur*, as I use it throughout this part, is not meant disparagingly, but may refer to both members of the scientific elite around the universities and the metropolitan geological societies (for my purpose primarily the lectures at Oxford and Cambridge and the sessions of the Geological Society of London) and those not part of the elite, such as women, local clergymen, physicians, lawyers, and other educated provincials. They share in common a taste for geology, which for the less socially affluent amateurs was often a dearly bought pastime because they lacked the financial means of the more gentlemanly geologists (women were not part of the inner circle even if they did have the necessary station and leisure). Another group of people who were not scientific professionals but who contributed to geology were the fossil collectors and the miners and quarrymen who hit upon fossils in their work. Since the term *amateur* today carries negative connotations, however, I follow the practice of putting it in quotation marks.

3. Edmonds, 1978.

4. See, for example, Portlock, 1857.

5. Buckland gave a minute itinerary of the Continental tour to Lady Mary Cole in a letter dated 3 April 1817 (NMW, Bu P, Box 5, Folder: Buckland, William [1.] May 1816–October 1819, 84.2OG. D144–151 [subsequently Folder 1], p. 146).

6. On personalities involved in securing the readership for Buckland, see, for example, NMW, Bu P, Box 5, Folder 1, p. 148: Letter from Buckland to Lady Mary Cole, 24 December 1818. On the development of a historical geology out of its eighteenth-century precedents, see, for example, Porter, 1977.

7. British Library, Add MS 58995 (subsequently BrL, Bu C), letter to Lord Grenville, 27 June 1822 (pp. 96–97); Rupke, 1983, pp. 15–20 and chap. 15, speaks of an "English School of (Historical) Geology"; Leroy Page, 1984, criticizes the notion. Certainly, the English geologists were no monolithic entity, and there was no simple opposition between Scottish and English schools, but the reconstruction of their self-portrayal as carried out by Rupke does not seem to imply so. On the divergence between the self-image of the Geological Society as a "happy band of

brothers of the hammer" and their internal quarreling, see Thackray, 2003 (1999), quote from p. ix.

8. BrL, Bu C, letters to Lord Grenville, pp. 64–68. That Buckland undertook many geological journeys and received specimens even from outside Europe (such as America, Canada, Madagascar, and India) is documented in letters (BrL, Bu C, to Lord Grenville, 16 and 23 August and 23 September 1819, pp. 72–77).

9. Letter of 9 November 1818, reproduced in Edmonds, 1979, pp. 39–41.

10. On the peculiarities of the universities of Oxford and Cambridge in contrast to the Scottish universities of the Enlightenment tradition, the French reformed institutions, and the Humboldtian German university model in the first half of the nineteenth century, see Brockliss, 1997; on criticism leveled at Oxford, see Briggs, 1997; on Oxford and geology in particular, see Rupke, 1983, 1997; on the Oxford of Buckland's time, see also Brock, 1997; on the wider impact of the early geology, see, for example, Thomas, 1947.

11. BrL, Bu C, p. 83; Buckland again wrote to Lord Grenville on 27 March 1823 that his lectures were very well attended, with the bishop of Oxford, as well as the warden of New College, among his students (pp. 102–103); on Buckland's lectures, see also NMW, Bu P, Box 5, Folder: Buckland, William (5.) February 1826–May 1831, 84.2OG D175–182 (subsequently Folder 5), pp. 180 and 182: Letters from Buckland to Henry De la Beche, 1 and 25 May 1831, in which he described the success of De la Beche's *Duria Antiquior* in entertaining his class.

12. NMW, Bu P, Box 5, Folder 4, p. 168: Letter to Lady Mary Cole, 3 April 1823.

13. Buckland, 1821; on the worldwide distribution of the "Instructions," see NMW, Bu P, Box 5, Folder 1, p. 151: Letter from Buckland to Lady Mary Cole, 29 October 1819, in which he wrote of boxes with animal remains from Russia and importations from all the British colonies through Lord Bathurst for his collection and as visual aids in his lectures, so that "my Treasures in Geology continue more than ever to accumulate." In the same letter, he informed Cole of his receipt of a diploma from the Moscow Society of Naturalists (founded in 1805), of which he had become a full member on 15 February 1818. Indeed, he was subsequently made a member of the Mineralogical Society of St. Petersburg (Hugh Torrens, personal communication; see also Folder: Buckland, William [2.] January 1820–September 1821, 84.2OG D152–159 [subsequently Folder 2], p. 152). Buckland was a corresponding member of the Museum of Natural History in Paris (BrL, Bu C, Buckland to Lord Grenville, 23 August 1819, 18 June 1821, pp. 74–75, 83; see also the archives of the Royal Society of London, Papers of British Churchmen, 1780–1940, Buckland correspondence, MS 251 [subsequently RS, Bu C], letters from and to Cuvier about visits and exchanges of specimens; Cuvier had visited Oxford in the 1810s, but by letter of 3 August 1830 he canceled a visit with his family to the Bucklands because of the political situation in France and wished that Buckland would be spared similar bouleversements). On the journey to the Grisons, the Tyrol, and Vienna with the "amateur" geologist Count Breunner and Greenough, see NMW, Bu P, Box 5, Folder 2, p. 153: Letter to Lady Mary Cole, 18 May 1820, and p. 155: Letter to Jane Talbot, 11 December 1820. Buckland found it altogether less adventurous than the last Continental tour, on which they had faced imprisonment and bandits. Nonetheless, Breunner had put some incidents into drawings. On his visit to the German caves, see NMW, Bu P, Box 5, Folder: Buckland, William (3.) October

1821–December 1822, 84.2OG D160–166 (subsequently Folder 3), p. 165: Letter to Lady Mary Cole, 24 December 1822.

14. Buckland, 1817 (1816), 1821 (1819); see also Buckland, 1821b, 1824 (1822).

15. Quote from letter from Buckland to Lady Mary Cole, 18 May 1820 (NMW, Bu P, Box 5, Folder 2, p. 153), with which he sent a copy of the inaugural lecture, i.e., Buckland, 1820. In an earlier letter to Lady Mary Cole, Buckland suggested that the inaugural lecture was a direct response to the claim that geology afforded no proof of the Deluge (NMW, Bu P, Box 5, Folder 2, p. 152, 5 January 1820; Whittacre's review article had appeared anonymously in the *Quarterly Review* of May 1819, pp. 14–66). Although Buckland's diluvialism was attacked by low as well as high church exponents, it enjoyed the patronage of a number of Anglican clergymen, and by 13 January 1820, he was relieved that the bishop of London and the Evangelical John Bird Sumner (1780–1862), later archbishop of Canterbury, approved of the lecture's contents (NMW, Bu P, Box 5, Folder 2, p. 154: Letter to Lady Mary Cole). On the inaugural lecture and Conybeare's influence, see Edmonds, 1991.

16. Buckland, 1823, pp. 171–228. On diluvialism and associated geologists, institutions, and developments, see Clark, 1961, pp. 86–94; Gillispie, 1996 (1951); Gould, 1983; Rupke, 1983.

17. Buckland, 1822, p. 224.

18. Buckland quoted from Benjamin Silliman's (1779–1864) *American Journal of Science* 14 (April 1828) (Oxford University Museum of Natural History, Papers of British Scientists, 1600–1940, Buckland Papers, NRA 42502 [subsequently OUMNH, Bu P], Notes for the Bridgewater Treatise/Notes by Subject [box 2], "Species change of Lamarck," folder 6i, p. 4).

19. OUMNH, Bu P, Lecture Notes (box 1), "Lecture notes on the Mosaic Deluge and on organic remains," folder 1, p. 2.

20. Buckland, 1824; on Buckland's discovery of and work on dinosaurs, see Delair and Sarjeant, 1975, pp. 12–21; Gordon, 1894, pp. 87–119; North, 1956, pp. 136–138; for accounts of Anning's life and work, see Goodhue, 2002; Tickell, 1996; Torrens, 1995.

21. In his paper read to the Geological Society in 1829, Buckland mentioned Mary Anning's discovery of many coprolites in connection with ichthyosaurs. This, however, happened only "since her attention has been directed to these bodies" (Buckland, 1835 [1829], p. 224). The description of Anning's plesiosaur fell to Conybeare, 1824 (see also De la Beche and Conybeare, 1821). A letter to De la Beche seems to suggest that a later paper on coprolites was rejected for publication by the Geological Society because of considerations of decency (NMW, Bu P, Box: DLB, Buckland, William, October 1831—undated to Carlisle 84.2OG D183–192 [subsequently Box 6], p. 196, 10 May 1847?;—De la Beche was president of the Geological Society in 1848–1849, so this letter might well have been written during that period).

In the course of this book, I will use different designations for kinds of evidence that I think can be distinguished in the work of the geologists and anthropologists under concern. There is the *vicarious evidence* used here, which generally refers to a *pars pro toto* relation or proxy. Then there is what I call *collateral evidence*, which means that evidence that allows making a certain inference thereby suggests further inferences that seem associated with this interpretation. I use *negative evidence* exclusively in the sense of the absence of evidence being equated with the evidence of absence, since although its use is ambiguous in my sources, this was the main meaning of the notion. To refer to the converse, i.e., the inference of something from nothing, I introduce the expression *invisible evidence* (for a more detailed discussion of the issues associated with these differentiations, see Sommer, 2004d).

22. Imperial College Library, London University, Papers of British Scientists, 1600–1940, B/Playfair, NRA 11556 Playfair (subsequently IC, B/Playfair), 1 and 6 January 1843 (pp. 89–90); on her engraving, see, for example, RS, Bu P, letter from Cuvier to Buckland, 20 June 1824; on Mary Morland/Buckland and contemporary women geologists, see Creese and Creese, 1994; Kölbl-Ebert, 1997a, 1997b, 2001; Gordon, 1894.

23. On the British Association for the Advancement of Science, see Morrell and Thackray, 1981.

24. OUMNH, Bu P, Lecture Notes (box 1), "Lecture notes on the Mosaic Deluge" (folder 5, p. 9).

25. Lyell, 1830–1833; on Lyell, see Rudwick, 1970; on the discussions about "the doctrine of modern causes" versus "diluvialism" during the sessions of the Geological Society, see Thackray, 2003 (1999); the terms *diluvialist* and *fluvialist* seem to have been coined by William Conybeare (p. 33); see also Rupke, 1983, chap. 7, for the controversy with the fluvialists, and chap. 20 for more information on the Oxford or Tractarian movement.

26. Buckland, 1836.

27. Woodward, 1883, p. 229; on these discussions at the Geological Society meetings, see also Thackray, 2003 (1999), pp. 95–102, and Woodward, 1907, pp. 138–144.

28. Reproduced in Thackray, 2003 (1999), p. 124.

29. Buckland, 1842a, 1842b; on the project of a new edition of *Reliquiae*, see Buckland, 1830, p. 393, and Clark, 1961, p. 92.

30. Ramsay in his diary about meetings of 22 March 1848 and 25 February 1846; Edward Forbes (1815–1854) to Ramsay, 20 June 1846; Forbes to Roderick Impey Murchison (1792–1871), 31 January 1848; Gideon Algernon Mantell (1790–1852) in diary about 2 February 1848; Ramsay in diary about 5 April 1848; reproduced in Thackray, 2003 (1999), pp. 155, 135, 140, 153, 154.

31. On Buckland's life and work, see also Boylan, 1967, 1981, 1984, 1997; Frank Buckland, 1858; Cannon, 1970; Edmonds, 1978; Gordon, 1894; Sommer, 2003a, 2004b; Spencer, 1997a; Unt, n.d.

32. Buckland's as well as Sedgwick's way of doing geology are thus characterized by Martin Rudwick, 1982, as concrete, i.e., focusing on the order of strata and their fossil content, as opposed to, for example, the rather abstract style of work and sense of time of Lyell, who was more of a theoretician. A letter from Mantell to Buckland from 16 June 1825 contains a drawing of a kind of seal and coat of arms of the Geological Society of London with a banner saying "Facts Not Theories" (RS, Bu C). Cuvier, whose influence on Buckland was considerable, argued against the eighteenth-century tradition of grand theorizing. The accumulation of empiric data was preferable (Rudwick, 1976 [1972], p. 103; 1997, pp. 176–177). In an undated letter, Buckland replied to a request for reading suggestions. De la Beche's and Phillips's works were recommended first. Then he went on: "Sir Lyell's works are excellent but too full of shifting hypothesis and [hot?] theory for a beginner." Mantell's books were all good, and he himself in his Bridgewater Treatise had

given a general overview "avoiding technical detail as much as possible" (Bodleian Library, Papers of British Churchmen, 1833–1849, MS Eng lett d 5, pp. 251–252; the letter must date from after Lyell was knighted in 1848).

2. Man among the Cannibalistic Hyenas?

1. On early pictorial representations of the prehistoric world, see Rudwick, 1992, chaps. 2–3; on the distribution of the image in particular, see pp. 38–39. Keble, John. 1840 (1827). *The Christian Year*, 2nd Am. ed., Philadelphia, p. 27, in Cannon, 1978, p. 11.

2. On the "romantic cave" in poetry, geology, folklore, and tourism, see Sommer, 2003b; on geology and English romantic poetry in general, see Heringman, 2004; on poetry inspired by Buckland's person and science, see O'Connor, 1999; for poems on and by geologists, see Daubeny, 1869, and letters from Buckland to Lady Mary Cole, in which he reproduced several poems touching on his person and work (NMW, Bu P, Boxes 5 and 6). Buckland generally believed in "the close Connexion which subsists between geological Speculations, and not the ideas only but also the language of complete Poetry" (NMW, Bu P, Box 5, Folder 2, p. 158: Letter from Buckland to Lady Mary Cole, 25 July 1821, which contains transcriptions of William Conybeare's "Ode to a Professor's Hammer" and Duncan's "Specimen of a Geological Lecture" that were later published in Daubeny, 1869, pp. 78–79 and 84–87).

3. Buckland, 1823, p. 98.

4. BrL, Bu C, 24 January 1822, pp. 87–88.

5. OUMNH, Bu P, "Miscellaneous geology," box 1, folder 4, p. 47.

6. Buckland, 1822. Carlo Ginzburg, 1983 (1979), regards the reconstruction of the past through retrospective divination by means of traces as characteristic of paleontology, which thus shares with human sciences the "evidential paradigm." Cuvier (1812), in *Recherches sur les ossemens fossiles*, referred to Voltaire's *Zadig* (1747), this master in the reconstruction of past events by present signs, when claiming that he could reconstruct an animal from its track. Huxley, 1915 (1880), who saw in the method nothing but the application of trained common sense as opposed to the mysterious way of producing knowledge by a proclaimed caste of magi, called the method of retrospective prophecy common to history, archeology, geology, and paleontology *Zadig's method*. This definition of the historical sciences and their practitioners was at variance with the self-image, if not the actual practice, of the early paleontologists such as Cuvier and Buckland, and Huxley undertook a considerable amount of rewriting of Cuvier's self-perception. At the same time, that Huxley retained the notion of prophecy in combination with an evidence- and analogy-based method of acute observation seems well suited to Buckland's rude harmony between the image of romantic divination and the practicality of geology (on the notion of rude harmony, see Sommer, 2003b, p. 200 in particular).

7. Buckland reasoned that some animals would have died long before the Flood, and thus their bones would have been much further decayed than the bones of those who died shortly before or even during the catastrophe. This meant that in the same cave, at the same position, bones of the same species and of the same geological age could be found in very different states of decomposition. The consequence of this is

clear: No inference can be made from the state of a bone to its ante- or postdiluvian age, a consequence relevant to the following discussion of potentially fossil human bones.

8. Buckland mentioned Cuvier's agreement in a letter to Lord Grenville (BrL, Bu C, 21 July 1822?, pp. 92–94). De Luc, 1779; Cuvier, 1812; see also Rudwick, 1976 (1972), pp. 135–136, and Rudwick, 2005, pp. 622–632, on Kirkdale Cave.

9. Buckland, 1822; 1823, pp. 1–51, 162–163; on various criticisms of Buckland's combination of geology with religion, see BrL, Bu C, 10 December 1824?, pp. 111–112 (on Granville Penn, a staunch advocate of scriptural geology); letters to Lady Mary Cole in NMW, Bu P, Box 5, pp. 152, 153, 154, 165, 171 (on criticism/acceptance of *Vindiciae Geologicae*, the Kirkdale interpretation, and *Reliquiae Diluvianae*). In his review of *Reliquiae Diluvianae*, Edward Copleston, 1823, defended diluvialism against Huttonianism, on the one hand, and biblical literalism, on the other hand, thereby illustrating the precarious situation Buckland found himself in. For a discussion of the critiques against diluvialism from within geology, including catastrophists, and particularly from the Scottish geologist John Fleming, see also Page, 1969, pp. 267–271; on the literalist critique, see also Boylan, 1997; Millhauser, 1954; North, 1942, pp. 93–99, 119–123; Rupke, 1983, chaps. 2–3.

10. Eastmead, 1824, pp. 30–40; Young, 1822; on the Whitby museum, see Knell, 2000, chap. 8, notes (and personal correspondence with the author).

11. Eastmead, 1824, pp. 13–14.

12. For an analysis of the social structure of early nineteenth-century geology revealed through the practice of collecting (and its analogies to practices of gift giving), see Knell, 2000. Anne Secord, 1994, analyzes the interactions (mainly correspondence) between artisan and gentleman naturalists and brings to light the fact that not only the gentlemen but also the artisan community was organized in networks. David Allen, 1994 (1976), p. 138, points out that it was unseemly for a gentleman to engage in manual labor of any kind or even to carry tools, which was a constant source of embarrassment to the gentleman geologists, which they met by hiring "boys," bringing along menservants, and hiding tools and fossils from public view.

13. Frere, 1800 (1797). Fossils became recognized as the remains of organisms, and flint implements as human-made tools, in the course of the seventeenth century. Before contemporary meaning, they were grouped together with stones and gems under the term *fossil*, which simply meant "dug up." Stone instruments, in addition to being viewed as naturally formed regular stones, appeared in folklore as thunderbolts or elf bolts, missiles that fairies and witches threw after cattle and humans. On the history of interpretations of fossils, see Rudwick, 1976 (1972); on the history of "ceraunia," with a main focus on the eighteenth century, see Goodrum, 2002.

14. OUMNH, Bu P, Letters, box 2/P 3, 28 January 1821.

15. Buckland, 1822, pp. 216–217. In *Reliquiae Diluvianae*, 1823, p. 100, Buckland claimed that Rosenmüller classified the best-preserved bones of animal species, including *Homo*, as postdiluvian. On Esper and the history of excavations at Gailenreuth, see also Heller, 1972.

16. Quoted from Buckland, 1823, p. 168 and pp. 169–170; see also Buckland, 1822, pp. 167–170; Weaver, 1823, in general. According to a modern radiocarbon dating, Buckland's doubts were justified (Grayson, 1983, p. 97; on the excavations at Gailenreuth and in the Elster Valley in general, see also pp. 89–98).

17. Buckland, 1822, p. 226; see also Buckland, 1823, pp. 164–165. Today, the skeleton from Burrington Cave in the Mendips is considered to be of Mesolithic or older origin. After Buckland, the University of Bristol Spelaeological Society fully investigated the cave from 1914 onward. Human remains of at least ten individuals were unearthed. Some of these were in formal burials accompanied by burial goods of an Upper Paleolithic type (Jackson, 1953, p. 199).

18. Buckland, 1823, p. 166; on Dillwyn, see Howes, 1988, pp. 298–299. The Swansea Philosophical and Literary Society was named the Royal Institution of South Wales after it received a royal charter from Queen Victoria in 1938.

19. Buckland, 1822, p. 227.

20. On the problem of reconciling the concept of plentitude with the idea of the extinction of species, see Grayson, 1983, chap. 4; on Buckland's harmonization of the abundance of carnivora with a lovingly designed world, see Buckland, 1836, pp. 129–134; see also Rupke, 1983, pp. 248–254.

21. Eastmead, 1824, pp. 25–27.

22. Ibid., p. 30; on the hyena as avatar in Eastern mythology, see Dart, 1956.

23. Buckland, 1822, p. 217.

24. Reproduced in Murchison, 1868, pp. 576–577.

25. Lyell to Mantell, reproduced in Brooke, 1979, p. 48; Rupke, 1983, pp. 71–72.

26. Emerson, 1969.

27. RS, Bu P, letter from Humphry Davy to Buckland, 21 November 1822.

28. NMW, Bu P, Box 5, Folder 3, p. 165, 24 December 1822.

29. That the concepts traveled both ways is shown by John Keats's *Endymion*: "While yet our England was a wolfish den;/Before our forests heard the talk of men" (1818:IV.5–6), and Shelley's *Prometheus*: "what a low yet dreadful groan/Quite unsuppressed is tearing up the heart/Of the good Titan, as storms tear the deep,/And beasts hear the sea moan in inland caves" (1818–1820:I.578–581). The stanza and line numbers for the quotes from Byron, Keats, and Shelley refer to Byron, 1980–1993; Keats, 1900–1901; Shelley, 2002 (1977). On Byron and geology, see, for example, O'Connor, 1999, and Rupke, 1983, p. 76; see Rupke, 1990, and Sommer, 2003b, 2005b, on the romantic appeal of the cave in geology and poetry in general.

3. The Romance of the Witch of Paviland

1. Shelley, *The Witch of Atlas*, 1820:I.55–56.

2. Reproduced in Porter, 1977, p. 141; on the Buckland anecdotes, see Gordon, 1894, p. 36; on the early history of cave symbolism, see Weinberg, 1986; for an in-depth discussion of the adaptation of the romance genre, with the geologist as on a chivalric quest into the cave for visions of the past, for which he as hero or knight appeared to be particularly gifted, see Sommer, 2003b, 2005b.

Ammonites were not only important stratigraphic indicators because they were geographically widely distributed and extended through the entire series of fossiliferous strata, with the numerous species differing according to their age. To Buckland, their symmetry and delicacy were also a beautiful manifestation of the Almighty's benevolent providence in the contrivance of each creature (see Buckland, 1836, vol. 1, pp. 333–357). Ammonites were found in great numbers at Lyme Regis, and a species was named after Buckland (*Arietites bucklandi*). The importance of these fossils to geological instruction is evidenced in Figure 1.1. (*Ammon* may also refer to the Egyptian king of the gods or to a member of the Semitic tribe whose principal city was Rabbath Ammon, in Palestine.)

3. Letter written in 1823, quoted in Gordon, 1894, p. 78.

4. "The Wonders of the Antediluvian Cave!" 1822, p. 492.

5. Letters from Buckland to Lady Mary Cole and Jane and Mary Talbot, NMW, Bu P; quote from Box 5, Folder 1, p. 147: Letter to Lady Mary Cole, 14 July 1817.

6. According to F. J. North, 1966, the cave had been discovered by the brothers Davies of Reynoldston, a surgeon and a curate. On the Coles/Talbots, see also North, 1942, pp. 93, 102–104, and Gordon, 1894, pp. 15–23.

7. NMW, Bu P, Box 5, Folder 3, p. 166: Letter from Buckland to Mary Talbot, 31 December 1822 (reproduced in Howes, 1988, pp. 301–302, and in North, 1942, pp. 103–104); see also p. 165: Letter from Buckland to Mary Cole, 24 December 1822, in which he asked for confirmation of the rumor that there was a new cave and for a box of the best-marked teeth and bones. Buckland made the time he needed to prepare his lectures responsible for the delay of *Reliquiae Diluvianae*'s appearance in print (BrL, Bu C, letter to Lord Grenville, 27 March 1823, pp. 102–103).

8. Buckland, 1822, p. 227.

9. Notebook entry by Murchison on the Geological Society session of 17 June 1825, reproduced in Thackray, 2003 (1999), pp. 9–10.

10. Buckland, 1823, pp. 87–89, 95–97. That it was Buckland who found the skeleton we know only from his own account in *Reliquiae Diluvianae*, 1823, p. 87. Dillwyn does not mention the find in his diary (Howes, 1988). However, since Buckland spent the morning of 21 January alone in Paviland Cave, where he was only later joined by the rest of the party to depart for other caves, this possibility seems the most likely.

11. NMW, Bu P, Box 5, Folder 4, p. 167: Letter from Buckland to Lady Mary Cole, 15 February 1823.

12. Giraldus Cambrensis, 1908 (1188), p. 82. Giraldus accompanied the archbishop of Canterbury on his mission to Wales in 1188, which aimed to win support for the Third Crusade. However, besides being an itinerary of the preaching tour, the book provides a history of Wales and wonderful insight into the customs and beliefs of the inhabitants they encountered (on his life and work, see Giraldus Cambrensis, 1978 (1188), introduction by Lewis Thorpe, pp. 9–62).

13. NMW, Bu P, Box 5, Folder 4, p. 167: Letter from Buckland to Lady Mary Cole, 15 February 1823. The idea of a witch might not have seemed as far-fetched in Buckland's time as it would today. When I was checking the relevant newspapers for entries about the excavation at Paviland Cave, I came across a remarkable article in the *Cambrian* (Swansea, Wales) of 1 February 1823. A heading under "Varieties" read "Belief in Witchcraft," and the article told of a woman and her three daughters in Milverton, Somersetshire, who cut and maimed an old woman, whom they took for a witch. They accused the old woman, who sustained herself by collecting rags, of having bewitched one of the daughters who suffered from fits. Furthermore, the Devonshire conjurer confirmed their prejudice and recommended the drawing of the witch's blood (Newspaper Library of the British Library). In fact, in the 1908 edition of *The Itinerary of Archbishop Baldwin through Wales*, W. Llewelyn Williams observes in a note that the custom of scapulomancy was still practiced there (Giraldus Cambrensis, 1908 [1188], p. 80).

14. Giraldus Cambrensis, 1908 (1188), pp. 145–146.

15. Buckland, 1836, vol. 1, p. 1.

16. NMW, Bu P, Box 5, Folder 4, p. 167: Letter to Lady Mary Cole, 15 February 1823; see discussion in the introductory chapter to Part I.

17. OUMNH, Bu P, Letters: Eyles Colln., p. 24, my emphasis.

18. Giraldus Cambrensis, 1908 (1188), p. 80.

19. Buckland, 1823, p. 90.

20. Ibid., pp. 90–91.

21. Hoare, 1812, 1819; Strabo, 1854, bk. 4, p. 298. For Hoare, archeology meant the study and excavation of ancient remains, which, even though the engagement with such antiquities as field monuments and artifacts as historical documents became more widespread in the late seventeenth century, was not self-evident in the eighteenth century, when *archeology* still meant "ancient history." Until well into the nineteenth century, inquiries into the early history of humanity or of nations were predominantly made through studies of religious and other historical texts and, from the middle of the eighteenth century, through comparative philology (Haycock, 2002). It is interesting to observe at this point that the early Paleolithic archeologists, whom A. Bowdoin Van Riper has termed *geological archeologists* and who will concern us in Part II, allied themselves with anthropologists and geologists, with whom they mingled in the Royal Society, the Geological Society, the Society of Antiquaries (1707), and most importantly the Ethnological Society (1843) (and later the Anthropological Institute, 1863), while historical archeology, which was organized within the Society of Antiquaries, the British Archaeological Association (1843), and the Archaeological Institute, continued to be mostly interested in Roman and later remains (Van Riper, 1993, p. 190).

22. NMW, Bu P, Box 5, Folder 4, p. 167: Letter to Lady Mary Cole, 15 February 1823.

23. See, for example, Hoare, 1829, pp. 11–12. Hoare, who was well acquainted with Wales in general, had in fact found some of these traces of Ancient British habitations there (Hoare, 1983 [1810]).

24. OUMNH, Bu P, Lecture Notes (Box 1), "Lecture notes on the Mosaic Deluge and on organic remains" (folder 1, pp. 5 and 11).

25. Buckland, 1823, p. 83.

26. NMW, Bu P, Box 5, Folder 4, p. 171: Letter to Lady Mary Cole, 3 December 1823, reproduced in North, 1942, pp. 112–113.

27. Gordon, 1894, p. 78.

28. The understanding of this technological sequence is evidenced in William Borlase's (1695–1772) work *Antiquities Historical and Monumental of the County of Cornwall* (1754).

29. Daniel, 1950, pp. 29–31; Daniel and Renfrew, 1988 (1962), pp. 10–16; Kendrick, 1950. Already Giraldus described how the Welsh traced their genealogy to Brutus and beyond to the sons of Noah. In Tudor and Stuart Britain, Gomer, the eldest son of Japhet, was linked to the Cymraeg, the Welsh, so that by the early eighteenth century, the peoples that then came to be called the Celts were identified as the descendants of Gomer (Piggott, 1989, p. 59).

30. This picture of the development of the meanings of *Celts* and *Celtic* is obviously simplified, and the reader is referred to Morse, 2005, for an adequately complex reconstruction of their history. Simon James, 1999, pp. 47–48, has pointed out

that the very year the Celtic-language theory was introduced into Britain, the Treaty of Union between England and Scotland was signed (1707). Thus at the moment the term *Britons/British*, formerly reserved to non-English peoples, was usurped for the new state, the alternative *Celtic* became available to designate the independent roots of areas such as Scotland and Wales. Similarly, the Celtic revival coincided with the incorporation of Ireland into the United Kingdom (1801), threatening not only the political but also the cultural autonomy of the Irish. These observations also hint at the fact that there was a counternarrative, that of Anglo-Saxonism, the history of which from the sixteenth to the twentieth century is reconstructed in MacDougall, 1982.

31. Kendrick, 1950, pp. 123–125 and figs. XII, XIII, XV, XVI; see also Piggott, 1989, pp. 73–86. On the refinement of knowledge on stone implements in eighteenth-century antiquarianism and natural history, and the role of ethnological comparison therein, see Goodrum, 2002.

32. Strabo, 1854, bk. 4, pp. 296–300. It seems as though Strabo structured his books 3 and 4 of the *Geography* along a west-east axis, as well as along a gradient from civilized to barbaric. There were several reasons for Strabo to see the British as representatives of the barbaric end: The British Isles belonged to the lesser-known regions of Europe, were situated toward the north and far from Rome, were an island, and, most of all, were perceived as insufficiently Romanized, which to Strabo was the civilizing agent (Thollard, 1987).

33. Haycock, 2002, pp. 253–255, on Browne.

34. This, at any rate, was John McEnery's (1796–1841) conjecture (Pengelly, 1869, p. 446; on McEnery, see Chapter 4 of this volume).

35. Hoare, 1829, pp. 5–6.

36. Buckland, 1845 (1844), p. 108.

37. Stukeley had been a pioneer in the Druidic interpretation of British antiquities. Like other antiquaries of his time, he had been caught in the climate of reformation and national unity (in the 1707 Act of Union England and Wales had been united with Scotland). He saw the British church as derived from a universal religion and as prefigured in the beliefs and practices of the ancient Druids. This secured a pre-Roman Catholic foundation for the Church of England. He accordingly interpreted stone circles and henges as Druid temples. Again, Hoare, who relied entirely on classical sources without regard to biblical evidence or grand theory, disagreed (Haycock, 2002, pp. 249–250, on Hoare in particular).

38. Hoare, 1812; on this, see also Piggott, 1989, p. 153.

39. For a reappraisal of Hoare, see Morse, 2005, pp. 87–94. That Buckland and the geologists of his time did not hold the highest opinions of the antiquarian craze is expressed by Whewell in his poem *Nugae Bartlovianae*, which addresses the disputes on prerogatives of access to and interpretation of antiquities between the geologist and the antiquarian, referring to the antiquaries as "robbers" (reproduced in Daubeny, 1869, pp. 213–224).

40. NMW, Bu P, Box 5, Folder 4, p. 171: Letter to Lady Mary Cole, 3 December 1823. Buckland left the question of the cause by which nearly the entire right half and skull of the human skeleton had been removed open, simply referring to the previous operations, "whatever they were" (Buckland, 1823, p. 88).

41. NMW, Bu P, Box 5, Folder 4, p. 168: Letter to Lady Mary Cole, 3 April 1823, reproduced in Gordon, 1894, p. 51, and Edmonds and Douglas, 1976, pp. 153–154.

At an earlier date, Cole had sent two boxes full of Paviland bones to him in London, where Buckland spent the main part of two days inspecting them, as he informed Cole in the letter dated 15 February 1823 (NMW, Bu P, Box 5, Folder 4, p. 167). We cannot know whether the Red Lady's bones were among them.

42. Prichard, 1813; see also Stocking, 1973; on Buckland favoring Prichard as Thomas Arnold's (1795–1842) successor, see Rupke, 1983, p. 203.

43. König, 1814a, p. 488; König, 1814b. In an address to the Archaeological Institute, Mantell stated that the human and animal bones deposited in the same stratum would be in a similar state of mineralization. However, he warned the antiquaries or archeologists—terms that he used interchangeably—that there were many ways in which human bones might become associated with older strata, become fossilized, or become sealed in rock without being of a geological age. The Guadeloupe skeleton which had been encased in limestone and impregnated by carbonate of lime within the short time of about 150 years, was one such case. This specimen and the other skeletons found with it were the remains of a tribe of Gallibis, slaughtered by the Caribs (Mantell, 1850, pp. 337–339, on the Guadeloupe find; König noted that *Calibi* was simply another way of spelling *Caribee*; on early interfaces between archeology and geology, see Torrens, 1996).

44. Blumenbach, 1790–1828; on the anatomical distinction between male and female skeletons in eighteenth- and early nineteenth-century comparative anatomy, see Schiebinger, 1989, chap. 7; on the female pelvis size as sex marker, see pp. 208–209 in particular.

45. Schaaffhausen, 1872 (1869), pp. 116–117.

46. NMW, Bu P, Box 5, Folder 4, p. 167: Letter to Lady Mary Cole, 15 February 1823.

47. On Buckland's blue bag, see also Thomas Sopwith in his diary, reproduced in Thackray, 2003 (1999), pp. 75–77.

48. Duncan had coined another verse that combined a geological phenomenon with the notion of blushing as a symbol for innocence and its loss: "Of yore tis said a heavenly Red/The Cheeks of modest Maids o'erspread/Some say with Innocence it fled/But where it went no Man could know/The Truth our modern Trav'lers shew/It went to dye the Arctic Snow" (NMW, Bu P, Box 5, Folder 1, p. 148: Letter from Buckland to Lady Mary Cole, 24 December 1818; red snow was a puzzle recorded by the polar expedition). On a cultural history of red hair, see Roach, 2005; on the meanings of *scarlet* and *red* at Buckland's time, see *The Oxford English Dictionary Online* (Oxford University Press, 2006). Michael Taylor has directed my attention to the fact that *scarlet lady* or *woman* was also a designation for the pope and papal or pagan Rome, as well as for the anti-Christian world.

49. NMW, Bu P, Box 5, Folder 4, p. 167: Letter from Buckland to Lady Mary Cole, 15 February 1823.

50. On gendered humor as an instrument of oppression and maintenance of existing power structures, see Gaut, 1998, p. 51. In a group that shares similar social status and ideology, humor that targets members of a group with whom it does not empathize will increase social cohesion in the group and at the same time solidify the exclusion of members of the other group (Robinson and Smith-Lovin, 2001, pp. 125–127, 139). Jane Browne, 1992, has analyzed students' magazines in Britain around 1830, and emphasizes the importance of scientific humor as a social cement in the formation of an identity for "college men."

51. Reproduced in Gordon, 1894, p. 123. This made the attendance of Mary Somerville impossible (Kölbl-Ebert, 1997b, p. 35). At the third meeting at Cambridge, organized by Whewell in 1833, women could attend the general meetings in the Senate House but not the sectional meetings. Although the ban from the sections was ignored, women were only officially allowed to attend the majority of sections—not the delicate subject of zoology—in 1838, a development that was mainly motivated by monetary considerations. This last formal exclusion was dropped for the 1840 meeting in Glasgow (Morrell and Thackray, 1981, pp. 148–157).

52. Buckland, 1823, pp. 83, 85, 86–87.

53. The first quote represents Buckland's reaction to Duncan's verse on the Red Lady (NMW, Bu P, Box 5, Folder 4, p. 167: Letter to Lady Mary Cole, 15 February 1823); the second quote is from Buckland, 1823, p. 92.

54. Cuvier's "Preliminary Discourse" in Rudwick, 1997, text 19, pp. 183–252; on the question of human fossil bones in particular, see pp. 232–234. The Swiss physician and natural theologian Scheuchzer had explained all fossils as relics of the biblical Flood that took place two thousand years ago. He had thus been an opponent of the concept of fossils as *Naturspiele*. The most famous piece of his collection (Scheuchzer's "herbarii diluviani") was the so-called *Homo diluvii testis*, which he discovered in the stone quarry of Öhningen in 1725, and which to him was a testimony of that cursed human race that drowned in the Deluge. Cuvier later described the fossil as a huge salamander. In 1838, Johann Jacob Tschudi renamed the find *Andrias scheuchzeri*, "Scheuchzer's Man" (Cuvier, 1809; Scheuchzer, 1726, 1984 [1731]).

55. Buckland, 1823, p. 231. Cuvier's chronology of the animal remains initially differed from Buckland's (ibid., pp. 229–231); BrL, Bu C, letters from Buckland to Lord Grenville, June and August 1823; on the *Mosasaurus*, see Gordon, 1894, p. 211.

56. OUMNH, Bu P, Lecture Notes (box 1), "Lecture notes on the Mosaic Deluge" (folder 5, p. 11). He found support for this conjecture in Edward Stillingfleet's *Origines Sacrae* (1663), which also limited antediluvian humans to the East (OUMNH, Bu P, Lecture Notes [box 1], "Lecture notes on the Mosaic Deluge and on organic remains" [folder 1], "No pecora till the Deluge gravel" [p. 4].

57. OUMNH, Bu P, Lecture Notes (box 1), "Lecture notes on the Mosaic Deluge and on organic remains" (folder 1), "No pecora till the Deluge gravel" (pp. 3–16); see also Rupke, 1983, pp. 91–92. This is the first usage of one of my categories of evidence that illustrates the difficulties associated with them. Although I use collateral evidence here, one might be more justified in referring this usage to vicarious evidence, accepting Buckland's notion that the Pecora must be regarded as an integral part of the human way of life, that they have no existence outside the human realm and therefore may function as a substitute for human remains.

4. Buckland Accused

1. Boylan, 1967, p. 250.
2. Pengelly, 1869, p. 214.
3. Ibid., p. 205.
4. Ibid., p. 214; on inscriptions, see also pp. 274–275, 314, 426, 459.

5. Ibid., p. 207.

6. On Northmore's work in Kent's Cavern, see Blewitt, 1832, pp. 110–138, where two of his letters are published.

7. Buckland to Trevelyan, 4 May 1825, reproduced in Pengelly, 1873b, p. 53.

8. Letter dated 16 March 1823, reproduced in Pengelly, 1873b, pp. 56–57.

9. Pengelly, 1873b, p. 59.

10. *The Monthly Magazine; or British Register* (London), 1825, 59:190–191; reproduced in Pengelly, 1869, p. 194. In McEnery's account, it was he who discovered the implement (Pengelly, 1869, p. 443).

11. Pengelly, 1869, p. 208.

12. See ibid., pp. 293, 301, 329, for examples.

13. Ibid., 1869, p. 210.

14. For a detailed discussion of experiences in the process of cave excavation, see Sommer, 2003b, 2005b.

15. Evans, 1872, pp. 441, 443; Sollas, 1913a, p. 328; see also Pengelly, 1866, pp. 542–543; 1869, pp. 327–338.

16. L. K. Clark has attributed McEnery's delay of the publication of his manuscript to the fact that he was afraid to disagree with his professional and financial mentor, Buckland, besides his own declining health (Clark, 1961, pp. 33–34; see also, for example, Daniel, 1950, p. 37). Plates that were meant to accompany McEnery's publication were prepared at Buckland's expense (Boylan, 1970, p. 352, see also 1967, pp. 243–244). On the other hand, Edward Vivian in the preface of *Cavern Researches* (1859) informs the reader that the project of publishing the manuscript was abandoned because of financial problems (in Pengelly, 1869, pp. 196–197).

17. Pengelly, 1869, p. 332.

18. Ibid., pp. 226–227.

19. Ibid., p. 338; see also pp. 249, 226–227.

20. Ibid., pp. 327–334, 337, 435–438. Buckland reasoned that because of the confusion of languages and the subsequent quarreling of the nonkindred tribes (i.e., those that did not share a language), in combination with the postdiluvian sterility of Asia, the human race spread much faster across the globe after the Flood (OUMNH, Bu P, Lecture Notes [box 1], "Lecture notes on the Mosaic Deluge" [folder 5], "Dispersion of human race," 1820 [p. 13], and Lecture Notes [box 2], "Lecture notes on palaeontology" [folder 5, p. 34]).

21. Blewitt, 1832, pp. 107–108. On the natural philosopher as sage, to whom my geologist as conjurer seems related, see Knight, 1998 (1967), 1998 (1996), who mainly develops the example of Humphry Davy, and Schaffer, 1990, who discusses the notion of the "genius" in romantic science and art.

22. Blewitt, 1832, p. 108. The chemist Andrew Ure integrated Buckland's cave model into his theory of the exchange of dry land and sea by sparing some land from having been at the bottom of the ocean before it (*New System of Geology*, 1829).

23. Buckland was informed in correspondence about human remains found on the Continent and in Britain (NMW, Bu P, Box 5, Folder 4, p. 167: Letter to Lady Mary Cole, 15 February 1823; OUMNH, Bu P; see, for example, Lecture Notes [box 1], "Lecture notes on Mosaic Deluge" [folder 5, p. 10], and Letters: Eyles Colln., letter from Stephen Hensted [p. 20], 8 February 1825?); see also, for example, the distinguished French geologist Ami Boué, 1829, who reported bones from an open site rather than a cave.

24. Buckland, 1830. Buckland here reiterated the opinion stated in the *Edinburgh New Philosophical Journal* (1829), where Pierre Cordier's (1777–1861) reading to the Academy of Science of part of the memoir addressed to him by de Christol in 1829 was summarized. Cordier considered the facts announced by de Christol more conclusive with regard to the coexistence of humans with the extinct fauna than either those from the cave at Bize brought forward by Marcel de Serres (1780–1862), professor of mineralogy and geology at the Faculty of Science in Montpellier, who was a friend and admirer of Buckland, or those presented by Paul Tournal (1805–1872) (" 'Fossil Antediluvian Animals Mingled with Human Remains in the Caves of Bize,' 'Caves Containing Human Remains,'" 1829).

25. De Serres had examined the human bones for the French Academy and found them to be of the same age as those of the hyenas and not at all in a similar state to that of those from an ancient Gaulish graveyard. Nonetheless, in the last edition of the *Recherches sur les ossemens fossiles*, which appeared posthumously in 1834, Cuvier brushed the finds from southern France off in one sentence (Cuvier, 1834, p. 216). On de Christol and similar finds from France, see also Howorth, 1902; Rupke, 1983, pp. 93–94; Grayson, 1983, chap. 6.

26. Desnoyers, 1834, p. 306.

27. Letter of 9 November 1818, reproduced in Edmonds, 1979, pp. 39–41.

5. Man as a Crocodile Superior?

1. James Secord, 2000, pp. 299–320; on radical evolutionism, see also Bowler, 1990, pp. 20–27.

2. Desmond, 1987, 1989.

3. Lamarck, 2002 (1809), chaps. 6–7 in particular; on Lamarck's transmutationism, see also Corsi, 1988 (1983); Lefèvre, 1984, chap. 2.

4. For translations of and commentaries on Cuvier's works, see Rudwick, 1997; for the arguments summarized here, see Cuvier's "Preliminary Discourse" (*Ossemens fossiles*), text 19, pp. 183–252, in particular pp. 232–234 and pp. 226–232; see also Grayson, 1983, pp. 51–52.

5. Grayson, 1983, chap. 4; on the differences between Cuvier's and Lamarck's ideas, see also Rudwick, 2005, pp. 388–399, 445–456. The classical study on the history of the *scala naturae* concept, which will reappear throughout this book, is Lovejoy, 1964 (1936).

6. On Geoffroy Saint-Hilaire, see Corsi, 1988 (1983), pp. 230–268.

7. OUMNH, Bu P, Notes for the Bridgewater Treatise/Notes by Subject (box 2), "Species change of Lamarck" (folder 6i, p. 5, and 6ii, pp. 15 and 28A in particular); on Geoffroy Saint-Hilaire, see Bourdier, 1969, pp. 45–51.

8. OUMNH, Bu P, Notes for the Bridgewater Treatise/Notes by Subject (box 2), "Species change of Lamarck" (folder 6ii, p. 26).

9. Ibid., p. 29; see also folder 6i, pp. 6, 7, 11. That Buckland regarded every creature as perfect becomes wonderfully obvious from his defense of the sloth, which, too, was perfectly fitted for its niche (Buckland, 1837, pp. 17–27). On the notion of progressionism, see Rupke, 1983, chap. 12.

10. OUMNH, Bu P, Notes for the Bridgewater Treatise/Notes by Subject (box 2), "Species change of Lamarck" (folder 6ii, p. 26).

11. Ibid., p. 29; see also Rupke, 1983, pp. 174–176. The words that finally entered the Bridgewater Treatise against the speculative philosophy of species transmutation were more docile: see Buckland, 1836, vol. 1, pp. 53–54, on the argument of the absence of plant or animal life in the primary strata; the note on pp. 107–108 on the notion of increasing perfection in the sense of overall complexity of organization; pp. 114–116 in particular on the nonexistence of an orderly succession from lower to higher organisms, with pp. 254–255 on the crocodile, pp. 293–294 on fishes, and pp. 395–396 on the trilobites; and pp. 585–586 in particular on the necessity of repeated direct creative intervention to account for the successive systems of organic life in earth history (see also the note on p. 293, in which Buckland insisted on keeping the terms *development* and *transmutation* apart; while the first was used to denote ontogenetic development, as well as the progression of fossil genera and species in the gradual advancement of the system of creation, Lamarck had illegitimately expanded its meaning to include the notion of the transmutation of species, as if this last mechanism would logically follow from the first two observations).

12. Buckland, 1836, vol. 1, p. 21, pp. 94–95, note. Patrick Boylan, 1970, p. 353, argues that Buckland must have had serious doubts about the universal flood theory, as well as about scriptural chronology, as early as 1825. In lecture notes dated April 1833, however, one still encounters the concept of the universal Deluge (OUMNH, Bu P, Lecture Notes, box 1, folder 5, p. 18). In his lecture notes on the Mosaic Deluge originating from the 1820s, Buckland strongly argued against the possibility of a quiet Deluge. He reasoned that even a broken dam induced visible effects. How should it be possible that the Deluge did not produce gravel, as even rivers did, and where were the remains of the animals it drowned? To the Buckland of the 1820s, a quiet Deluge was a contradiction in terms (OUMNH, Bu P, Lecture notes [Box 1], "Lecture notes on the Mosaic Deluge and on organic remains" [folder 5, pp. 3–4, and folder 1, p. 12]).

13. Buckland, 1836, vol. 1, p. 15.

14. Ibid., vol. 1, p. 103; for a synopsis of Buckland's stance toward human antiquity in relation to his theoretical development, see Grayson, 1983, pp. 63–72, and Rupke, 1983, pp. 91–95.

15. Schmerling, 1833–1834, vol. 2, p. 179, my translation; on the flint tools and worked bone in general, see chap. 10.

16. Blumenbach, 1790–1828.

17. All of this was later contradicted by Huxley, who found the forehead neither narrow nor receding and judged that the brow ridges were well but not excessively developed: "It is, in fact, a fair average human skull, which might have belonged to a philosopher, or might have contained the thoughtless brains of a savage" ("On Some Fossil Remains of Man" [1863], in Huxley, 1894, p. 204; see also pp. 167–168).

18. Schmerling, 1833–1834; on human bones, see vol. 1, chap. 3. The last edition of Cuvier's *Recherches sur les ossemens fossiles*, which appeared posthumously in 1834, maintained the claim that none of the human bones so far discovered were fossil (Cuvier, 1834, p. 216).

19. Morren, 1838, p. 147, my translation. As a physician, Schmerling had been interested in deformed fossil bones to determine from what kind of diseases the antediluvian animals might have suffered.

20. Buckland, 1845 (1844), p. 107.

21. De Beaumont, 1869, p. 1213, my translation; the report of the meeting is given in "Erste Sitzung am 19. September," 1836.

22. Blumenbach, 1790–1828, vol. 4, p. 9.

23. Schaaffhausen, 1861 (1858), p. 159; on the event, see also Schaaffhausen, 1872 (1869), pp. 116–117, where he quoted from August Friedrich Walchner's *Handbuch der gesammten Mineralogie* (Karlsruhe, 1832), who drew attention to the fact that the tongue test was sometimes admitted as a reliable method to determine the age of bones and in other cases, such as the human bones from southern France, was excluded.

24. Buckland, 1836, vol. 1, p. 598, note to p. 106; see also Buckland, 1837 (1836), vol. 1, p. 602. In contrast, the professor of botany of the University of Liège, Charles Morren, 1838, presented a paper in honor of Schmerling to the Royal Academy of Science in Brussels in which he applauded his work. Although 1838 seems an early date, the same year, Etienne Geoffroy Saint-Hilaire wrote a letter to the Académie des Sciences in Paris in which he reported that the members of the University of Liège accepted Schmerling's human remains as antediluvian. For Geoffroy Saint-Hilaire, the Engis skull represented *Homo troglodytes*. In fact, the elongated shape of the cranium, which had been underrepresented in Schmerling's illustration, confirmed his belief that the pre-Adamic cavemen had had elongated heads (Geoffroy Saint-Hilaire, 1838, p. 13; on Geoffroy Saint-Hilaire's *Homo troglodytes*, see Bourdier, 1969, pp. 58–59). Huxley later confirmed that the skull was somewhat longer ("On Some Fossil Remains of Man" [1863], in Huxley, 1894, p. 164). The fact that Schmerling did not give adequate descriptions to his plates, and that some of them were imprecise, is surprising, considering that he himself had to work with plates in the absence of a large-enough collection of fossil species for comparison (on the central role of visualizations of specimens, or proxy images, as "immutable and combinable mobiles," see Rudwick, 2000). Although Schmerling's adult skull (Engis 1) dates from the young Paleolithic period, the second skull is that of a Neanderthal child, as Charles Fraipont (1883–1946) recognized about one hundred years later (Fraipont, 1936).

25. Buckland to Henry Brougham, 26 November 1838, reproduced in Rupke, 1983, p. 164. In fact, in 1837, Édouard Lartet (1801–1871) described the jaw of a small ape similar to the gibbon, subsequently named *Pliopithecus*, from the late Tertiary deposits of southern France (at Sansan, Gers), and in 1856 he discovered remains of *Dryopithecus*, a more advanced Miocene primate of southern France.

26. Gillispie, 1996 (1951), pp. 73–79; Leguebe, 1976, pp. 25–26, 33; Rupke, 1983, chap. 3. Leroy Page, 1969, makes the same observation with regard to those geologists who worked with a geological revolution, i.e., whether or not they agreed that the traces of the last inundation were proof of the Flood accounted for in the Bible, they shared the belief in the recency of humankind.

27. The radical French popular writer, botanist, and geologist Pierre Boitard (1789–1859) took the side of Fossil Man and organic transmutation as early as 1838, the year of the coronation of Queen Victoria, in his article "L'Homme fossile" in the *Magasin universel* (Boitard, 1838). This article, which drew on human remains claimed to be fossil, such as those discovered by Tournal, de Christol, and Schmerling, was accompanied by a visualization of an ape-"Negro" chimera representing Fossil Man. He swings a hand ax modeled on Schmerling's finds. Boitard made fun

of the orthodox geologist who was unmoved by evidence, explicitly mocking the deceased Cuvier and his conservative ghost that still haunted the Academy of Science. To herald the English infidel paper *Oracle of Reason*'s use of science to undermine religion, the image opened the "Theory of Regular Gradation" series in 1841 (see James Secord, 2000, pp. 312–313).

28. Darwin, 1859.

29. Godwin-Austen, 1842, pp. 444–446; on his work at Kent's Cavern, see also Godwin-Austen, 1840.

30. *Athenaeum* and *Literary Gazette*, 14 August 1841, reproduced in Clark, 1961, pp. 44–45.

31. Pengelly, 1869; Vivian, 1848; on Kent's Cavern, see also Clark, 1961, pp. 43–58.

32. On the glacial theory episode, see Woodward, 1883, p. 229, quoted in Chapter 1; on this kind of humor, see Robinson and Smith-Lovin, 2001, p. 125; on Buckland's abhorrence of controversy, see OUMNH, Bu P, Lecture Notes (box 1), "Lecture notes on the Mosaic Deluge" (folder 5, p. 9).

6. The Holy Order of Nature and Society

1. On the BAAS, see Morrell and Thackray, 1981, pp. 224–245; on the establishment of provincial societies, see also Knell, 2000.

2. Morrell and Thackray, 1981, chap. 1. On Buckland's time as part of the age of improvement, see Briggs, 2000 (1959), particularly chaps. 3–7. Broad Church members of the Church of England understood its formularies and doctrines in a comparatively liberal sense, advocating for a church that would tolerate a certain variety of opinion in matters of dogma and ritual. The designation came into use about 1848 in analogy to the older phrases *High Church* and *Low Church*.

3. Desmond, 1989, p. 2.

4. See Morrell and Thackray, 1981, pp. 226–227; on the readings and appropriations of Buckland's Bridgewater Treatise, see Topham, 1998, pp. 250–261.

5. Sedgwick, Adam. 1845. "Natural History of Creation." *Edinburgh Review* 82:1–85, on p. 3; James Secord, 2000; on Whewell's and Sedgwick's responses, see pp. 227–247; on responses from exponents of Oxford University, see pp. 253–260. Among the latter, the strategies of coping with the claims brought forward in *Vestiges* varied from straightforward attack and defense of natural theology to an agnostic stance toward its scientific claims with an emphasis on religion's independence from science and to attempts at rendering transmutationism compatible with Christian doctrine.

6. Brooke, 1979, p. 50; Morrell and Thackray, 1981, p. 245. Brooke, 1979, argues that the design argument of the natural theology of the time had a mediating function between different religious traditions. It was thus advantageous in the unification of men of different Christian sects under the roof of the BAAS. On Buckland's reactions, see Rupke, 1983, p. 179; James Secord, 2000, pp. 209, 226.

7. Gordon, 1894, pp. 241–246.

8. The sermon is reproduced in Gordon, 1894, pp. 242–244; see also Gillispie, 1996 (1951), pp. 201–202.

9. Young, 1954, p. 30. On utilitarianism in this context, see also Brooke, 1979, p. 52; Gillispie, 1996 (1951), chap. 7. That Buckland as a geologist took advantage

of the various British colonies to get hold of specimens can be learned from letters (see, for example, NMW, Bu P, Box 5, Folder 1, p. 151: Letter from Buckland to Lady Mary Cole, 29 October 1819).

10. Gordon, 1894, pp. 150–172; on Buckland's engagement with agriculture and farming, see also IC, B/Playfair.

11. *Illustrated London News*, 22 September 1849, p. 89, in Williams, 1990, p. 106.

12. On the happenings in Dudley Caverns, see Gordon, 1894, pp. 79–83; see also James Secord, 1982, pp. 439–440. That the naming of geological strata and their geographical expansion were associated with nationalistic competition is supported by a letter from Buckland to Roderick Murchison in which he approved of Murchison's *Silurian* and warned him against the possible anticipation of the system's naming by de Beaumont (NMW, Bu P, Box 6, p. 185, 17 July 1835).

13. Boucher de Perthes, 1847–1849, chaps. 3, 12, 23, note 38. Among the sites where Boucher de Perthes excavated were Menchecourt (Abbeville), L'Hôpital (Abbeville), St. Acheul (Amiens), Moulin-Quignon (St. Gilles), and Champ de Mars (Abbeville). On Boucher de Perthes, see Cohen and Hublin, 1989, particularly chaps. 6, 7. His later editions were more in line with the current scientific views and were better received (Boucher de Perthes, 1857, 1864).

14. Boucher de Perthes, 1847–1849, chaps. 21, 23.

15. Buckland, 1845 (1844), quote from p. 107.

16. Pengelly, 1873a; Pengelly et al., 1873.

17. Pengelly, 1866, p. 535.

18. Reproduced in Murchison, 1868, p. 597. Falconer and his niece Grace Milne McCall visited Abbeville on their way to Sicily, where Falconer discovered human flint implements in the ossiferous caves of the north coast (Falconer, 1860b).

19. However, Jacob W. Gruber, 1965, pp. 383–384, makes the point that Prestwich, who was a wine merchant and was appointed professor of geology at Oxford only at the age of sixty-two in 1874, as well as Pengelly, who was a local schoolteacher in Devon, were in that sense rather part of the "amateur" community, typical of the early proponents of antediluvian humans, than of the kind of Buckland, Lyell, Murchison, Owen, or Cuvier abroad, who opposed the concept of a high human antiquity. Even Evans's main occupation was that of a paper manufacturer.

20. Lartet, 1860; Lyell, 1860 (1859); Prestwich, 1860 (1859), 1861 (1859). On Brixham Cave and the related events, see Clark, 1961; Cohen and Hublin, 1989, chap. 9; Evans, 1872, pp. 477–478; Grayson, 1983, pp. 179–190; Lubbock, 1869 (1865), pp. 332–335; Sommer, 2004c; Van Riper, 1993, chap. 4.

21. Reproduced in Murchison, 1868, pp. 576–577. In fact, true Neanderthal skulls had already been found at Engis by Schmerling, and at Forbes' Quarry, Gibraltar, in 1848. Bones of the same type had been discovered near Düsseldorf in 1856, and, first of all, Buckland had exhumed the Upper Paleolithic Red Lady. But none of these genuine finds was immediately universally accepted.

22. Reproduced in ibid., p. 576.

23. Prestwich, 1860 (1859), p. 57. For a confirmation of Prestwich's prophecy, see Evans, 1872, chap. 22, on the discoveries of Paleolithic implements in caves and river drifts; Jackson, 1953, for a summary of the great number of caves examined only in the British Isles; and Lubbock, 1869 (1865), chaps. 10, 11.

24. As A. Bowdoin Van Riper, 1993, pp. 118–127, shows, neither did the general acceptance of human coexistence with the extinct Pleistocene fauna mean the end of skepticism, including that from biblical literalists.

25. Ruse, 1999 (1979), pp. 250–263.

Conclusion: The Red Lady Is No Fossil Man

1. Martin Rudwick, 1992, makes this point with respect to the early visualizations of scenes from deep time.

2. Gillispie, 1996 (1951), pp. 73–79; Leguebe, 1976, pp. 25–26, 33; Rupke, 1983, chap. 3.

3. See also Van Riper, 1993, p. 131, on this point.

4. On Lyell's troubled engagements with the possibility of human evolution, see Bartholomew, 1973; on this point in particular, see also Trautmann, 1991, p. 383, and Grant, 1979, pp. 32–37.

5. Portlock, 1857, p. xxvi, in his obituary of Buckland, described him as always having been ready to abandon a theory for one that explained the existing evidence better.

II. William Sollas, a Cro-Magnon Man, and Issues of Human Evolution

1. On the institutionalization of anthropology, see Schlette, 1990, for Germany; Keuren, 1982, for Britain; and Spencer, 1981, 1982a, for the United States; on the centrality of skeletal measurement and racial hierarchy in early anthropology, see Stepan, 1982, chap. 4; on the development and importance of instruments in the history of physical anthropology, see Bowles, 1974, pp. 175–177, and Hoyme, 1953.

2. On different ways of appropriating the new insights into human prehistory, see Grayson, 1983, chap. 9; for a concise history of modern attempts at racial classification and theories of racial origin(s) and diversification, as well as the early institutionalization of physical anthropology, see Sommer, 2005b; on the continuation of polygenist thinking, see Stocking, 1968, chap. 3.

3. Gould, 1996 (1981), pp. 105–106.

4. Schaaffhausen, 1861 (1858), pp. 159–160.

5. Lartet, 1861; Lartet and Christy, 1864; see also Lartet, 1860; on relative dating, see Oakley, 1964b, p. 107. For a discussion of the dating methods of paleoanthropology of a couple of years later, see Schaaffhausen, 1872 (1869), where, apart from dendrites and furrows that might have been caused by gnawing insects or plant roots, and the traces on bones rolled by water, gnawed by animals, or worked by humans, he mentioned the chemical composition of bones as indicative of their age. As we have already seen for Buckland, the decomposition of bones was seen not only as due to time but also as dependent on the access of air and water to the place where they were buried. Schaaffhausen explained that fossilized bones had integrated new mineralogical elements such as carbonated chalk (*kohlensaurer Kalk*) or siliceous earth (*Kieselerde*). The amount of animal matter (nitrogen) was

seen as only a relative dating tool, as when the content of human bones could be compared with that of animal bones of the same stratum. Like the organic matter, so too did the calcium carbonate (*kohlensaurer Kalk*) decrease, while the calcium phosphate (*phosphorsaurer Kalk*) and fluorine content increased. As Hugh Torrens has found out, the former principal of the college at Agra, Bengal, J. Middleton, experimented on the relative content of fluorine in bones as a key to determine their age as early as the 1840s (Middleton, 1844; Torrens, 1996, p. 27).

6. Lubbock, 1865.

7. Falconer, 1860a, pp. 490–491; see also Murchison, 1868, p. 522. Falconer based his judgment, as Buckland had done before him, on Strabo's *Geography*. For a similar reiteration of Buckland's interpretation of the Red Lady, see Mayer, 1864.

8. Lartet and Christy, 1875 (1865–1875), "Description of the Plates—Bone Implements, etc.," pp. 93–94.

9. Pruner-Bey, 1875 (1865–1875), pp. 86–87; on the history of Cro-Magnon discoveries and their acceptance as a fossil race, see also Cohen and Hublin, 1989, chap. 12.

10. Broca, 1875 (1865–1875), pp. 120–122; see also Broca, 1868. The dolicho- and brachycephalic indexes referred to length-breadth ratios of the skull, after Anders Adolf Retzius (1796–1860). The dolichocephals, or longheaded as opposed to roundheaded types, were considered more advanced. *Prognathism* (and its counterpart *orthognathism*) referred to the facial angle, after Petrus Camper (1722–1789). A more protruding face was seen as a maker of primitiveness.

11. Quatrefages, 1875 (1865–1875), p. 125.

12. Evans, 1872, p. 441; see also pp. 440–442.

13. Boyd Dawkins, 1874, pp. 232–263.

14. Quatrefages, Hamy, and Formant, 1882, p. 44.

15. Henry Howorth, 1902, p. 22, also classified as Paleolithic the human material from Burrington described by Buckland. However, the Red Lady continued to be from time to time described in print as of doubtful age (see, for example, Zittel, 1891–1893, p. 718). The third race described in *Crania ethnica* was that of Furfooz (Belgium), based on human crania from the Furfooz cave, excavated, among other Belgian sites, by Édouard Dupont (1841–1911).

16. From *Man's Place in Nature* (1863), in Huxley, 1894, p. 205; see also pp. 168–207.

17. On the Neanderthal discoveries and controversial interpretations, see, for example, Campbell, 1956. The Neanderthal controversy took its time, and Virchow still called the Neanderthal and Canstadt races *Fantasiegebilde* ("creatures of the imagination") at a meeting of the German Anthropological Association in the 1890s (Fraipont, 1893).

18. See, for example, Prichard, 1848, on the various branches of knowledge converging in the reconstruction of human history (which he subsumed under ethnology).

19. Quatrefages, Hamy, and Formant, 1882, pp. 98–146.

20. In his paper on the troglodytes from the Périgord (Les Eyzies), Broca, 1868, p. 374, explicitly used the dolichocephaly of the skulls from Les Eyzies to contradict Retzius's claim that the autochthonous Europeans had consisted exclusively of a brachycephalic race. He reiterated again and again that his analysis proved that they belonged to a different race from the one whose remains had been discovered

in the Belgian caves, and also from any of the living races (pp. 351–352, 363, 373, 384, 388, 390–391). While Pruner-Bey thought that the autochthonous Europeans had been Mongoloid and others described them as Negroid, Broca emphasized the fact that there had been more than one type (Broca, 1875 [1865–1875]; Pruner-Bey, 1875 [1865–1875]).

21. On the Aryan theory, see, for example, Bowler, 1986, pp. 56–58; Erickson, 1974, pp. 114–119; Peake, 1940, pp. 105–111.

22. On the history of Anglo-Saxonism, see MacDougall, 1982.

23. See, for example, Boyd Dawkins, 1880, p. 247.

24. The earlier race of the River-drift men, whose cultural remains were discovered at open sites, was identified by Boyd Dawkins with de Mortillet's Chellean/Acheulean. When it became clear that some of the Mousterian tool types (associated with Boyd Dawkins's early caveman) could be found in the river gravel, as well as in the lower cave strata, a third division was introduced that resulted in a partition of the Paleolithic into Lower (drift only), Middle (drift and cave), and Upper (cave only) (Boyd Dawkins, 1880, pp. 199, 230–233).

25. Ibid., pp. 229–243, 495.

26. Boyd Dawkins referred de Mortillet's Paleolithic cultures Mousterian, Solutrean, and Magdalenian all to the same race of caveman, who, like the Eskimos today, showed local tribal variation in material culture (ibid., pp. 202–203, 206–207, 229).

27. Tylor, 1881, p. vi. Although the succession scenario drew on philological models, George Stocking interprets Tylor as a successor to the eighteenth-century belief in European civilization as the culmination of a progress from a savage state that was essentially followed by all human social groups. He further regards the attempt of Tylor to synthesize the Enlightenment heritage of the uniformity of human reason and universal cultural progress with evolutionism partly as a reaction to the rival anthropological approaches of polygenism and degenerationism ("The Dark-Skinned Savage: The Image of Primitive Man in Evolutionary Anthropology," in Stocking, 1968, chap. 6).

28. Tylor, 1881; see particularly chaps. 8 and 16. On Tylor, see also "Matthew Arnold, E. B. Tylor, and the Uses of Invention" and "'Cultural Darwinism' and 'Philosophical Idealism' in E. B. Tylor," reproduced in Stocking, 1968, as chaps. 4 and 5.

29. Lubbock, 1865, p. viii.

30. Ibid., p. 256.

31. Ibid., pp. 330–332, treated the Engis skull discovered by Schmerling and the Feldhofer Neanderthal skull, mainly following Huxley's interpretation. Where the age of humankind on earth was concerned, Lubbock considered it likely that a Miocene hominid would be found in a tropical region (p. 334; see also pp. 473–484).

32. Mortillet, 1867, pp. 186–187 in particular.

33. Mortillet, 1883, p. 29, my translation.

34. Ibid., pp. 2, 16, 103–106, 628–629 in particular.

35. Haeckel, 1898 (1868), p. 726.

36. Ibid., "28. Vortrag. Wanderung und Verbreitung des Menschengeschlechts. Menschenarten und Menschenrassen," pp. 729–765.

37. Ernst Haeckel, *The History of Creation: or, the Development of the Earth and Its Inhabitants by the Action of Natural Causes*, trans. E. Ray Lancaster, 2 vols. (New

York: D. Appleton & Co., 1876). On Haeckel's life and work, see Di Gregorio, 2005; on human evolution, see pp. 227–239.

7. William Sollas

1. Imperial College Library, London University, HMC MS Papers of British Scientists, 1600–1940, 1982 (hereafter IC), Notes of Ramsay's Geology Lectures, 906/Ka 8, Geology Ramsay R.M.S. (1869), 37th lecture (29 April: Bone Caves), quote from pp. 562–563.

2. Ibid., pp. 571–572. Sollas might also have heard Huxley's view on the Aryan question as later explained in "The Aryan Question and Prehistoric Man" (*Nineteenth Century*, November 1890; reproduced in Huxley, 1894, pp. 271–328). Although Huxley, as outlined later, criticized the philological model as too simplistic, he, too, believed in an Aryan race that he aligned with the blond longheads who had once spread across Europe and into Asia from eastern Europe. They resembled the present Frisians and Danes, who again showed similarities to the Neanderthals. (So did the Australian Aborigines, so that the present races might have evolved from Quaternary Neanderthal Man.)

3. IC, Notes on Huxley Lectures, 924 KB8, vols. 1 and 2 (1869); see in particular vol. 2, 29th lecture, 12 November 1869, and 30th lecture, 15 November 1869. *Steatopygia* referred to "excessive fatness of the buttocks," as Europeans had for example met with in the famous Hottentot Venus.

4. IC, Notes on Huxley Lectures, 924 KB8, vol. 2, 31st lecture, 16 November 1869.

5. Smith Woodward and Watts, 1938, p. 265.

6. On Huxley's crusade for the inclusion of science in general education and for his emphasis on practical training, see, for example, Allen, 1994 (1976), pp. 162–163, 181–182.

7. Reproduced in Huxley, 1893, pp. 38–95, on p. 45.

8. Huxley, 1915 (1880).

9. White, 2003; for a comprehensive contextual biography, see Desmond, 1997 (1994).

10. Daston and Galison, 2002; 2007, chap. 4. Although the differences between Buckland's and Huxley's constructions of a self-image seem to accord well with these observations, with Buckland arguably in a transition phase that led to a divergence between image and practice, Keith, for example, was suspicious of the excessive reliance on instruments and measurements in anthropology and advocated the methods of intuition and tacit knowledge, as in the identification of races by the unaided eye, which he continued to defend against the introduction of the statistical method (see, for example, Keith, 1931b, pp. 394–395). Smith Woodward is also said to have made fun of Sollas's soft spot for finicky measurements and his invention of new apparatus associated with his obsession for precision (on the persistence into the present of intuition as a tool for establishing phylogenies, see Tattersall, 1995, p. 168).

11. Huxley, "On Emancipation—Black and White" (1865), reproduced in Huxley, 1893, pp. 66–75.

12. See, for example, Sollas, 1896–1897, 1898, 1900, 1901–1902, and American Philosophical Society, William Johnson Sollas Papers, B:So4, Rawlins Autograph

Collection, APS B:R199 (subsequently Am Phil Soc), letter from Henry Carlin, 18 July 1888; letter from Alfred John Jukes-Browne, 30 March 1891; letter from George Bidder, 14 February 1892; letters from Cara David, 5 and 19 December 1897.

13. Sollas, 1904 (1903).

14. Smith Woodward and Watts, 1938, p. 271.

15. See Sollas, 1914; Sollas and Sollas, 1912, 1914 (1913), 1916; Sollas, 1918, 1920; Sollas and Sollas, 1904.

16. University College London, MS Add 152, T. G. Bonney Correspondence, 98719 (hereafter UCL), letter to Sollas, 2 June 1912.

17. British Library, Add MS 55222, Correspondence with Macmillan and Co., 1909–1934 (hereafter BrL), Sollas to Macmillan, 17 October 1911.

18. Suess, 1904–1924; on Sollas's career, see also Smith Flett, 1938, who provides a complete bibliography of Sollas's publications, and Spencer, 1997b, pp. 966–967.

19. Symonds, 2000, pp. 690–691; on the Rhodes Scholarships, see Darwin, 1993, pp. 347–352; Williams, 2000.

20. However, colonial officials were suspicious toward "professional" anthropologists (Kuklick, 1991, pp. 194–209, and personal communication), and anthropology had encountered particular difficulties at Oxford because of the unorthodoxy of "the science of man," as well as its infringement on the territory of the classics (see Keuren, 1982, pp. 216–241); on Oxford and empire, see Symonds, 1986, pp. 10–19 and chaps. 7–9 in particular; 2000, pp. 690–691 in particular; see also Darwin, 1993, pp. 345–346, on the establishment of an Indian Institute in 1875, with Indian studies as a part of Oriental studies for Indian Civil Service probationers. The academic study of forestry, too, entered Oxford as training for the Indian forest service.

21. On Lankester, see also Lester, 1995, chap. 10 for his time at Oxford, chaps. 13 and 14 for his struggle to better the place of science in British society.

22. Symonds, 1986, pp. 125–128; on Sollas's work with Stein, see Bodleian Library (hereafter BL), Correspondence with Sir Aurel Stein, 1924–1927, 107, fols. 175–216; and Stein, 1928; on the journey to India, see MS Eng lett d 329, pp. 6–7, letter to Sollas, 4 February 1925?.

23. Sollas, 1910, p. 56.

24. UCL, Bonney to Sollas, 24 May 1916 and 21 October 1917; on the rifle episode, see Morrell, 1997, p. 12.

25. On the women question and Oxford, see Brittain, 1960, chap. 6, and Howarth, 1994.

26. On Oxford University during World War I, see Green, 1993, pp. 70–71; and Winter, 1994; on the sciences at Oxford, see Darwin, 1993, pp. 357–359; Howarth, 2000; and Morrell, 1994; 1997, p. 217.

27. For correspondence with Broom, see Findlay, 1972, pp. 44–45.

28. Three years after the death of his wife Helen, Sollas married Amabel Nevill, widow of the anatomy professor at Oxford.

29. Douglas recorded these and other anecdotes meant to give testimony to Sollas's eccentricities on tape in 1976 (Oxford University Museum of Natural History, Geological Collection, NRA 42536, HMC MS papers of British scientists, 1600–1940, 1982, William Johnson Sollas, 1849–1936 [subsequently OUMNH, So P], box 20, "Sollesiana" typescript [folder 17]); see also Morrell, 1997, pp. 220–225, who reproduces the quote on p. 224.

8. Ancient Hunters and Their Modern Representatives

1. Sollas, 1908 (1907), p. 337.

2. Dubois, 1894; Sollas, 1895; for a closer analysis of this article and a discussion of the development of Sollas's anthropology in the context of imperialism and war, see Sommer, 2005a; on the general reception of *Pithecanthropus*, see Shipman and Storm, 2002, pp. 110–111. Many anthropologists did not readily accept *Pithecanthropus* because of its unexpected combination of a small cranium and an upright posture. Haeckel, for one, had expected his hypothetical *Pithecanthropus* to be the opposite, not fully upright but big brained. It was widely assumed that brain expansion had preceded bipedalism in the split of the hominid from the pithecoid line.

3. Schwalbe, 1906; Sollas, 1908 (1907); on Schwalbe, see also Spencer and Smith, 1981, p. 436.

4. Sollas, 1908 (1907), pp. 337–339.

5. Sollas again used the method for comparative cranial analyses of *Homo sapiens* and *neanderthalensis*, of the Chancelade skull and the Eskimo, and of the Taung child, *Pithecanthropus*, *Sinanthropus*, several Neanderthal specimens, great apes, and modern human races (Sollas, 1922, 1927, 1933). He applied the method in his detailed study of *Australopithecus africanus* (Sollas, 1926 [1925]), or, more accurately, he made Broom apply it to the Taung child for him and send him the paper proxy. In the case of the australopithecine, the method provided "ample confirmation of Prof. Dart's conclusions. *Australopithecus* is doubtless generically distinct from all known Apes, and in those important characters by which it differs from them it makes a nearer approach to the Hominidae" (Sollas, 1926 [1925]), p. 10). Raymond Dart (1893–1988), who could use some support for his widely rejected idea that australopithecines were the hominid ancestors closest to the apes, was enthralled by the article on his find and asked Sollas to send him the one on the general method (OUMNH, So P, Letters, box 1, 1882–1935: Dart to Sollas, 16 September 1927).

The Deutsche Gesellschaft für Anthropologie wanted to standardize anthropometric measurements, an endeavor that culminated in the Frankfort Agreement in 1882. The Frankfort horizontal, which fixed the ear-eye plane, was defined as the standard line to orient skulls for comparative measurements and illustrations. Other standardizations in craniometry comprised angles, indexes, and volumes (on the attempts at standardization in German anthropology and their meaning, see Zimmerman, 2001, pp. 86–94).

6. Sollas, 1908 (1907), p. 336.

7. See Appendix B, Model A.

8. Sollas, 1910.

9. Sollas, 1909.

10. Sollas, 1911, p. vii.

11. The year before Sollas's publication, Boyd Dawkins had again publicized his race-succession scenario for Europe, which explained the progress throughout the Paleolithic, as well as in the Neolithic, by means of invasions into Europe of races or tribes in possession of more advanced cultures. He reasoned that if they represented an evolutionary sequence, the transformations from one industry or one race into the next had taken place outside Europe. In opposition to the Tylorian

Notes to Pages 160–165

school, he explicitly argued against the explanation of cultural similarities between peoples in different regions and/or times by independent invention along parallel lines (Boyd Dawkins, 1910).

12. See Appendix A, which gives an overview of the geological periods and prehistoric cultures with associated fauna and hominids according to Sollas's *Ancient Hunters* (1911).

13. Sollas, 1911, p. 170.

14. Ibid., p. 161.

15. Ibid., p. 87; see also pp. 170, 161. For comparison, Otto von Bismarck's (1815–1898) brain was often taken to be representative of the other end of the spectrum of human cranial capacity. His brain was listed by Sollas as of 1,965 cc (Sollas, 1910, table on p. 64). According to Sollas, the average German might have had a cranial capacity of about 1,500 cc (Sollas, 1910, fig. 6, p. 70).

16. The woman referred to as Truganini (ca. 1812–1876) is generally considered to have been the last "pure" Tasmanian Aboriginal in Tasmania. On the appropriations of indigenous Australians in the anthropology of the first decades of the twentieth century, in particular with regard to the totemism debate, see Kuklick, 2006, where the differentiation between the sociocultural anthropologists' and physical anthropologists' perceptions is discussed.

17. Both Sollas and Boyd Dawkins theorized the Eskimo-European Paleolithic link before the discovery of the Chancelade remains (Boyd Dawkins, 1880, pp. 233–242; Sollas, 1880 [1879]). The argument from morphology was brought in by Testut, 1890, and later tested in detail by Sollas, 1927. In fact, by 1910, Boyd Dawkins no longer claimed that the cavemen and the Eskimos were of the same race, but ascribed their close cultural similarity to contact (Boyd Dawkins, 1910, pp. 259–262; on the history of the Magdalenian-Eskimo link, see also Laguna, 1932).

18. Sollas, 1911, pp. 382–383. On the Middle Paleolithic and the Australian Aborigines, see ibid., chaps. 6–7; on the Aurignacian Age and the Bushmen, see chaps. 8–9; on the Solutrean Age, Magdalenian Man, and the Eskimo, see chaps. 10–12. It is confusing that Sollas in this conclusion equates the Bushman with the Paleolithic Solutrean culture rather than, as in the preceding text, with the Aurignacian stage. This kind of inconsistency often makes it hard to extrapolate his exact views.

19. On the racial-succession model as antievolutionary, see Brace, 1997.

20. Sollas, 1911, pp. 48–50.

21. Ibid., p. 50. This is a verbal expression of a model of hominid evolution one might visualize schematically along the lines of Appendix B, Model B. As we will see, the strategy of projecting fossils back in time onto the main line leading to modern humans, so to speak, was one widely employed as different, but roughly contemporary, hominid fossils turned up and needed to be accommodated.

22. Sollas, 1911, p. 50.

23. Boyd Dawkins, 1925; Sollas, 1925.

24. Lankester, 1912b.

25. Sollas to Broom, July 1925, reproduced in Findlay, 1972, p. 53. On Sollas's quarrels, see also Gould, 1981a, p. 16, on Smith Woodward; and Johanson and Edey, 1981, p. 46, on his troubles with Keith.

26. On book sales, see BrL, Sollas to Macmillan, 18 January 1912; on Bonney's response, see UCL, Bonny to Sollas, 29 May 1912; see also Brown, 1912. Sollas

received more than twenty cuttings of English, Scottish, Irish, imperial, German, and French newspapers, magazines, and journals from Macmillan with reviews of *Ancient Hunters* (OUMNH, So P, box 19, folder 1). They generally applauded Sollas for his achievement of a really comprehensive outlook on the Paleolithic in the English language. Some juxtaposed his stance in the eoliths debate with the well-publicized beliefs of Lankester (see Chapter 12). The *Naturwissenschaftliche Wochenschrift* (13 October 1912) was one of the more critical because the author did not agree with Sollas's racial classification, and the *Revue d'Ethnographie et de Sociologie* (1912, by A. van Gennep) disagreed with him on racial identification and the new theory of migration.

27. Hazzledine Warren, 1912, p. 204. This critique was again raised against the succeeding editions (again Hazzledine Warren, 1916; and the U.S.-based anthropologist Robert H. Lowie, 1915, pp. 575–576, who was also a vehement critic of Elliot Smith's "hyperdiffusionism" (see Chapter 11); see also the Yale anthropologist George Grant MacCurdy, 1915, p. 135; *"Ancient Hunters and Their Modern Representatives* [Review]," 1925, pp. 630–631; P., 1925).

28. Brice, 1955, p. 9.

9. The Red Lady Is a Cro-Magnon Man

1. BrL, Sollas to Macmillan, 14 September 1913; see also Sollas to Macmillan, 9 May 1912, 19 October 1912.

2. Boyd Dawkins, 1910, still did not mention the Red Lady and claimed that no Cro-Magnon remains had been found in Britain (pp. 258–259); Sollas, 1911, pp. 212–215, was indeed cautious about the Red Lady's ascription to the Aurignacian and deemed this possible only because of the absence of ornaments, which he at this point associated with the Magdalenian.

3. Sollas communicated with Colonel William Llewyn Morgan, president of the Royal Institute of South Wales, which ran the Swansea Museum. When he visited the museum with Breuil, Morgan was absent, but the curator showed them the specimens from Paviland Cave and other caves in Gower. He also assisted in their visit to the cave itself. Subsequently, Sollas asked Morgan for permission to loan and to describe the flint and bone implements in his paper on Paviland. Breuil and Sollas also visited Bacon Hole, where there were ten nearly horizontal broad bands in vivid red on the wall that they took to be the first discovered Paleolithic wall art in the British Isles, and Sollas would have liked to chemically analyze a sample. It later turned out to be the work of a vandal (the original letters are at the West Glamorgan Archive Service, County Hall, Swansea, and copies can be found in OUMNH, So P, Letters, box 1, 1882–1935: 8 October 1912, 11 October 1912, 17 November 1912; see also press cutting from *Illustrated London News* [21 December 1912, "science jottings"] on Sollas's and Breuil's discovery of "first cave art in Britain" in Bacon's Hole [box 18, folder 11]).

4. Breuil, 1913 (1912); MacCurdy, 1915, 1926.

5. According to Sollas's accounting, the expenditures (£819) included three weeks and four days of hotel accommodation, food, and drink for Balfour, Marett, and himself (OUMNH, So P, box 18, folder 5). Sollas put a Harry Long in charge of excavations, who reported mishaps such as flooding of the cave overnight or the

collapsing of the floor (OUMNH, So P, Letters, box 1, 1882–1935: Long to Sollas, 13 January and 3, 9, and 23 May 1913).

6. Sollas, 1913a, p. 339.

7. Ibid., p. 353.

8. Ibid., p. 364.

9. The Mentone or Grimaldi caves, close to the French border in Italian Liguria, had first been excavated by the French paleoanthropologist Emile Rivière (1835–1922) in the 1870s. After his spectacular finds of human burials, Boule, Breuil, Cartailhac, and Verneau carried out a systematic excavation of the Grotte des Enfants at the beginning of the twentieth century.

10. Sollas, 1913a, p. 365, note 2.

11. See references in ibid., p. 368, notes 1 and 2: Manouvrier, Léonce-Pierre. 1893. "La Détermination de la taille d'après les grands os des membres," *Mémoires de la Société d'Anthropologie de Paris*, 2nd ser., 4:347–402; Pearson, Karl. 1899. "Mathematical Contributions to the Theory of Evolution. V. The Reconstruction of the Stature of Palaeolithic Races," *Philosophical Transactions of the Royal Society of London* 192:169–244.

12. Sollas, 1915 (1911), fig. 133, p. 293.

13. Sollas, 1913a, p. 365. This specimen was later labeled Paviland 2 and will reappear in Part III.

14. Ibid., p. 325.

15. Ibid., p. 372.

16. Ibid., p. 373.

17. Ibid., p. 372.

18. On the role of hunting in male Anglo-Saxon self-definition, see, for example, Bederman, 1995; on hunting as an imperial practice, see MacKenzie, 1988. In fact, *Country Life* (6 January 1912), most likely inspired by the title, reviewed *Ancient Hunters* and welcomed the Bushmen as ideal sportsmen.

19. Again, the classification of the evidence used by Sollas is not straightforward. One might argue here that he did not work with invisible evidence but rather with collateral evidence if the findings of Aurignacian art on the Continent and the fact that the Red Lady was ritually buried are accepted as such.

20. On the Cro-Magnon race as spanning the Aurignacian and Magdalenian and as possibly ancestral to existing Europeans, see also Sollas, 1915 (1911), p. 388. In the final edition of *Ancient Hunters*, Sollas seems to have reverted to the Cro-Magnon–Native Americans relation, with some Cro-Magnon traces also in present France (Sollas, 1924 [1911], p. 599).

10. Human Evolution as a Trunkless Tree

1. On changes in theories of human evolution during the first decades of the twentieth century, see also Hammond, 1988, and Bowler, 1986, particularly pp. 87–104. On the affair of the La Chapelle-aux-Saints Neanderthal, see Sommer, 2006, which analyzes the newspaper-clippings collection Boule made on the media coverage of the discovery and his reconstruction of the find. The collection of articles and excerpts allows tracing the back-and-forth between the French press and Boule with the certainty that Boule was aware of the affair. The article situates the

media coverage and Boule's reactions to it in the cultural and scientific contexts of turn-of-the-century France (the collection can be found at the Bibliothèque Centrale du Muséum National d'Histoire Naturelle [hereafter MNHN], Fonds Boule, B 34: Documents relatifs à la découverte de l'Homme de la Chapelle-aux-Saints; extraits de presse, correspondance).

2. Bouyssonie, Bouyssonie, and Bardon, 1908; on the priests Bouyssonie and Breuil, see Albarello, 1987, chaps. 2–3, and Houghton Brodrick, 1963, particularly pp. 127–142; on de Mortillet, see Hammond, 1980, and Cohen, 2001.

3. Boule, 1908; 1911, 1912, 1913, 1914 (1912); for a summary of Boule's findings and conclusions, see Boule, 1923 (1921a), chap. 7. For a more detailed analysis of how Boule arrived at his results, see also Trinkaus and Shipman, 1993, pp. 190–194. The reason for this brutish reconstruction has been attributed partly to the fact that the bones of the Old Man of La Chapelle-aux-Saints had been afflicted with osteoarthritis. Although Boule was aware of the deforming illness (Boule, 1923 [1921b], p. 241), it seems that he did not take it sufficiently into account in his reconstruction. Straus and Cave, 1957, have shown that although the pathology of the Old Man from La Chapelle-aux-Saints may well have forced him into something of a stoop, classic Neanderthal in a healthy state was fully human in posture.

4. Boule, 1908, p. 525, my translation.

5. Hammond, 1982; see also Hammond, 1988, pp. 118–120. Alternatively, C. Loring Brace, 1964, has interpreted Boule's work as being mainly in the Cuvierian tradition, which he identifies with antievolutionism and catastrophism. Boule's delegation of the Neanderthals to a dead-ending side branch and their sudden replacement by anatomically modern humans would thus represent such an instance of catastrophic change.

6. MNHN, Fonds Boule, B 34: Documents relatifs à la découverte de l'Homme de la Chapelle-aux-Saints; extraits de presse, correspondance, *Le Radical*, 19 May 1909, "Fortes têtes," by Albert Gorey, my translation.

7. MNHN, Fonds Boule, B 34: Documents relatifs à la découverte de l'Homme de la Chapelle-aux-Saints; extraits de presse, correspondance, *Le Temps*, 19 December 1908, "L'homme fossile de la Chapelle-aux-Saints," by Henry de Varigny.

8. MNHN, archives Henri Breuil, Cachet 2 AP7 A3 A.1. La Quina, Feuille GN 818.9, A5/A16, Quina, ZM, "Notre plus vieil ancêtre" (newspaper clipping), my translation. In fact, the traumatic experience with the Le Moustier specimen led to the first law protecting the national prehistoric heritage (Albarello, 1987, p. 59).

9. Boule, 1914, p. 576, my translation.

10. Ibid., p. 578, my translation.

11. Ibid., p. 578, my translation.

12. British Academy research professor Robin Dennell has drawn my attention to the fact that the article "La guerre" is followed by the obituary of Joseph Déchelette (1862–1914), killed on 5 October, most likely in one of the vain charges the French made that fall against German lines. This must have seemed a tragic and futile loss of the famous French archeologist at the beginning of an industrialized war that demanded casualties on an unexpected scale, which might well have contributed to Boule's anger (personal correspondence, 7 November 2006).

13. King, 1864. Breuil began to teach at the institute as professor of prehistoric archeology in 1910 and was professor at the Collège de France from 1929 to 1947. Obermaier was at the institute from 1911 up to World War I.

14. On the Praesapiens theory, see Vallois, 1954; Bowler, 1986, pp. 75–111.

15. BrL, Sollas to Macmillan, 2 July 1914; Fleure, 1920. Fleure later became professor of geography at Manchester University (1930–1944). He was president of the Royal Anthropological Institute from 1945 to 1947. His research interests included biological studies, physical anthropology, and human geography. For another example of an article that is largely based on Sollas's writings, see Taylor, 1919.

16. The literature on the Piltdown forgery is extensive, and some of the most famous anthropologists of the time have been proposed as guilty of the crime; even Sollas at one time figured among the suspects (see Gould, 1981a; Halstead, 1979; Hammond, 1979; Spencer, 1990a, 1990b; Straus, 1954; Tobias, 1992; Walsh, 1996; Washburn, 1979).

17. On Piltdown, compare Sollas, 1911, pp. 48–50, and Sollas 1915 (1911), pp. 56–57; compare also 1911, pp. 168–169, and 1915 (1911), p. 204; on *Pithecanthropus*, see Elliot Smith, 1914; Sollas, 1915 (1911), p. 38, note 2; on Neanderthal, see Sollas, 1915 (1911), p. 197, and fig. 91, p. 198. The only hint at a possible change in outlook was the shift from the phrase "a fresh discovery [*H. heidelbergensis*] was made which adds another link to the chain of human descent" (Sollas, 1911, p. 40) to "a fresh discovery was made which adds another branch to man's family tree" (Sollas, 1915 [1911], p. 41). In the context of the entire book, and in view of the fact that Sollas's model of human evolution was at that point still an indirect linear one, along the lines of Appendix B, Model B, the significance of this minor change is ambiguous.

18. Sollas, 1924 (1911), p. 246; BrL, Sollas to Macmillan, 5 January 1924?, 20 January 1924, 23 October 1925, 19 November 1926, 9 November 1933.

19. Sollas, 1924 (1911), p. 44.

20. Ibid., pp. 192–193.

21. Gregory, 1920, p. 690.

22. Sollas, 1924 (1911), p. 63; on his revision of the Piltdown section, see also OUMNH, So P, box 18, folder 15, which contains a card to Elliot Smith (11 December 1922) on the topic with pencil notes about Gregory and Keith. Sollas might therefore be seen as having finally adopted a model along the lines of Appendix B, Model C.

23. On *Pithecanthropus*, see Boule, 1923 (1921a), pp. 108–110. With regard to national communities, Schwalbe, for example, continued to believe in the ancestral status of the Neanderthals (Schwalbe, 1913), and Klaatsch developed an idiosyncratic theory in which he aligned the modern races with different anthropoid apes (Klaatsch, 1910). On the impact of national conflict on the relations between the anthropological communities, see, for example, Hammond, 1982, p. 26.

11. The "Evolutionary" versus the "Historical" Model

1. Sollas, 1911, p. vii.

2. Sollas, 1915 (1911), p. ix.

3. See Breuil, 1913 (1912), on his emphasis on migration and contact; at the same congress, Boule restated his ideas on the Neanderthals and *Pithecanthropus* (Boule, 1914 [1912]; see also Delisle, 2000). Later, in his inaugural lecture at the

Collège de France, Breuil enumerated the Paleolithic European races and cultures in a way reminiscent of Sollas's views, once again stressing that the Negroid Grimaldi, white Cro-Magnon, Eskimoid Chancelade, and Nordic Laugerie Basse types had all migrated into Europe: "Nothing allows to think that they were autochthones, and it is without doubt far toward different points in the Orient that one would have to search for their origin, which is today unknown" (Bibliothèque Collège de France [hereafter CDF], CDF 16/8, Henri Breuil, 7A, "La Préhistoire: Leçon d'ouverture de la chaire de Préhistoire au Collège de France," extrait de la *Revue des Cours et Conférences* [30 December 1929], Emmanuel Grevin, Imprimerie de Lagny, 1930, p. 9, my translation).

4. On diffusionism in general, see Elliot Smith, 1911, 1915, 1929, 1933.

5. Elliot Smith, 1924, chap. 2, "Primitive Man," first published 1916, p. 111.

6. See in particular Elliot Smith, 1933, chap. 4, which treats Tylor, 1871. For a defense of the ethnologists, see A. A. Goldenweiser, 1916, who was rightly annoyed by the simplistic picture Elliot Smith drew of the theories of such "evolutionists" as Tylor, who in fact had been well aware of the phenomenon of diffusion through contact and migration.

7. Elliot Smith, 1924, chap. 2, "Primitive Man," first published 1916, pp. 97–108.

8. For praise of Sollas, see Elliot Smith, 1929, p. 98; for the phylogenetic "tree," see Elliot Smith, 1924, fig. 1; on racial characteristics and cultural diffusion between races, see Elliot Smith, 1924, chap. 1, "The Evolution of Man," first published 1912, p. 35; 1929, pp. 122–130.

9. Elliot Smith, 1924, p. 11.

10. For a later critique of the Aryan theory and of the concept of evolution in ethnology, but also of a purely cultural definition of race, see Elliot Smith, 1935. He particularly criticized the fallacy committed by the Germans when they attributed their defeat in World War I to the myth that the Aryan race of original Germans had been diluted and weakened by the blood of inferior races (pp. 202–203).

11. Elliot Smith, 1924, chap. 2, "Primitive Man," first published 1916, p. 52; see also pp. 51–56, 97–98. William Ripley's influential *The Races of Europe* (1899) also suggested that the Cro-Magnon race had evolved into the Nordic and Mediterranean types.

12. Elliot Smith, 1929, pp. 58–59. Although Elliot Smith already adhered to the model of successive immigrations into Europe in 1916, he still had some doubts about the question of the Neanderthals being a separate species from modern humans.

13. Ibid., p. 60.

14. This kind of model of human evolution might be abstracted as in Appendix B, Model D. For another example, see Smith Woodward, 1935, p. 404.

15. Elliot Smith, 1929, p. 85. For his ideas on the evolution of the brain, see, for example, Elliot Smith, 1924, 1928.

16. Smith Woodward, 1925, p. 18.

17. Ibid., pp. 18–19. Although the idea of Asia as the cradle of humankind, which has already been a topic in Part I, had a long history, the central Asia hypothesis drew specifically on William Diller Matthew's "Climate and Evolution" (Matthew, 1915 [1911]; see especially pp. 209–214). Prominent supporters were to be found in Europe and the United States, such as the American paleontologist Henry Fairfield Osborn (1857–1935), who was Matthew's superior at the American Museum

of Natural History (see, for example, Osborn, 1928). Boule, 1923 (1921), p. 468, considered Matthew's theory a good working hypothesis for the paleoanthropological data then available but did not want to exclude Africa as the possible scene of human origin. Breuil, on the other hand, came to favor the central Asia hypothesis (see, for example, CDF 16/8, Henri Breuil, 7A, "La Préhistoire: Leçon d'ouverture de la chaire de Préhistoire au Collège de France," extrait de la *Revue des Cours et Conférences* [30 December 1929], Emmanuel Grevin, Imprimerie de Lagny, 1930, in particular p. 15).

The change in outlook in physical anthropology from a linear to a diversified model indeed had its parallel in Paleolithic archeology, where it became increasingly apparent that there was no clean series of cultures throughout the industrial layers, but that the picture was more complex. As we have seen, there were proponents of these more complex pictures as early as the first decade of the twentieth century, and in the late 1920s and early 1930s, the concept of parallel industries in the early Paleolithic became popular (see O'Connor, 2005; on these developments in cultural anthropology, see Stocking, 1995, pp. 208–220). That the diffusionist model of cultural evolution could also be associated with a model of physical evolution that inferred from similar structures a common ancestry rather than parallel evolution is discussed in Keith, 1934, pp. 40–42, which provides an overview of phylogenetic "trees."

18. See Elliot Smith, 1924, chap. 1, "The Evolution of Man," first published 1912, p. 40, and chap. 2, "Primitive Man," first published 1916, pp. 80–81, where Darwin's view of Africa as the birthplace of the first hominids is viewed as supported by the gorilla and the chimpanzee, the "unenterprising relatives" of man that were satisfied with their undemanding life in the lush tropical forest, where they stagnated in their evolution; in Elliot Smith, 1924, the foreword and chap. 2 (1916) deal with hominid fossils and the evolution of modern races and their cultures; Elliot Smith, 1929, pp. 46–48 and chap. 2, treats the wanderings of man and his relations; see also Elliot Smith, 1931 (on Elliot Smith, see also Bowler, 1986, pp. 167–173, 217–218; Dart, 1974; Elkin and Macintosh, 1974; and Dennell, 2001, p. 50, on his ideas on the place of human origin).

19. Elliot Smith, 1924, chap. 2, "Primitive Man," first published 1916, p. 131; 1929, chaps. 5–7.

20. This was a notion shared by other paleoanthropologists; see, for example, Osborn, 1928 (1927), p. 82.

21. Eriksen and Nielsen, 2001, p. 36; Sowerwine, 2001, chaps. 5–7. On discourses of degeneration around the turn of the twentieth century, see particularly Pick, 1989.

12. The Moral Authority of Nature

1. As outlined in the introduction to this part, de Mortillet called to life a hypothetical shaper for his eoliths, *Anthropopithecus* or *Homosimius*, which he subdivided into species according to eolith cultures in Europe: *bourgeoisii* after the abbé Louis Bourgeois, who found Oligocene eoliths in Thénay, southern France, in 1867; *ribeiroi* after Carlo Ribeiro, who discovered Upper Miocene eoliths near Madrid in 1871; and *ramesii* after J. B. Rames, who also claimed to have unearthed Upper Miocene tools at Puy Courny, near Aurillac, France, in 1877. In Britain, the sup-

posed tools came from below the presumably Upper Pliocene marine deposit referred to as Red Crag found in Suffolk, Norfolk, and northeast Essex (Lankester, 1907, 1912a, 1920 [1919], 1921; Reid Moir, 1916, 1921, 1927, 1935; Reid Moir and Peake, 1922). On the exhibition of eoliths, see Osborn, 1928 (1927), p. 33; on the eolith controversy as discussed here, see Sommer, 2004d; on the British context, see also O'Connor, 2003; Spencer, 1988.

2. Sollas, 1913b, p. 790.

3. Sollas, 1915 (1911), p. 77; BL, MS Eng lett d 329, Lankester to Sollas, 16 July ? (p. 12) and 19 September ? (p. 14) (the year must be 1912); see also BrL, Sollas to Macmillan, 16 April 1912, and Am Phil Soc, Lankester to Sollas, 1 May ? (before 1915). For résumés of the eolith issue, see also Boule, 1923 (1921a), chap. 5, and Sollas, 1915 (1911), chap. 3.

4. UCL, Bonney to Sollas, 15 November 1920. On Foxhall Hall, see Reid Moir, 1917. In 1920, Reid Moir supported his eolith theory with an additional workshop site in a Pliocene deposit of Cromer, Norfolk, where he found rostrocarinates and their supposed derivatives, Early Chellean implements, side by side (Reid Moir, 1921, especially p. 413). Breuil's visit seems to have taken place in 1920 (Sollas, 1924 [1911], pp. 96–105). On Breuil's cautious stance in the eolith debates, see also CDF 16/8, Henri Breuil, 7A, "La Préhistoire: Leçon d'ouverture de la chaire de Préhistoire au Collège de France," extrait de la *Revue des Cours et Conférences* (30 December 1929), Emmanuel Grevin, Imprimerie de Lagny, 1930. In contrast, Reid Moir and Peake, 1922, p. 54, claimed that Breuil also accepted the Pliocene tools from the sub-Crag deposit. That Boule could not be convinced is also evidenced in letters (OUMNH, So P, Letters, box 1, 1882–1935: Boule to Sollas, 15 February 1928: Boule still thought that the eoliths from Puy Courny and Puy de Boudieu, Cantal, were of geological origin). On eoliths in general, see also Hazzledine Warren, 1923.

5. BrL, Sollas to Macmillan, 27 November 1920.

6. Sollas, 1924 (1911), p. 106.

7. Ibid., p. xi, chaps. 2–3.

8. Reid Moir and Keith, 1912, pp. 346–348, 359–377. The Galley Hill find had been described by Newton, 1895. On the events surrounding the Moulin-Quignon jaw, see Boylan, 1979; Falconer, Busk, and Carpenter, 1863. For a critique of Reid Moir's and Keith's interpretation of the Ipswich remains, see Irving, 1914. The Ipswich Man affair was carried into the open and found entrance into the *East Anglian Times* and *Morning Post*.

9. Keith, 1911, pp. 26 and 73. For Keith's deconstruction of Sollas's Chancelade-Eskimo link, see also Keith, 1925. Most interestingly, Keith here questioned Sollas's craniological measurements as the epistemic foundation of a claim to racial identification. In Keith's view, it was the naked eye of the well-experienced anthropologist that was most able to classify people on the street, as well as a Pleistocene skeleton.

10. Keith, 1911; see Appendix B, Model B.

11. Keith, 1912, 1915a. In fact, Piltdown Man did not fit Keith's expectation. In agreement with his theory of the great antiquity of modern human anatomy, he disagreed with Smith Woodward's reconstruction of the skull and Elliot Smith's analysis of the endocranial cast, producing an alternative skull reconstruction that suggested a more modern brain. He also interpreted the jaw and teeth as more hu-

manlike. This renders speculations that Keith had been involved in the Piltdown forgery very unlikely. Keith essentially dedicated the second volume of *The Antiquity of Man*, or chapters 18 to 27 in the single-volume edition, to the Piltdown discoveries and reconstruction (see Elliot Smith, 1916, for a defense of his reconstruction).

12. See, for example, Boule, 1923 (1921a), p. 124, note.
13. Keith, 1915a, pp. 510–511, on eoliths as evidence and the comparative strategy.
14. See Reid Moir, 1913, pp. 173–175, and Osborn, 1930, for other uses of the comparison strategy. The success or failure of intentional eolith reproduction by humans also entered the arguments. For a history of experimental flint knapping, see Johnson, 1978.
15. Reid Moir, 1927, p. 162.
16. Boule, 1905, pp. 266–267, my translation.
17. Keith, 1915a, p. 270.
18. Keith, 1915a, 1924.
19. Hrdlicka, 1927; on Hrdlicka's theory, see also Spencer and Smith, 1981. There had been earlier detractors of a Neanderthal phase in human ancestry (see, for example, Verneau, 1924).
20. Reid Moir, 1917, pp. 386–391, 411–412; Reid Moir, 1928.
21. Sollas, 1924 (1911), p. 75.
22. Ibid., p. 74; see, for example, the lemuroid theory of the American paleontologist Edward Drinker Cope (1840–1897), in whose model hominids (as part of the Anthropomorpha) bypassed the monkeys proper since they sprang directly from Eocene lemuroids (Cope, 1893, pp. 324 and 326; for later expressions, see, for example, Osborn, 1927; Osborn and Reeds, 1922; Wood Jones, 1919, 1929; on the tarsioid theory and similar models, see also Bowler, 1986, chap. 5). Sollas had been an early critic of the theory (see, for example, Sollas, 1895, p. 151). Gregory, Osborn's assistant at the American Museum of Natural History (1899) and later a full science staff member and professor of zoology at Columbia University, became another outspoken critic of such notions. He attacked his former boss Osborn's combination of a center of evolution and dispersal in central Asia with an enormous antiquity of the hominid line (Oligocene or even Eocene) and the associated dissociation of the anthropoid from the hominid lineage. Rather, Gregory recognized the close phylogenetic relationship between modern humans and the apes (Gregory, 1927, 1928).
23. Sollas, 1924, p. 190.
24. Keith was therefore developing an ever-stronger parallelism as expressed in Appendix B, Model D. See Keith, 1925 (1915), especially vol. 2, pp. 725–728, and fig. 263, p. 714; 1931b, pp. 27–28; 1946; and 1948, especially "tree," pp. 158–159. His work in the 1930s with the Berkeley paleoanthropologist Theodore D. McCown (1908–1969) on the Mount Carmel Neanderthals, finally convinced him that the evolution of the Caucasian type had gone through a Neanderthal phase after all (Keith, 1950, p. 599). Other paleoanthropologists also came to accept a so-called pre-Neanderthal or a Neanderthal-phase-of-man theory (see, for example, Smith Woodward, 1935; see Bowler, 1986, pp. 105–111). An early model of parallel evolution that even aligned different recent races with different ape genera was developed by Hermann Klaatsch (1863–1916), associate professor of anatomy

and anthropology at Breslau University (Klaatsch, 1920; *Der Werdegang der Menschheit* was published posthumously, since Klaatsch had died in 1916).

25. Keith, 1950; see chaps. 7–8 on Siam.

26. Ibid., p. 378.

27. Ibid., p. 415.

28. Keith had always regarded himself as a staunch Darwinian, and in the later years of his life he felt thrilled by the thought of living in Darwin's former home village of Downe, Kent, where after his retirement in 1933 he was master of the Buckston Browne Research Station (of the Royal College of Surgeons) near Down House. He also venerated Herbert Spencer (1820–1903) and was enthusiastic about being chosen as one of three trustees to carry out his will in 1917 (Keith, 1950, pp. 427–428). Keith began reading Spencer while in Siam and was strongly influenced by his theory of evolution. In particular, he drew on Spencer's codes of amity and enmity for his theory of racial diversification, which supported his belief that cohesive emotions and behaviors would arise within, and disruptive emotions and behaviors between, human groups, tribes, or nations (see Spencer, 1892, vol. 1, pp. 313–318, 471; for Darwin's use of intertribal rivalry and intratribal cohesion, see Darwin, 1874 [1871], chap. 5). Besides drawing on both Spencer and Darwin, he also seems to have undergone a Haeckelian influence. He greatly admired Haeckel as a scientist, as well as a personality (Keith, 1934).

29. Keith, 1915b, 1916, 1919b, 1919c, 1920 (1919); on Keith's thoughts on race, see also Brace, 2002, pp. 226–233.

30. Keith, 1925 (1915), vol. 2, pp. 724–725; 1927, p. 204. Keith's theory of racial conflict culminated in *A New Theory of Human Evolution* (1948). Before moving on to the larger book, Keith published the smaller *Ethnos—The Problem of Race, Today and Tomorrow* (1931a), which in turn was essentially based on the lecture published as "The Evolution of Human Races, Past and Present" in *Early Man* (1931c), and on *New Discoveries Relating to the Antiquity of Man* (1931b) as an update on fossil hominids.

31. Keith, 1936, pp. 14–15; for a very similar passage, see Keith, 1931b, p. 30.

32. Keith, 1925 (1915), vol. 2, pp. 725–728. Keith here followed Osborn's ideas on orthogenesis (e.g., Osborn, 1917).

33. See communication with Broom, reproduced in Findlay, 1972, p. 45.

34. Sollas, 1911, pp. 405–406.

35. On the persistence of "Lamarckism" and the role of racial conflict in theories of evolution, see also Bowler, 1986, pp. 186–209, 223–237; 1993, pp. 65–73; Stocking, 1994.

36. Sollas, 1924 (1911), pp. 599–600; for a similar line of argument, see, for example, Keith, 1931c, p. 54.

37. Daston and Vidal, 2004.

38. Some indeed saw God's will manifested in the development of ever-higher intellect and morals in the course of evolution, as well as in the course of the individual human life (see, for example, Morgan, 1928; Osborn, 1928 [1927]).

39. Bowler, 1986, chaps. 2 and 4, discusses nineteenth-century anthropological progressionism and the turn toward the so-called Praesapiens model within anthropology; on connections between imperial ideology, the experiences and opportunities provided by the British and French empires, and Paleolithic archeology,

see also Dennell, 1990; on national rivalries between the German, English, and French communities and on the role of nationalism in the reception of fossils and in theories of morphological and cultural progress through migration and struggle, see Hammond, 1988, pp. 127–130.

40. On the conflation of *nation* and *race*, see Stocking, 1994.

41. However, many anthropologists who speculated about the evolutionary consequences of war did not follow Keith in his rationalization of war as evolutionarily positive. Within a group-selectionist approach, the greatest fear associated with World War I might have been due to the fact that it was seen as representing a conflict between the master nations or races rather than the simple extermination of an inferior race. Also on the assumption of individuals as selective units, the war appeared as a double-edged sword. On the one hand, the war effort seemed to rob the nation of its most able and evolutionarily fit young men, while the invalid in mind and body remained safe and sound at home, handing their defects on to the next generation. On the other hand, war could be seen as evolutionarily positive in the sense that those who would survive would no doubt have been selected under enormous pressures, and thus their fitness would be above average (Vergata, 1994; for a pre–World War I example, see Ripley, 1899, pp. 86–89).

42. Clearly, if World War I was rationalized within an evolutionary framework at all, it was associated with concerns about the fitness of the nation or the race. The specter of the possible degeneration of the British race had been raised, and in his presidential address to the Anthropological Institute in 1917, Keith reprimanded the government for its lack of support for survey projects such as the establishment of a Bureau of Ethnology, the institutionalization of an anthropometric survey of Britain and empire, and an anthropometric survey of schoolchildren. Keith was optimistic that World War I would make the authorities perceive the administrative value of anthropology and that the anthropometric survey would in due course become a government measure (Keith, 1917, pp. 27–30). Indeed, the same year Keith became the anthropologist on a medical board that was appointed by the minister of the National Service to advise recruiting boards and to examine the criteria used to group recruits into grades. The outcome was considered worse than expected: low-grade men were very numerous, especially in the industrial populations of northern England, where undernourished and underdeveloped bodies abounded (Keith, 1919a). Keith, who was strongly influenced by Francis Galton (Keith, 1950, p. 399), thus provides one of the missing links between the paleoanthropologists' reconstructions of hominid evolution as discussed here and the larger issues of eugenics, to which, however, I have no room to do justice (for an analysis of Keith's interfering roles as paleoanthropologist, state anatomist and anthropometrist, and public voice, see also Sawday, 1999).

Conclusion: Turbulent Times for an "Old Lady"

1. See also Hammond, 1979, p. 53. Piltdown Man finally gave the English hominid relics as old as, if not older than, those found on the Continent and diverted attention from the French caves to the English gravel pit (Lankester, 1915, pp. 288–289; *Diversions of a Naturalist* is one of several collections Lankester

published of his weekly popular science articles from "Science from an Easy Chair" [1907–1914] and "More Science from an Easy Chair" [1919] in the *Daily Telegraph;* see also Lankester, 1934).

2. Sollas, 1926 (1925).
3. Sollas, 1933, p. 394.
4. See, for example, Sollas, 1924 (1911), p. 371.

III. An Interdisciplinary Team, an Early Upper Paleolithic Shaman, and a Definitive Report

1. Oakley, 1948, 1963, 1964a, 1980; Oakley and Montagu, 1949; Oakley and Vries, 1959; on Oakley, see Sommer, 2007a; for the unmasking of Piltdown, see also Clark, Weiner, and Oakley, 1953; Weiner et al., 1955.

2. On the history of the australopithecines, see, for example, Broom, 1950; for a recent account of the role of the australopithecines in the development of a new physical anthropology, see Gundling, 2005, chaps. 4–5 in particular.

3. On the old versus the new systematics, see, for example, Mayr, 1982 (1942), pp. 6–8; for linear models of hominid evolution by "the makers of the synthesis," see, for example, Dobzhansky, 1944, 1962, and Mayr, 1950; on the influence of the synthesis on paleoanthropology, see also Delisle, 1995; Gould, 1980, on the role of George Gaylord Simpson (1902–1984). The Oxford anatomist Wilfrid Le Gros Clark's (1895–1971) work, for example, is indicative of the transitions in the history of paleoanthropology summarized here. Initially working with parallelism and the australopithecines as fossil apes, he was eventually involved in the exposure of the Piltdown forgery and the acceptance of the australopithecines as hominids, as well as the introduction of a biological concept of species and the new systematics. He became aware of the importance of modern evolutionary and genetic concepts for paleoanthropology and the significance of functional anatomical studies. He also followed the trend of simplifying nomenclature and of reducing taxonomic diversity (see in particular Clark, 1934, 1947, 1950, 1955, 1967; on Clark, see Sommer, 2007b). However, my account is grossly simplified, and although the notion of a Darwinian synthesis as such has been subjected to critical inquiry, the descriptive-historical approach of typology has not disappeared entirely from physical anthropology (see, for example, Armelagos, Carlson, and Gerven, 1982; Lovejoy, Mensforth, and Armelagos, 1982).

4. Keith, 1948; another example was Reginald Ruggles Gates (1882–1962), a geneticist who presented a model with multiple centers of origin and parallelism in the evolution of the different racial types, which for him had species status (Ruggles Gates, 1948). The Harvard professor Carleton S. Coon (1904–1981), although working within the single-species framework and with gene flow, thought that the five human subspecies lines (Australoid, Capoid, Congoid, Mongoloid, and Caucasoid) had evolved through the grades of polymorphic *Homo erectus,* (Neanderthaloids) and *Homo sapiens* in parallel and at different paces in the respective geographical regions of the world (Coon, 1962; see particularly pp. 305–309, 332–337, 656).

5. See, for example, Tattersall and Eldredge, 1977; Tattersall, 1995, chap. 12. On the impact of cladistics on hominid taxonomy and its differences from adaptationism, see also Cartmill, 2001; Delisle, 2001.

6. Sarich and Wilson, 1967. There still remained the problem of importing one date from paleoanthropology into the molecular scheme to set the clock. Sarich and Wilson had originally used an assumption of thirty million years ago for the Old World monkeys-apes split (on the history of molecular anthropology up the 1980s, see, for example, Goodman and Cronin, 1982).

7. Mitochondrial DNA (hereafter mtDNA) is the circular chromosome found in the intracellular organelles called mitochondria. For the purposes of molecular anthropology, those who work with it see several advantages over nuclear DNA. It is transmitted exclusively through the mother; it contains only a few thousand base pairs; it does not recombine (although this has been questioned); and it has a much higher mutation rate due to the absence of an effective error-checking system. Furthermore, there is a so-called control region of a few hundred base pairs that contains about one-third of the variation found.

8. Cann, Stoneking, and Wilson, 1987; see also Stoneking and Cann, 1989, which was published as part of *The Human Revolution* volume (Mellars and Stringer, 1989) composed after "Origins and Dispersal of Modern Humans" (Cambridge, 1987), the international conference that brought together specialists from human evolution, archeology, and molecular genetics to discuss the revolutionary new methods developed in the latter field and the meaning of their results when applied to human evolution; for a discussion of the sexism and racism betrayed by some of the reactions against the African Eve theory, see Cann, 1997.

9. Wolpoff also questions the selective neutrality of mtDNA (see, for example, Wolpoff, 1989; Wolpoff and Thorne, 2003).

10. The current multiregional-evolution hypothesis especially draws on precedents such as the German anatomist Franz Weidenreich (1873–1948) (but also Coon), who was an influential proponent of a unilinear multiregional phylogeny on the basis of gene flow and within the single-species framework (see, for example, Weidenreich, 1946, particularly chaps. 1–4). The proponents of the out-of-Africa theory usually refer it to the Harvard anthropologist William White Howells. Reminiscent of Grafton Elliot Smith's distinction between an evolutionary and a historical model, Howells identifies evolutionists (multiregional-continuity model of modern human origins) and migrationists among the contemporary anthropologists. Although he favors the migrationist model of a single origin of *Homo sapiens* and subsequent migrations, he does not exclude gene flow and absorption of other forms, as does the extreme version of the migratory model based on genetic data, to which he refers as the Noah's Ark hypothesis (Howells, 1976).

11. Wolpoff, 1989, p. 98.

12. That the out-of-Africa scenario with a human racial history of less than 50,000 years can nonetheless be combined with the belief in significant differences between the races (morphologically, behaviorally, and mentally, but not genetically speaking) is exemplified by Sarich and Miele, 2004. One is again reminded that theories are adjustable to a range of ideologies (the matter is complicated by the fact that Sarich started out with a belief in the great antiquity of the human races). Sarich and Miele reaffirm the notion of a link of brain size to intelligence that in their view establishes a hierarchy from African to white to Asian. They also make use of the racism accusation strategy when in an idiosyncratic appropriation of history they link the out-of-Africa scenario to the pre-Darwinian monogenist view, whereas the multiregional hypothesis is referred to polygenist scenarios (p. 129).

However, the post-Darwinian migrationist frameworks discussed in Part II could be seen as having a polygenist tendency, while the nineteenth-century linear evolutionists tended to be "monogenists." Then again, the multilinear variant of the multiregional model as promoted by Keith and Ruggles Gates shows affinities to the polygenist tradition. Even this complicated picture is a gross simplification, however, as discussed in previous chapters and further examined in following ones, not least because the successive "migrationist" as well as "local continuity" schemes are not comparable in a straightforward way.

13. Oakley, 1968; 1980, pp. 20–22. See also Godwin, 1970, and Harris, 1987, on the impact of radiocarbon dating on British archeology.

14. Campbell, 1977, pp. 144–145; Molleson, 1976; Molleson and Burleigh, 1978; see also Aldhouse-Green, 2001b. The Gravettian is a period of the (Middle) Upper Paleolithic in Europe from about 28,000 to 22,000 BP (also known as the Perigordian), succeeding the Aurignacian. It is named after the type site of La Gravette in the Dordogne region of France.

15. Jacobi, 1980, chap. 2; on the Holocene date, see Spencer, 1997b, p. 691.

16. Jackson, 1953, p. 211.

17. Ibid., pp. 170–174, 210–211. The Creswellian is a British Upper Paleolithic culture named after the type site of Creswell Crags in Derbyshire (approximately 12,500 to 12,000 BP). It was followed by a Mesolithic culture.

18. Swainston and Brookes, 2000, p. 33.

19. *The Oxford English Dictionary Online* (Oxford University Press, 2006).

13. The Paviland Project and Its Results

1. Cadw (Welsh for "to keep") is the historic-environment agency within the Welsh Assembly Government with responsibility for protecting, conserving, and promoting the historic environment of Wales. Aldhouse-Green is an expert on the Welsh Paleolithic, on which he has published abundantly (Aldhouse-Green, 1998, 2001a, 2001c, 2004a; Aldhouse-Green et al., 1992, 1995; Aldhouse-Green and Walker, 1991; Green, 1984; Green et al., 1989). On how the story began, see also Aldhouse-Green, 2001b, p. 23.

2. The Red Lady, for example, appears as part of the comparative sample of Upper Pleistocene anatomically modern humans in Trinkaus, 1984. Trinkaus is currently professor of anthropology at Washington University in St. Louis, Missouri.

3. Adrienne Mayor, 2000, shows how already the ancient Greeks searched for and quarreled over the bones of giant heroes (large fossil mammals) that were claimed as patrons of city-states. Often the oracle at Delphi unraveled the identity of a particular set of bones. Local and founder heroes, such as Asterios, Pelops, Hector, Orestes, and Theseus, strengthened the ties to the land and conferred religious and political power on the governing people. Some of them were also thought to bring financial wealth (see in particular pp. 72–75, 95–98, 110–114). In medieval Britain, the bones of giants (most likely mammoths) were discovered that supported the legendary role of giants in the peopling of Britain as offered by Geoffrey of Monmouth in the twelfth century (Piggott, 1989, pp. 48–53). As Paul Semonin, 2000, describes, the mastodon also functioned as a human giant in the early eighteenth century in America, before the animal was turned into a symbol for the new nation.

4. Aldhouse-Green, 2000c, preface, p. xxiii.
5. See, for example, Roebroeks et al., 2000.
6. The field assessment was published before the *Definitive Report* in Aldhouse-Green, 1997.
7. Aldhouse-Green, 2004b.
8. Aldhouse-Green, 2000b, pp. 232–233.
9. Lowe, 2000, p. 58; Young, 2000.
10. Swainston and Brookes, 2000; see also the accompanying appendix on the history of the museum collections by Elizabeth Walker, 2000.
11. Swainston, 1999. For reasons that have not become clear to me, Swainston never received a PhD for her extensive work on the project.
12. On how the project was funded and on its participants, see also Aldhouse-Green, 2000c, summary on pp. xxxi–xxxiii.
13. Obviously, this is a subjective judgment that, although it mirrors in particular the division of popular attention between hominid and faunal material, is by no means intended to devalue the work carried out on the Paviland fauna (such as that summarized in Turner, 2000).
14. Trinkaus, 2000. Archeologists tended to refer to Upper Paleolithic material older than about 20,000 years as Early Upper Paleolithic, as opposed to the Late Upper Paleolithic, which was the more recent time period. When they began to realize over the last decade or two that there are fairly significant differences between what is more precisely referred to as the Aurignacian and the Gravettian in Europe, the term Middle Upper Paleolithic was introduced to differentiate the Gravettian from the older Early Upper Paleolithic cultures such as the Aurignacian, on the one hand, and from the younger Late Upper Paleolithic cultures, on the other hand (the distinction was introduced by Roebroeks et al., 2000).
15. Trinkaus, 2000, p. 191.
16. Holliday, 2004; Trinkaus, 2004.
17. Trinkaus, 2000, p. 145.
18. Sykes, 2000.
19. Ibid., p. 75.
20. Holliday, 2000.
21. His MUP sample consisted mostly of Gravettian material, but also proto-Magdalenian, 20,000 to 30,000 years BP, and some Aurignacian, pre–30,000 years BP, including data from the Cro-Magnon site. Body proportions are measured as tibia length/femur length × 100; the relative femoral head size is taken as the index for body linearity; the ratio of femoral length to bi-iliac breadth indicates relative body breadth.
22. Holliday, 2000, p. 199.
23. Ibid., p. 204.
24. Bull, 2000; Debenham, 2000; Ivanovich and Latham, 2000; Mourne and Case, 2000.
25. Richards, 2000.
26. Pettitt, 2000. The dates from the radiocarbon dating program were published beforehand in Aldhouse-Green and Pettitt, 1998.
27. Bowen, 2000.
28. Aldhouse-Green, 2000a.
29. Swainston, 2000.

14. There Is Magic at Work

1. Aldhouse-Green, 2000a.
2. Aldhouse-Green, 2000b, pp. 235–241.
3. Ibid., p. 241.
4. Ibid., p. 233.
5. In fact, Aldhouse-Green considers the possibility that the Red Lady's dead body was transported in a frozen state to the venerated site from much farther south (Aldhouse-Green and Aldhouse-Green, 2005, p. 36). This idea originated in the 1970s (in communication with Molleson) and accorded with the notion of a Paviland at the verge of the glaciers (Green, 1989; and personal commentary by Aldhouse-Green, 26 January 2006).
6. Aldhouse-Green, 2000b, pp. 241–242.
7. Ibid., pp. 236–238.
8. Aldhouse-Green, 2000a, p. 117.
9. The possibility of the disappearance of wall art once present at Paviland is supported by Lowe, 2000, pp. 58–59. The claim of ancient shamanism has mainly been based on cave art and its imagery (for a recent example, see Clottes and Lewis-Williams, 1996).
10. Aldhouse-Green, 2000b, p. 244.
11. Ibid., p. 242. Experiments in neuropsychology have shown that in altered states of consciousness a particular set of images appears in front of the mind's eye. Some of these, the geometric figures and the so-called entoptics, such as therianthropes, have been identified in Paleolithic cave art (see, for example, Clottes, 2000). Because of this insight and the pervasiveness of shamanism in hunter-gatherer societies, it has been argued that (European) Upper Paleolithic cave art can be explained as shamanistic (Lewis-Williams, 1997, 2002).
12. Aldhouse-Green, 2000b, p. 244; *peregrinatio pro numinum amore* is a "pilgrimage for the love of the spirits" in analogy to the medieval pilgrimage for the love of God.
13. Jones, 2000, p. 264.
14. Freud, 1913.

15. Visualizing Paviland Cave

1. Figure 15.1 is reproduced in Aldhouse-Green, 2000c, pl. 11.5. The question of public appeal is not unimportant for museum curators, as well as scientists, who ultimately depend on the fact that their area of research is considered relevant. For example, the Prehistoric Society has identified the promotion of the Paleolithic and Mesolithic in Britain and Ireland, through museums and other public venues such as site displays, as one of its goals (*Research Frameworks for the Palaeolithic and Mesolithic of Britain and Ireland. A Report by the Working Party for the Palaeolithic and Mesolithic Annual Day Meeting and the Council of the Prehistoric Society*, 1999). Because actually reconstructing prehistoric people, places, and happenings is impossible first and foremost because of lack of evidence, it has been suggested to replace the term *reconstruction* by *simulation*. However, besides carrying depreciative connotations, *simulation* is a term already employed in science and technology for

practices different from those covered by the term *reconstruction* (discussed in James, 1997, p. 22).

2. Aldhouse-Green and Walker, 1991, figs. 4 and 5, p. 14.

3. In visualizations of the lives of our prehistoric ancestors, women traditionally constitute a minority. If shown, they tend to be close to the home base, often seated or on all fours at the margin of the scene, and passive or engaged in stereotyped occupations such as child care. They may also be ornamental or eroticized, inviting the male gaze (on the gendering of prehistoric reconstruction paintings, see, for example, Gifford-Gonzalez, 1993; Moser, 1993; see p. 77 for male dominance in reconstructions of burial scenes).

4. Aldhouse-Green and Walker, 1991, p. 17.

5. Wood and Caselli, 1976. See in particular the double-page reconstruction paintings of life scenes of australopithecines, *Homo habilis, Homo erectus, Homo neanderthalensis,* and Paleolithic "races" of *Homo sapiens*. The beautifully colored paintings, which incorporate many then-current scientific ideas, also betray the gender bias of the time. Although in most of the scenes females are present, they are often sitting and engaged with offspring, while the males are the tool users and hunters who are shown standing and in action, coming and going between hunting forays and home base. Caselli also illustrated *The Evolution and Ecology of the Dinosaurs* (L. B. Halstead, 1975), which contributed to his renown.

6. The popular *Flashman* novels by George McDonald Fraser tell the contrived story of Sir Harry Flashman, a cowardly, ignorant, and unprincipled no-good turned Victorian hero. The character is inspired by Tom Hughes's nineteenth-century novel *Tom Brown's Schooldays*. Flashman's memoirs document his scandals, misadventures, and career in the military and are as politically incorrect as it gets. As befits Fraser's parody, D'Achille's cover images are in the style of the trash adventure novel.

7. D'Achille, 2005.

8. Although illustrators intend to fill the evidential gaps by informed guesses (or what Wiktor Stoczkowski, 1997, p. 250, calls deductive imagination), they may be unaware of the way in which they are constrained by visual traditions of depicting early and/or "primitive" humans that can be traced back to antiquity (Moser and Gamble, 1997). As a result, they in fact work with what Stoczkowski refers to as conditioned imagination.

9. D'Achille, 2005.

10. D'Achille's reconstruction has also been reproduced in Aldhouse-Green, 1996, pp. 6–7; 2001b, p. 24; Aldhouse-Green and Pettitt, 1998, fig. 3, p. 766; on the notion of idées fixes, see James, 1997.

11. Aldhouse-Green, 2004b.

12. This last point has been brought to my attention by Swainston, 2004a.

13. The inclusion of women and also children in hunting has been argued for the European Gravettian (Soffer, 2000).

14. Swainston, 2004a.

15. Swainston, 1999, p. 113.

16. Swainston, 2004b, 2005.

17. Photograph by Aldhouse-Green, reproduced in Aldhouse-Green, 2000c, pl. 11.4.

18. A brochure was produced to accompany the exhibit (Aldhouse-Green, 1996).

19. Aldhouse-Green, 2004b.

20. On the Gower Society's plans, see, for example, Turner, 2004b, 2004a, available online at http://icwales.icnetwork.co.uk/0100news/0200wales/tm_objectid=14235936&method=full&siteid=50082&headline=return--red-lady--skeleton-to-wales--english-urged-name_page.html; http://icwales.icnetwork.co.uk/0100news/0200wales/tm_objectid=15015937&method=full&siteid=50082&headline=campaign-to-bring--red-lady--back-to-swansea-after--180-years-name_page.html, the national website of Wales, accessed 4 November 2006. Other caves, such as Kent's Cavern, with similar historical appeal (see Part I, Chapter 4) have already become tourist attractions, with a tour through the cave that has been rendered wheelchair friendly, a museum shop, and a large restaurant that can seat busloads of people. Evocative of nineteenth-century cave spectacles, there is even a special ghost tour at night. In line with the growing demands for repatriation and reburial of ancient bones, the group Dead to Rights, led by a "Druid," demands that the Red Lady's bones be reburied in Paviland Cave; see http://news.bbc.co.uk/1/hi/wales/south_west/5372598.stm (23 September 2006), and http://news.bbc.co.uk/2/hi/uk_news/wales/south_west/6038026.stm (11 October 2006), at the website of BBC Wales accessed 1 November 2006. "The 'Red Lady' Is to Return Home" was broadcast on BBC Wales News, 11 October 2006.

21. See Colour Plate 7 in Aldhouse-Green and Aldhouse-Green, 2005; fig. 1.8 in Aldhouse-Green, 2004a.

22. However, the claim for cave art at Paviland, as part of the vision of sacred place, pilgrimage, and shamanism, again suggests that this vision is gendered and might not have been evoked by a female burial. In what seems to be a projection of contemporary stereotypes into the past, the creation of cave paintings (but also of statues and engravings) is most often envisioned as having been an exclusively male occupation. Already in the nineteenth century, when the erotic ideal had turned from being male to being female, the female imagery of prehistoric art was understood as the work of the male (Paleolithic) artist with females as subjects. By inference, where cave art has been presumed to be associated with shamanic ritual, the shaman has been envisioned as male. Correspondingly, those images that have been interpreted as representing a shaman are thought to show a male person (on these points, see Cohen, 1999, chap. 3; 2003, chap. 3; Conkey, 1997b).

23. Aldhouse-Green and Aldhouse-Green, 2005, p. 37.

24. Anne Leaver, personal communication, 22 February 2006.

25. See, for example, Aldhouse-Green and Walker, 1991, pp. 10–11, for the discussion of the Welsh "deficit" in the kind of large cult centers, including cave art, known from France. The fact that the earlier suggestion of murals at Paviland Cave has only been visualized in this latest reconstruction might be due to the recent discovery of decorative rock carving in British caves (Anne Leaver, personal communication, 22 February 2006).

26. On the role of reconstructions in the history of anthropology and geology, see, for example, Moser, 1992; Rudwick, 1989, 1992. Several authors have dealt with visual reconstructions of human origins, highlighting the fact that they betray anthropocentric, ethnocentric, and androcentric perspectives (Berman, 1999; Conkey, 1997b; Gifford-Gonzalez, 1993; Gould, 1989; Haraway, 1997, pp. 175–187; Moser, 1993, 1998; Stringer and Gamble, 1993, chap. 1; Wiber, 1998, chap. 3).

27. Gifford-Gonzalez, 1993.

28. Redknap, 2002.

29. The role of prehistoric reconstruction as research tool can, and in many cases does, go far beyond the expression of, and engagement with, changing interpretations of a site over time. Reconstructions may be an integral part of the excavation process, with new questions and possible answers being created in feedback between the two (James, 1997, pp. 27–33).

30. Brian Leigh Molyneaux has in this context spoken of a metaphorical reinforcement of scientific attitudes through images. However, as discussed earlier, and as pointed out by Molyneaux, images also reinforce power structures and political attitudes (Molyneaux, 1997, pp. 3–4).

31. Aldhouse-Green and Aldhouse-Green, 2005, p. 35.

16. The End of the Red Lady's Story

1. White, 2002, available online at *The Prehistoric Society Homepage*, www.ucl.ac.uk/prehistoric/reviews/02_10_paviland.html, accessed 4 November 2006.

2. Aldhouse-Green, 2000c, preface, p. xxiv.

3. In feminist epistemologies of science, the national, ethnic, gender, and disciplinary diversity of scientific communities as a prerequisite for a strong kind of objectivity has long been stressed (the expression *strong objectivity* has been introduced by Sandra Harding, 1991). Helen Longino, 1993, proposes to move away from the individual as an epistemic unit to the scientific community as the smallest unit of knowledge production. She believes that this move could account for the problems of observation being theory laden, as well as of theories being underdetermined by data, both of which, as we have seen, are notoriously true for paleoanthropology. Less biased knowledge is seen to be the outcome of critical dialogue between different points of view within a group and among groups, in the course of which the different points of view will make visible each other's background assumptions. For this democratic process to work optimally, the community must be diverse. However, Longino presents a list of very high-stakes prerequisites for this model of doing science: public critical engagement; openness to the different; democratic procedures; and consideration of social and environmental factors. Lynn Nelson, 1993, also sees the solution to the postmodern critique of positivism in an epistemology of science that takes the scientific community as the smallest epistemological unit, rather than the individual, because the problems of individual knowledge acquisition, such as limited sense organs and underdetermination of theories by evidence, are perceived as not applying to the community.

4. Krull, 2000, p. 265.

5. In fact, Wilhelm Krull's description of interdisciplinarity no doubt reminds the reader of the transdisciplinary way of doing contemporary science (and/or the way science should be done) as conceptualized in Helga Nowotny et al., 1994, chap. 1 (i.e., Mode 2, versus the traditional (inter-/multi) disciplinary Mode 1). Whereas interdisciplinarity is seen to work with a common method and terminology on diverse problems, transdisciplinarity involves the mutual interpenetration of disciplinary epistemologies in a temporary congregation of experts from different disciplines to tackle a specific problem. Here the theoretical approach may be crosscutting disciplines or may convey a plurality of approaches in a local

context. The problem at hand, as well as solution strategies, is defined within the context of application in a cooperative effort. To the contrary, multidisciplinarity is described by Nowotny et al. as the cooperation of autonomous disciplines that remain unaffected by the process. In a multidisciplinary approach, a problem or topic is addressed by several disciplines through the application of the knowledge, methods, and technologies that already exist within the individual fields. As we will see, this comes closest to an adequate description of how the Paviland project proceeded.

6. For a similar low-key definition of interdisciplinarity, see Thompson Klein, 2000, p. 8.

7. Trinkaus, for example, had this experience with one of his projects in southwestern Rumania. While the human remains from the site under investigation (Pestera cu Oase) are likely to be among the oldest modern humans so far discovered in Europe (35,000 years BP), national pride has sometimes turned them into the first humans in the Rumanian media (Trinkaus, 2004). See also the discussion in Chapter 13, note 3.

8. Aldhouse-Green, 2004b.

9. Aldhouse-Green, 2000c, p. xxv.

10. Ibid., p. xxv.

11. Swainston, 1999, p. 104.

12. Already in Aldhouse-Green and Walker, 1991, pp. 31 and 35, the Paviland site and the Red Lady are used to argue for the importance of Paleolithic Wales vis-à-vis the Continent, as well as Britain at large. Historically, nationalistic concerns have also entered stratigraphic geology. In the nineteenth century, the Silurian system was named after a Roman-British tribe that once inhabited the Welsh border. Correspondingly, the extension of the Silurian system was viewed as a great national achievement. The Cambrian system carries the ancient name of Wales. The creator of each system made an attempt to subsume the other formation under his *Cambrian* or *Silurian*, respectively (James Secord, 1982).

13. Morgan, 2000, p. xxix.

14. Holliday, 2004.

15. Aldhouse-Green, 2004b.

16. Trinkaus, 2004. In the analysis of Neanderthal DNA, the danger of contamination is seen in the fact that the preconception in this case is that the DNA is outside the modern human ranges of variation (archaic), so that modern DNA sequences might automatically be disqualified as contamination. In addition, the multiplication of ancient DNA by polymerase chain reaction (PCR) may produce "mutations" in the sequences, so there is no knowing if the sequences singled out as Neanderthal, since they are divergent from those of modern humans, are in fact genuine (Trinkaus, personal communication, 14 May 2004). To prevent flakes of skin from entering the laboratory at Oxford, Sykes has installed a filtered-air clean room (Sykes, 2001, p. 178; this does not exclude older contamination of the bones with recent DNA).

17. Trinkaus, personal communications, 13 and 14 May 2004; for a concise history of molecular anthropology and its difficult relation to paleoanthropology, see Schwartz, 2005.

18. The model of the seven daughters of Eve is explained in detail in one of Sykes's popular books on molecular anthropology (Sykes, 2001; on the method, see

also Sykes, 1999). Sykes here makes up life stories for the seven women, thereby putting faces on the mtDNA sequences to render the information provided by Oxford Ancestors of some value to the customers. However, the chapters tell mostly of the thrilling male hunts, the ingenuity of their toolmaking, and their scary trips into caves to perform ritual. In his second popular book, which can be regarded as a companion to the company's Y services, Sykes fully embraces sociobiology's genetic determinism and molecular anthropomorphism when he ascribes such stereotypical male attributes as greed, aggression, and promiscuity to the Y chromosome. Again, the determined inseminator is personalized by identifying successful Y chromosomes with mythic Viking warriors or the legendary Genghis Khan (Sykes, 2003). Indeed, Oxford Ancestors also offers a service to determine whether one carries a Genghis Khan Y chromosome. Two further volumes related to the Tribes of Britain™ service are *The Blood of the Isles* (London: Bantam, 2006) and *Saxons, Vikings, and Celts: The Genetic Roots of Britain and Ireland* (New York: W. W. Norton, 2006). A parallel reading of www.oxfordancestors.com and these popular books thus strongly suggests that although the technologies employed by Oxford Ancestors and Sykes in his work are innovative, the results from comparative mtDNA analyses need to be translated back into stories, and these have a very familiar ring.

19. Galton, 1892.

20. *Oxford Ancestors*, www.oxfordancestors.com/the-team.html (the earlier version of this site referred to was accessed on 17 May 2004).

21. For the establishment of the European lines, Sykes's team first collected blood samples in Welsh schools because Wales was considered to have a high proportion of long-established families, and because linking the collecting with class presentations on genetics meant that the teachers were likely to respond favorably. The children had to be sixteen to give legal consent. The mitochondrial base sequences did not fall into two clear and comparatively old clusters, which one would have expected with a mixed Neanderthal and Cro-Magnon descent (Sykes,.2001, chap. 9).

22. Samples from European regions other than Wales were collected by Sykes's team and helpers as the opportunity arose. To a lesser degree, they could draw on published data on mtDNA sequences. The complicated mutational patterns found in Europe were then arranged in seven clusters, which, on the assumption of one base mutation per 10,000 years, indicated that these lines reached back further in time than the date assumed for the Neolithic invasions from the Middle East, with the one exception of the Jasmine clan (to give an example, if along the 500-bases segment of the mtDNA control region, the average distance between any two people is 4.5 mutations within one cluster, the members of this clan are assumed to have a common female ancestor some 45,000 years ago) (Sykes, 2001, chap. 10).

23. The seven-daughters-of-Eve model was strongly corroborated by ancient mtDNA from Cheddar Man, radiocarbon dated to 9,000 years ago, whose sequence neatly fit one of the European mtDNA clusters (Sykes, 2001, chap. 12), and of course by the much older Red Lady. It is indeed remarkable that Sykes did not use the evidence from the Red Lady's mtDNA in his book, but as we have seen, the Red Lady data only partly fit the seven-daughters model. (Another interpretation is the rumor that one of his employees has a very similar sequence.) As in the critiques issued by multiregionalists against the promoters of the out-of-Africa hypothesis, the technique of the molecular clock takes center stage in the critical reactions to Sykes's view of the origins of modern Europeans (on the history of

molecular anthropology up the 1980s, see, for example, Goodman and Cronin, 1982; for a critique of the molecular clock as based on the assumption of a too-high mutation rate, see, for example, Wolpoff, 1989). Although the new molecular techniques have mostly been seen as supporting the replacement scenarios, Sykes's refutation of the "wave-of-advance" model shows that the same technology can also be used to disprove genetic replacement. Thus, in his case, the argument works the other way around. If it could be shown that the clock Sykes assumed as the basis of his calculations of European lineages ran too slow, then the lineages might still not be traced back beyond about 10,000 years, or the Neolithic invasion (Sykes, 2001, pp. 155–168).

24. Trinkaus, 2004.

25. See, for example, Trinkaus, 1984. Trinkaus also cooperated in a critical reconstruction of the history of interpretation of the Neanderthals (Trinkaus and Shipman, 1993).

26. Holliday, 2004. Holliday's more tentative stance with regard to the Neanderthal question seems evidenced in Churchill et al., 2000, pp. 36–39, where he interprets the body-shape comparisons between Early, Middle, and Late Upper Paleolithic Europeans as supporting either a replacement or an admixture model (but not local continuity). The argument from morphology depends on the presupposition that body proportions are to a large degree genetically controlled. The fact that contemporary African Americans do not have tropical limb proportions but have in a few hundred years changed to more European body proportions (through adaptation to new climate plus intermixture with Europeans and First Nations Peoples) puts this claim into perspective (Pat Shipman, adjunct professor of biological anthropology, Pennsylvania State University, personal communication, 14 May 2004).

27. In Aldhouse-Green and Walker, 1991, where Paviland Cave and the Red Lady constitute the most important Early Upper Paleolithic evidence on the British Isles, the modern humans who were the Neanderthals' contemporaries in Europe are seen as superior in every respect. The Neanderthals are thought to have been less articulate and less intelligent, to have had less agility, and to have died earlier. Indeed, modern humans would not have perceived them as of the same kind, which prevented mating (pp. 12–13). The Neanderthals "were beings very different from ourselves" (p. 13). Thus modern humans extinguished them or drove them to marginal and less hospitable regions, turning them into the "world's first refugees" (p. 17).

28. Sykes, 2001, p. 115.

29. Trinkaus, 2004; see also Trinkaus et al., 2000.

30. Aldhouse-Green, 2000b, pp. 235–241.

31. See, for example, Gere, 1999; Hager, 1997a, pp. 10–15.

32. See, for example, Kästner, 1995, chaps. 2.2 and 2.3; Lohrke, 2004. Denise Donlon, 1993, discusses the numerical sex bias in human remains, in particular for the Pleistocene and early to mid-Holocene Australian Aboriginal skeletal collections, considering the possibility of both a bias in the actual burials and a bias in the physical anthropologists' determinations of sex.

33. Roebroeks and Corbey, 2000, p. 79. Raymond Corbey has also dedicated a book to the question of the metaphysics of the animal-human boundary and has found similar mechanisms of boundary maintenance at work in both the human-ape and the modern-archaic human divide (Corbey, 2005).

34. Aldhouse-Green, 2000b, p. 245.

35. Swainston, 2004a.
36. Swainston, 1999, p. 104.
37. Swainston and Brookes, 2000, pp. 35–42.
38. White, 2002, available online at *The Prehistoric Society Homepage*, www.ucl.ac.uk/prehistoric/reviews/02_10_paviland.html, accessed 4 November 2006.
39. The question might be raised whether the Paleolithic ritual practices and cave veneration really stand at the beginning of the mythical, folkloristic, and scientific tradition of understanding caves as sacred places, or whether we do not rather find it easy to see magic at work in prehistoric caves because our passions are linked to the powerful tradition of this trope in our culture and its reliance on "living fossils"—there might be a certain danger of being duped by one's own magic (on cave symbolism, see Weinberg, 1986).
40. This again is an intricate use of evidence, since parietal art is first inferred by using the pigments from the cave, as well as the collateral evidence of other sites, and in a second step of inference the invisible wall paintings are used to support the notion of Paviland Cave as a sacred place of pilgrimage and shamanism, which traditionally has been associated with Paleolithic cave paintings.
41. See, for example, Conkey, 1997a. On the connection between altered states of consciousness, mind imagery, and cave art, see Chapter 14, note 11. The assumption of universal images created by the human mind in altered states of consciousness is also controversial and has been labeled neurological determinism (Helvenston and Bahn, 2002).
42. Swainston, 1999, p. 109. For a discussion of the history and problems but also the possibilities of the comparative approach, see Wylie, 2002, pp. 136–153.
43. Aldhouse-Green, 2004b.
44. Jones, 2000, p. 253.
45. White, 2002, available online at *The Prehistoric Society Homepage*, www.ucl.ac.uk/prehistoric/reviews/02_10_paviland.html, accessed 4 November 2006.
46. Jones, 2000, particularly pp. 253–254.
47. Swainston, 1999, p. 112.

Conclusion: An Unfinished Life

1. Aldhouse-Green, 2004b.
2. Aldhouse-Green, 2001b, p. 20; and personal commentary, 26 January 2006.
3. Weingart, 2000, pp. 26–27.

Postscript

1. Scott, 2005, pp. 39–45.
2. Structural analyses of scenarios of human evolution tend to be associated with the assumption of this kind of anthropological universals. Misia Landau, 1984, 1991, identifies key steps in hominization (such as terrestriality, bipedalism, encephalization, and socialization) found in theories of human evolution since Darwin. She also argues that evolutionary origin stories have a narrative structure akin to our folktales and hero myths and to the creation myths of many "primitive"

peoples (as well as our own). Peter Bowler, 2001, has criticized this universalizing and dehistorizing approach, as well as the similar tendencies found in Wiktor Stoczkowski, 2001, who claims the existence of a limited set of ideas in the history of archeology (some of pre-nineteenth-century origin), which are seen to be prompted by outer stimuli, such as new evidence, ethnological analogy, or sociocultural context, and newly combined according to an internal set of also preexisting rules to build "new" hypotheses, models, and scenarios. Bruno Latour and Shirley Strum, 1986, too, approach texts as different in content and context as, for example, Jean-Jacques Rousseau's *A Discourse on the Origin of Inequality* (1755), Sigmund Freud's *Totem and Taboo* (1913), and Richard Dawkins's *The Selfish Gene* (1976) with the same set of questions to allow for comparisons with regard to the logical coherence of their accounts of human social origins.

3. Darwin himself had hoped that the triumph of evolutionism would mean the end of polygenism (Darwin, 1874 [1871], pp. 272–273). On parallels between nineteenth-century evolutionism and polygenism, see "The Dark-Skinned Savage: The Image of Primitive Man in Evolutionary Anthropology," in Stocking, 1968, chap. 6.

4. I would like to thank British Academy research professor Robin Dennell for sharing with me his views on the dangers of this kind of simplification, among other things. He is acutely aware of the fact that encapsulations (such as "the migratory model") tend to be simplistic, overlooking the diversity of opinion diachronically as well as synchronically. Although it is possible to discern major strands and themes, there are at any point in history many contradictory and ambiguous arguments due to, among other things, the different age, background, and nationality of scientists that may cause idiosyncratic thinking (Dennell, 2001, and personal communication, 27 October 2005; on the view that the history of paleoanthropology is marked by a resuscitation of old theories under new labels, see also Wolpoff, 2007, which is a preface to a book that in contrast arrives at a progressive picture of the science through an emphasis on continuities and similarities between hypotheses in the history of the field, albeit under the exclusion of sociocultural setting and historical context).

5. Proctor, 2003, p. 225.

6. On the ambiguity of ethnological analogy in the aftermath of classical evolutionism, see, for example, Ascher, 1961. Although "naïve" application continued, some archeologists have shied away from analogy to living peoples altogether, while others have proposed various constraints on the choice of analogs.

7. Appleyard, 19 June 2005, available online at *Times online*, www.timesonline.co.uk/article/0,,2102-1653983_1,00.html, accessed 4 November 2006.

8. Reported in the *Independent*, 27 February 1999, reproduced in Collis, 2003, p. 230. For the deconstruction of the notion of Iron Age Celts and the histories associated with it that provoked Morgan's response, see James, 1999; for a critical history of the concept of Celts/Celtic, see also Morse, 2005.

9. For a history of Anglo-Saxonism, see, for example, MacDougall, 1982, chaps. 2–7.

References

Albarello, Bruno. 1987. *L'Affaire de l'Homme de La Chapelle-aux-Saints, 1905–1909.* Treignac: Editions "Les Monédières."

Aldhouse-Green, Miranda, and Aldhouse-Green, Stephen. 2005. *The Quest for the Shaman. Shape-Shifters, Sorcerers and Spirit-Healers of Ancient Europe.* London: Thames and Hudson.

Aldhouse-Green, Stephen. 1997. "The Paviland Research Project. The Field Assessment." *Archaeology in Wales* 37:3–12.

———. 1998. "The Archaeology of Distance. Perspectives from the Welsh Palaeolithic." In *Stone Age Archaeology. Essays in Honour of John Wymer,* edited by Ashton, Nick, Healy, Frances, and Pettitt, Paul. Oxford: Oxbow.

———. 2000a. "Artefacts of Ivory, Bone and Shell from Paviland." In *Paviland Cave and the "Red Lady." A Definitive Report,* edited by Aldhouse-Green, Stephen. Bristol: Western Academic and Specialist Press.

———. 2000b. "Climate, Ceremony, Pilgrimage and Paviland. The 'Red Lady' in His Palaeoecological and Technoetic Context." In *Paviland Cave and the "Red Lady." A Definitive Report,* edited by Aldhouse-Green, Stephen. Bristol: Western Academic and Specialist Press.

———. 2001a. "Ex Africa aliquid semper novi. The View from Pontnewydd." In *A Very Remote Period Indeed. Papers on the Palaeolithic Presented to Derek Roe,* edited by Milliken, Sarah, and Cook, Jill. Oxford: Oxbow.

———. 2001b. "Great Sites. Paviland Cave." *British Archaeology* 61:20–24.

———. 2001c. "The Colonisations and Visitations of Wales by Neanderthals and Modern Humans in the Later Pleistocene." In *Histories of Old Ages. Essays in Honour of Rhys Jones,* edited by Anderson, Atholl, Lilley, Ian, and O'Connor, Sue. Canberra: Pandanus.

———. 2004a. "The Palaeolithic." In *Gwent in Prehistory and Early History,* edited by Aldhouse-Green, Miranda, and Howell, Raymond. Cardiff: University of Wales Press.

———. 2004b. Interview, 16 July.

——— (ed.). 1996. *Explaining the Unexplainable. Art, Ritual and Death in Prehistory.* Cardiff: National Museums and Galleries of Wales.

——— (ed.). 2000c. *Paviland Cave and the "Red Lady." A Definitive Report.* Bristol: Western Academic and Specialist Press.

Aldhouse-Green, Stephen, and Pettitt, Paul. 1998. "Paviland Cave. Contextualizing the 'Red Lady.'" *Antiquity* 72 (278): 756–772.

Aldhouse-Green, Stephen, Scott, Katharine, Schwarcz, Henry, Grün, Rainer, Housley, Rupert, Rae, Angela, Bevins, Richard, and Redknap, Mark. 1995. "Coygan Cave, Laugharne, South Wales, a Mousterian Site and Hyaena Den. A Report on the University of Cambridge Excavations." *Proceedings of the Prehistoric Society* 61:37–79.

Aldhouse-Green, Stephen, and Walker, Elizabeth A. 1991. *Ice Age Hunters. Neanderthals and Early Modern Hunters in Wales.* Cardiff: National Museum of Wales.

Aldhouse-Green, Stephen, Whittle, A. W. R., Allen, J. R. L., Caseldine, A. E., Culver, S. J., Day, M. H., Lundquist, J., and Upton, D. 1992. "Prehistoric Human Footprints from the Severn Estuary at Uskmouth and Magor Pill, Gwent, Wales." *Archaeologia Cambrensis* 141:14–55.

Allen, David Elliston. 1994 (1976). *The Naturalist in Britain. A Social History.* 2nd ed. Princeton: Princeton University Press.

"*Ancient Hunters and Their Modern Representatives* [Review]." 1925. *American Historical Review* 30 (3): 630–631.

Appleyard, Bryan. 2005. Review of *The Tribes of Britain* (2005), by David Miles. *The Sunday Times*, 19 June.

Armelagos, George J., Carlson, David S., and Gerven, Dennis P. van. 1982. "The Theoretical Foundations and Development of Skeletal Biology." In *A History of American Physical Anthropology, 1930–1980*, edited by Spencer, Frank. New York: Academic Press.

Ascher, R. 1961. "Analogy in Archaeological Interpretation." *Southwestern Journal of Anthropology* 17:317–325.

Balick, Michael, and Cox, Paul Alan. 1996. *Plants, People, and Culture. The Science of Ethnobotany.* New York: Scientific American Library.

Barkan, Elazar. 1992. *The Retreat of Scientific Racism. Changing Concepts of Race in Britain and the United States between the World Wars.* Cambridge: Cambridge University Press.

Bartholomew, M. 1973. "Lyell and Evolution. An Account of Lyell's Response to the Prospect of an Evolutionary Ancestry for Man." *British Journal for the History of Science* 6 (23): 261–303.

Beaumont, Elie de. 1869. "Mémoires et communications." *Compte rendu des séances de l'Académie des Sciences* 69 (July–December) (séance du lundi, 6 décembre): 1211–1213.

Beche, H. T. de la, and Conybeare, W. D. 1821. "Notice of the Discovery of a New Fossil Animal Forming a Link between the Ichthyosaurus and the Crocodile." *Transactions of the Geological Society of London*, 1st ser., 5:559–594.

Bederman, Gail. 1995. *Manliness and Civilization. A Cultural History of Gender and Race in the United States, 1880–1917.* Edited by Stimpson, Catharine R. Women in Culture and Society. Chicago: University of Chicago Press.

Berman, Judith C. 1999. "Bad Hair Days in the Paleolithic. Modern (Re)constructions of the Cave Man." *American Anthropologist* 101 (2): 288–304.

Blewitt, Octavian. 1832. *The Panorama of Torquay. A Descriptive and Historical Sketch of the District Comprised between the Dart and Teign.* 2nd ed. London: Simpkin and Marshall.

Blumenbach, Johann Friedrich. 1790–1828. *Decas collectionis suae craniorum diversarum gentium illustrata.* 7 vols. Göttingen: Dietrich.

———. 2001 (1798). *Über die natürlichen Verschiedenheiten im Menschengeschlechte.* Translated by Gruber, Johann Gottfried. 3rd ed. Concepts of Race in the Eighteenth Century 5. Bristol: Thoemmes.

Boitard, Pierre. 1838. "L'Homme fossile, étude paléontologique." *Magasin universel* 5 (April): 209–240.

Booth, James. 1989. "Enlightenment, Sensibility and Romanticism." In *Bloomsbury Guide to English Literature*, edited by Wynne-Davies, Marion. London: Bloomsbury.

Boucher de Perthes, Jacques. 1847–1849. *Antiquités celtiques et antédiluviennnes. Mémoire sur l'industrie primitive et les arts à leur origine.* 3 vols. Vol. 1. Paris: Treuttel et Wurtz.

———. 1857. *Antiquités celtiques et antédiluviennnes. Mémoire sur l'industrie primitive et les arts à leur origine.* 3 vols. Vol. 2. Paris: Treuttel et Wurtz.

———. 1864. *Antiquités celtiques et antédiluviennnes. Mémoire sur l'industrie primitive et les arts à leur origine.* 3 vols. Vol. 3. Paris: Jung-Treuttel.

Boué, Ami. 1829. "Ossemens humains présumés fossiles." *Revue bibliographique pour servir de complément aux Annales des Sciences Naturelles* 18:150–151.

Boule, Marcellin. 1905. "L'Origine des éolithes." *L'Anthropologie* 16:257–267.

———. 1908. "L'Homme fossile de La Chapelle-aux-Saints (Corrèze)." *L'Anthropologie* 19:519–525.

———. 1911, 1912, 1913. "L'Homme fossile de La Chapelle-aux-Saints." *Annales de Paléontologie* 6, 7, 8:1–64, 65–208, 209–279.

———. 1914. "La guerre." *L'Anthropologie* 25:575–580.

———. 1914 (1912). "L'Homo Néanderthalensis et sa place dans la nature." In *Comptes rendus, 14e Congrès international d'anthropologie et d'archéologie préhistoriques, Genève.* Geneva: Kündig.

———. 1923 (1921a). *Fossil Men. Elements of Human Palaeontology.* Translated by Ritchie, James, and Elliot, Jessie. 2nd ed. Edinburgh: Oliver and Boyd.

———. 1923 (1921b). *Les hommes fossiles. Éléments de paléontologie humaine.* 2nd ed. Paris: Masson.

Bouquet, Mary. 1995. "Exhibiting Knowledge. The Trees of Dubois, Haeckel, Jesse and Rivers at the Pithecanthropus Centennial Exhibition." In *Shifting Contexts*, edited by Strathern, Marilyn. London: Routledge.

Bourdier, Franck. 1969. "Geoffroy Saint-Hilaire versus Cuvier. The Campaign for Paleontological Evolution (1825–1838)." In *Toward a History of Geology*, edited by Schneer, Cecil J. Cambridge, MA: MIT Press.

Bouyssonie, A., Bouyssonie, J., and Bardon, L. 1908. "Découverte d'un squelette humain Moustérien à la bouffia de La Chapelle-aux-Saints (Corrèze)." *L'Anthropologie* 19:513–518.

Bowen, David Quentin. 2000. "Calibration and Correlation with the GRIP and GISP2 Greenland Ice Cores of Radiocarbon Ages from Paviland (Goat's

Hole), Gower." In *Paviland Cave and the "Red Lady." A Definitive Report*, edited by Aldhouse-Green, Stephen. Bristol: Western Academic and Specialist Press.
Bowker, Geoffrey C., and Star, Susan Leigh. 1999. *Sorting Things Out. Classification and Its Consequences*. Cambridge, MA: MIT Press.
Bowler, Peter J. 1986. *Theories of Human Evolution. A Century of Debate, 1844–1944.* Oxford: Blackwell.
———. 1990. *Charles Darwin. The Man and His Influence*. Edited by Knight, David. Blackwell Science Biographies. Oxford: Basil Blackwell.
———. 1993. *Biology and Social Thought: 1850–1914. Five Lectures Delivered at the International Summer School in History of Science, Uppsala, July 1990.* Berkeley Papers in the History of Science 15. Berkeley: Office for History of Science and Technology, University of California at Berkeley.
———. 2001. "Myths, Narratives and the Uses of History." In *Studying Human Origins. Disciplinary History and Epistemology*, edited by Corbey, Raymond, and Roebroeks, Wil. Amsterdam: Amsterdam University Press.
Bowles, Francis P. 1974. "Measurement and Instrumentation in Physical Anthropology." *Yearbook of Physical Anthropology* 18:174–190.
Boyd Dawkins, William. 1874. *Cave Hunting. Researches on the Evidence of Caves Respecting the Early Inhabitants of Europe*. London: Macmillan.
———. 1880. *Early Man in Britain and His Place in the Tertiary Period*. London: Macmillan.
———. 1910. "The Arrival of Man in Britain in the Pleistocene Age." *Journal of the Royal Anthropological Institute of Great Britain and Ireland* 40:233–263.
———. 1925. "Late Paleolithic Art in the Cresswell Caves." *Man* 25:48.
Boylan, Patrick J. 1967. "Dean William Buckland, 1784–1856. A Pioneer in Cave Science." *Studies in Speleology* 1 (5): 236–253.
———. 1970. "An Unpublished Portrait of Dean William Buckland, 1784–1856." *Journal of the Society for the Bibliography of Natural History* 5 (5): 350–354.
———. 1979. "The Controversy of the Moulin-Quignon Jaw. The Role of Hugh Falconer." In *Images of the Earth. Essays in the History of the Environmental Sciences*, edited by Jordanova, Ludmilla J., and Porter, Roy S. Aberdeen: Rainbow.
———. 1981. "The Role of William Buckland (1784–1856) in the Recognition of Glaciation in Great Britain." In *The Quaternary in Britain. Essays, Reviews and Original Work on the Quaternary Published in Honour of Lewis Penny on His Retirement*, edited by Neale, John, and Flenley, John. Oxford: Pergamon.
———. 1984. "William Buckland, 1784–1856. Scientific Institutions, Vertebrate Palaeontology, and Quaternary Geology." PhD diss., University of Leicester.
———. 1997. "William Buckland (1784–1856) and the Foundations of Taphonomy and Palaeoecology." *Archives of Natural History* 24 (3): 361–372.
Brace, C. Loring. 1964. "The Fate of the 'Classic' Neanderthals. A Consideration of Hominid Catastrophism." *Current Anthropology* 5:3–43.
———. 1997. "Modern Human Origins. Narrow Focus or Broad Spectrum?" In *Conceptual Issues in Modern Human Origins Research*, edited by Clark, Geoffrey A., and Willermet, C. M. New York: Aldine de Gruyter.
———. 2002. *"Race" Is a Four-Letter Word. The Genesis of the Concept.* New York: Oxford University Press.

Breuil, Henri. 1913 (1912). "Les subdivisions du paléolithique supérieur et leur signification." In *Comptes rendus, 14e Congrès international d'anthropologie et d'archéologie préhistoriques, Genève*. Vol. 1. Geneva: Kündig, 165–238.
Brice, W. C. 1955. "The Study of Prehistory [Review]." *Man* 55:9.
Briggs, Asa. 1997. "Oxford and Its Critics, 1800–1835." In *The History of the University of Oxford*, vol. 6, *Nineteenth-Century Oxford*, pt. 1, edited by Brock, M. G., and Curthoys, M. C. Oxford: Clarendon Press.
———. 2000 (1959). *The Age of Improvement, 1783–1867*. 2nd ed. Harlow: Longman.
Brittain, Vera. 1960. *The Women at Oxford. A Fragment of History*. London: George G. Harrap.
Broca, Paul. 1868. "Sur les crânes et ossements des Eyzies." *Bulletins de la Société d'Anthropologie de Paris* 54 (2nd ser., 3): 350–392.
———. 1875 (1865–1875). "On the Human Skulls and Bones Found in the Cave of Cro-Magnon, near Les Eyzies." In *Reliquiae Aquitanicae; Being Contributions to the Archaeology and Palaeontology of Périgord and the Adjoining Provinces of Southern France*, edited by Lartet, Édouard, Christy, Henry, and Jones, Thomas Rupert. London: Williams and Norgate.
Brock, M. G. 1997. "The Oxford of Peel and Gladstone, 1800–1833." In *The History of the University of Oxford*, vol. 6, *Nineteenth-Century Oxford*, pt. 1, edited by Brock, M. G., and Curthoys, M. C. Oxford: Clarendon Press.
Brockliss, L. W. B. 1997. "The European University in the Age of Revolution, 1789–1850." In *The History of the University of Oxford*, vol. 6, *Nineteenth-Century Oxford*, pt. 1, edited by Brock, M. G., and Curthoys, M. C. Oxford: Clarendon Press.
Brooke, John Hedley. 1979. "The Natural Theology of the Geologists. Some Theological Strata." In *Images of the Earth. Essays in the History of the Environmental Sciences*, edited by Jordanova, Ludmilla J., and Porter, Roy S. Aberdeen: Rainbow.
Broom, Robert. 1950. *Finding the Missing Link. An Account of Recent Discoveries Throwing New Light on the Origin of Man*. London: Watts.
Brown, Robert M. 1912. "Ancient Hunters, and Their Modern Representatives [Review]." *Bulletin of the American Geographical Society* 44 (9): 702.
Browne, Janet. 1992. "Squibs and Snobs. Science in Humorous British Undergraduate Magazines Around 1830." *History of Science* 30:165–197.
Buckland, Frank (ed.). 1858. "Memoir of the Very Rev. William Buckland, D.D., F.R.S., Dean of Westminster." In *Geology and Mineralogy Considered with Reference to Natural Theology*. The Bridgewater Treatises VI, On the Power, Wisdom, and Goodness of God, as Manifested in the Creation. 3rd ed. London: George Routledge.
Buckland, William. 1817 (1816). "Description of a Series of Specimens from the Plastic Clay near Reading, Berks: with Observations on the Formation to Which Those Beds Belong." *Transactions of the Geological Society of London*, 1st ser., 4:277–304.
———. 1820. *Vindiciae Geologicae; or, The Connexion of Geology with Religion: Explained in an Inaugural Lecture Delivered before University of Oxford, May 15, 1819*. Oxford: Oxford University Press.
———. 1821a. "Instructions for Conducting Geological Investigations, and Collecting Specimens." *American Journal of Science* 3:249–251.

———. 1821b. "On the Structure of the Alps and Adjoining Parts of the Continent, and Their Relation to the Secondary and Transition Rocks of England." *Annals of Philosophy* 1:450–468.

———. 1821 (1819). "Description of the Quartz Rock of the Lickey Hill in Worcestershire, and of the Strata Immediately Surrounding It; with Considerations on the Evidences of the Recent Deluge Afforded by the Gravel Beds of Warwickshire and Oxfordshire, and the Valley of the Thames from Oxford downwards to London; and an Appendix, Containing Analogous Proofs of Diluvian Action. Collected from Various Authorities." *Transactions of the Geological Society of London* 5:506–544.

———. 1822. "Account of an Assemblage of Fossil Teeth and Bones of Elephant, Rhinoceros, Hippopotamus, Bear, Tiger, and Hyaena, and Sixteen Other Animals; Discovered in a Cave at Kirkdale, Yorkshire, in the Year 1821: With a Comparative View of Five Similar Caverns in Various Parts of England, and Others on the Continent." *Philosophical Transactions of the Royal Society of London* 112:171–236.

———. 1823. *Reliquiae Diluvianae; or, Observations on the Organic Remains Contained in Caves, Fissures, and Diluvial Gravel, and on Other Geological Phenomena, Attesting the Action of an Universal Deluge.* London: Murray.

———. 1824. "Notice on the Megalosaurus, or Great Fossil Lizard of Stonesfield." *Transactions of the Geological Society of London* 1:390–396.

———. 1824 (1822). "On the Excavation of Valleys by Diluvian Action, as Illustrated by a Succession of Valleys Which Intersect the South Coast of Dorset and Devon." *Transactions of the Geological Society of London,* 2nd ser., 1:95–102.

———. 1830. "Extract of a Letter from Dr. Buckland to Prof. Hitchcock of Amherst, Dated February 1, 1830." *American Journal of Science* 18:393–394.

———. 1835 (1829). "On the Discovery of Coprolites, or Fossil Faeces, in the Lias at Lyme Regis, and in Other Formations." *Transactions of the Geological Society of London* 3:223–236.

———. 1836. *Geology and Mineralogy Considered with Reference to Natural Theology.* 2 vols. The Bridgewater Treatises VI, On the Power, Wisdom, and Goodness of God, as Manifested in the Creation. London: William Pickering.

———. 1837. "On the Adaptation of the Structure of the Sloths to Their Peculiar Mode of Life." *Transactions of the Linnean Society* 17:17–27.

———. 1837 (1836). *Geology and Mineralogy Considered with Reference to Natural Theology.* 2nd ed. 2 vols. The Bridgewater Treatises VI, On the Power, Wisdom, and Goodness of God, as Manifested in the Creation. London: William Pickering.

———. 1840. *Address Delivered at the Anniversary Meeting of the Geological Society of London, on the 21st of February, 1840; and the Announcement of the Award of the Wollaston Medal and Donation Fund for the Same Year.* London: Richard and John E. Taylor.

———. 1842a. "On the Evidences of Glaciers in Scotland and the North of England, First Part." *Proceedings of the Geological Society of London* 3:332–337.

———. 1842b. "On the Evidences of Glaciers in Scotland and the North of England, Second Part." *Proceedings of the Geological Society of London* 3:345–348.

———. 1845 (1844). "Report of Dr. Buckland's Observations, at the Canterbury Meeting of the British Archaeological Association." *Report of the Proceedings of the British Archaeological Association:* 106–113.

Bull, Peter. 2000. "Paviland Cave. A Scanning Electron Microscope Investigation of Quartz-Rich Deposits." In *Paviland Cave and the "Red Lady." A Definitive Report*, edited by Aldhouse-Green, Stephen. Bristol: Western Academic and Specialist Press.

Byron, George Gordon Noel. 1980–1993. *Lord Byron. The Complete Poetical Works.* Edited by McGann, Jerome J. 7 vols. Oxford: Clarendon Press.

Campbell, Bernard. 1956. "The Centenary of Neandertal Man." Pts. 1 and 2. *Man* 56 (November, December): 156–158, 171–173.

Campbell, J. B. 1977. *The Upper Paleolithic of Britain. A Study of Man and Nature in the Late Ice Age.* 2 vols. Vol. 1. Oxford: Clarendon Press.

Cann, Rebecca L. 1997. "Mothers, Labels, and Misogyny." In *Women in Human Evolution*, edited by Hager, Lori D. London: Routledge.

Cann, Rebecca L., Stoneking, Mark, and Wilson, Allan C. 1987. "Mitochondrial DNA and Human Evolution." *Nature* 325:32–36.

Cannon, Susan Faye. 1978. *Science in Culture. The Early Victorian Period.* New York: Science History Publications.

Cannon, Walter F. 1970. "William Buckland." In *Dictionary of Scientific Biography*, edited by Gillispie, Charles Coulston. New York: Scribner's.

Cartmill, Matt. 2001. "Taxonomic Revolutions and the Animal-Human Boundary." In *Studying Human Origins. Disciplinary History and Epistemology*, edited by Corbey, Raymond, and Roebroeks, Wil. Amsterdam: Amsterdam University Press.

Chambers, Robert. 1844. *Vestiges of the Natural History of Creation.* London: Churchill.

Chiarelli, B. 1992. "The Development of Physical Anthropology and Human Evolutionary Studies in Western Europe since World War II and After." *Human Evolution* 7 (1): 55–65.

Churchill, Steven A., Formicola, Vincenzo, Holliday, Trenton W., Holt, Brigitte M., and Schuman, Betsy A. 2000. "The Upper Palaeolithic Population of Europe in an Evolutionary Perspective." In *Hunters of the Golden Age. The Mid Upper Palaeolithic of Eurasia, 30,000–20,000 BP*, edited by Roebroeks, Wil, Mussi, Margherita, Svoboda, Jiri, and Fennema, Kelly. Leiden: University of Leiden Press.

Clark, L. K. 1961. *Pioneers of Prehistory in England.* Edited by Hodgson, P. E. Newman History and Philosophy of Science Series. London: Sheed and Ward.

Clark, Wilfrid E. Le Gros. 1934. *Early Forerunners of Man. A Morphological Study of the Evolutionary Origin of the Primates.* Baltimore: William Wood. New York: AMS Press, 1979 reprint.

———. 1947. "Pan-African Congress on Prehistory. Human Palaeontological Section." *Man* 47:101.

———. 1950. "Hominid Characters of the Australopithecine Dentition." *Journal of the Royal Anthropological Institute of Great Britain and Ireland* 80 (1/2): 37–54.

———. 1955. *The Fossil Evidence for Human Evolution. An Introduction to the Study of Paleoanthropology.* Edited by Bruyn, Peter P. H. de. The Scientist's Library, Biology and Medicine. Chicago: University of Chicago Press.

———. 1967. *Man-Apes or Ape-Men? The Story of Discoveries in Africa.* London: Holt, Rinehart and Winston.

Clark, Wilfrid Edward Le Gros, Weiner, Joseph S., and Oakley, Kenneth Page. 1953. "The Solution of the Piltdown Problem." *Bulletin of the British Museum (Natural History),* Geology Series, 2:141–146.

Clottes, Jean. 2000. "Art between 30,000 and 20,000 BP." In *Hunters of the Golden Age. The Mid Upper Palaeolithic of Eurasia, 30,000–20,000 BP,* edited by Roebroeks, Wil, Mussi, Margherita, Svoboda, Jiri, and Fennema, Kelly. Leiden: University of Leiden Press.

Clottes, Jean, and Lewis-Williams, David. 1996. *Les chamanes de la préhistoire. Transe et magie dans les grottes ornées.* Paris: Seuil.

Coe, Sophie D., and Coe, Michael. 1996. *The True History of Chocolate.* New York: Thames and Hudson.

Cohen, Claudine. 1999. *L'Homme des origines. Savoirs et fictions en préhistoire.* Paris: Seuil.

———. 2001. "Searching for the 'Missing Link' at the Turn of the 19th Century. Mortillet's 'Anthropopithèque' and Bourgeois' 'Tertiary Eoliths.'" Paper read at the Summer Academy of the Max Planck Institute for the History of Science on Human Origins, Berlin, 20–24 August.

———. 2003. *La femme des origines. Images de la femme dans la préhistoire occidentale.* Paris: Belin-Herscher.

Cohen, Claudine, and Hublin, Jean-Jacques. 1989. *Boucher de Perthes (1788–1868). Les origines romantiques de la préhistoire.* Un Savant, une Époque Series. Paris: Belin.

Collis, John. 2003. *The Celts. Origins, Myths and Inventions.* Stroud: Tempus.

Conkey, Margaret W. 1997a. "Beyond Art and between the Caves. Thinking about Context in the Interpretive Process." In *Beyond Art. Pleistocene Image and Symbol,* edited by Conkey, Margaret W., Soffer, Olga, Stratmann, Deborah, and Jablonski, Nina G. San Francisco: California Academy of Sciences.

———. 1997b. "Mobilizing Ideologies. Palaeolithic 'Art,' Gender Trouble, and Thinking about Alternatives." In *Women in Human Evolution,* edited by Hager, Lori D. London: Routledge.

Conybeare, William D. 1824. "On the Discovery of an Almost Perfect Skeleton of a Plesiosaurus." *Transactions of the Geological Society of London,* 2nd ser., 1:380–387.

Coon, Carleton S. 1962. *The Origin of Races.* Borzoi Books. New York: Alfred A. Knopf.

Cope, Edward D. 1893. "The Genealogy of Man." *American Naturalist* 27:316–335.

Copleston, Edward. 1823. "*Reliquiae Diluvianae; or, Observations on the Organic Remains Contained in Caves, Fissures, and Diluvial Gravel, and on Other Geological Phenomena, Attesting the Action of an Universal Deluge.* By the Rev. William Buckland, F.R.S. &c. London. 1823." *Quarterly Review* 23: 138–165.

Corbey, Raymond. 2005. *The Metaphysics of Apes. Negotiating the Animal-Human Boundary.* Cambridge: Cambridge University Press.

Corbey, Raymond, and Roebroeks, Wil. 2001. "Does Disciplinary History Matter? An Introduction." In *Studying Human Origins. Disciplinary History and Epistemology,* edited by Corbey, Raymond, and Roebroeks, Wil. Amsterdam: Amsterdam University Press.

Corsi, Pietro. 1988 (1983). *The Age of Lamarck. Evolutionary Theories in France, 1790–1830*. Translated by Mandelbaum, Jonathan. 2nd ed. Berkeley: University of California Press.

Creese, Mary R. S., and Creese, Thomas M. 1994. "British Women Who Contributed to Research in the Geological Sciences in the Nineteenth Century." *British Journal for the History of Science* 27:23–54.

Cros, Hilary du, and Smith, Laurajane (eds.). 1993. *Women in Archaeology*. Occasional Papers in Prehistory 23. Canberra: Department of Prehistory, Research School of Pacific Studies, Australian National University.

Cunningham, Andrew, and Jardine, Nicholas (eds.). 1990. *Romanticism and the Sciences*. Cambridge: Cambridge University Press.

Cuvier, Georges. 1809. "Sur quelques quadrupèdes ovipares fossiles conservés dans des schistes calcaires. Article II. Sur le prétendu Homme Fossile des carrières d'Oeningen, décrit par Scheuchzer, que d'autres naturalistes ont regardé comme un Silure, et qui n'est qu'une Salamandre, ou plutôt un Protée, de taille gigantesque et d'espèce inconnue." *Annales du Muséum d'Histoire Naturelle* 13:411–420.

———. 1812. *Recherches sur les ossemens fossiles de quadrupèdes, où l'on rétablit les caractères de plusieurs espèces d'animaux que les révolutions du globe paroissent avoir détruites*. 4 vols. Paris: Deterville.

———. 1813. *Essay on the Theory of the Earth, with Mineralogical Notes by Professor Jameson*. Translated by Kerr, Robert. Edinburgh: William Blackwood.

———. 1834. "Discours sur les révolutions de la surface du globe, et sur les changemens qu'elles ont produit dans le règne animal." In *Recherches sur les ossemens fossiles, où l'on rétablit les caractères de plusieurs animaux dont les révolutions du globe ont détruit les espèces*. Vol. 1. 4th ed. Paris: Dufour and d'Ocagne.

Cuvier, Georges, and Brongniart, Alexandre. 1808. "Essai sur la géographie minéralogique des environs de Paris." *Annales du Muséum d'Histoire Naturelle* 11:293–326.

D'Achille, Gino. 2005. Interview, 10 October.

Daniel, Glyn E. 1950. *A Hundred Years of Archaeology*. London: Duckworth.

Daniel, Glyn E., and Renfrew, Colin. 1988 (1962). *The Idea of Prehistory*. 2nd ed. Edinburgh: Edinburgh University Press.

Dart, Raymond A. 1956. "The Myth of the Bone-Accumulating Hyena." *American Anthropologist* 58 (1): 40–62.

———. 1974. "Sir Grafton Elliot Smith and the Evolution of Man." In *Grafton Elliot Smith. The Man and His Work*, edited by Elkin, A. P., and Macintosh, N. W. G. Sydney: Sydney University Press.

Darwin, Charles. 1859. *On the Origin of Species by Means of Natural Selection; or, The Preservation of Favoured Races in the Struggle for Life*. London: Murray.

———. 1871. *The Descent of Man, and Selection in Relation to Sex*. London: Murray.

———. 1874 (1871). *The Descent of Man, and Selection in Relation to Sex*. 2nd ed. London: John Murray, 1906 reprint.

Darwin, John. 1993. "The Growth of an International University." In *The Illustrated History of Oxford University*, edited by Prest, John. Oxford: Oxford University Press.

Daston, Lorraine. 1998. "Die Kultur der wissenschaftlichen Objektivität." In *Naturwissenschaft, Geisteswissenschaft, Kulturwissenschaft. Einheit, Gegensatz, Komplementarität?* edited by Oexle, Otto Gerhard. Göttingen: Wallstein.

———. 2001a. *Eine kurze Geschichte der wissenschaftlichen Aufmerksamkeit.* Themen 71. Munich: Carl Friedrich von Siemens Stiftung.

———. 2001b. *Wunder, Beweise und Tatsachen. Zur Geschichte der Rationalität.* Frankfurt am Main: Fischer.

———. 2004a. "Speechless." In *Things That Talk. Object Lessons from Art and Science,* edited by Daston, Lorraine. New York: Zone Books.

——— (ed.). 2000. *Biographies of Scientific Objects.* Chicago: University of Chicago Press.

——— (ed.). 2004b. *Things That Talk. Object Lessons from Art and Science.* New York: Zone Books.

Daston, Lorraine, and Galison, Peter. 1992. "The Image of Objectivity." *Representations* 40:81–128.

———. 2002. "Das Bild der Objektivität." In *Ordnungen der Sichtbarkeit. Fotografie in Wissenschaft, Kunst und Technologie,* edited by Geimer, Peter. Frankfurt am Main: Suhrkamp.

———. 2007. *Objectivity.* New York: Zone Books.

Daston, Lorraine, and Vidal, Fernando (eds.). 2004. *The Moral Authority of Nature.* Chicago: University of Chicago Press.

Daubeny, C. G. B. (ed.). 1869. *Fugitive Poems Connected with Natural History and Physical Science.* Oxford: James Parker.

Debenham, Nick. 2000. "Thermoluminescence Dating of Stalagmite and Sediment from Hound's Hole and Goat's Hole Caves, Paviland." In *Paviland Cave and the "Red Lady." A Definitive Report,* edited by Aldhouse-Green, Stephen. Bristol: Western Academic and Specialist Press.

Delair, Justin B., and Sarjeant, William A. S. 1975. "The Earliest Discoveries of Dinosaurs." *Isis* 66:5–25.

Delisle, Richard G. 1995. "Human Palaeontology and the Evolutionary Synthesis during the Decade 1950–1960." In *Ape, Man, Apeman. Changing Views since 1600. Evaluative Proceedings of the Symposium Ape, Man, Apeman. Changing Views since 1600, Leiden, the Netherlands, 28 June–1 July, 1993,* edited by Corbey, Raymond, and Theunissen, Bert. Leiden: Department of Prehistory, Leiden University.

———. 2000. "The Biology/Culture Link in Human Evolution, 1750–1950. The Problem of Integration in Science." *Studies in the History and Philosophy of the Biological and Biomedical Sciences* 31 (4): 531–556.

———. 2001. "Adaptationism versus Cladism in Human Evolution Studies." In *Studying Human Origins. Disciplinary History and Epistemology,* edited by Corbey, Raymond, and Roebroeks, Wil. Amsterdam: Amsterdam University Press.

———. 2007. *Debating Humankind's Place in Nature, 1860–2000. The Nature of Paleoanthropology.* Upper Saddle River, NJ: Pearson.

Dennell, Robin W. 1990. "Progressive Gradualism, Imperialism and Academic Fashion. Lower Palaeolithic Archaeology in the 20th Century." *Antiquity* 64:549–558.

———. 2001. "From Sangiran to Olduvai, 1937–1960. The Quest for 'Centres' of Hominid Origins in Asia and Africa." In *Studying Human Origins. Disciplinary History and Epistemology,* edited by Corbey, Raymond, and Roebroeks, Wil. Amsterdam: Amsterdam University Press.

Desmond, Adrian. 1987. "Artisan Resistance and Evolution in Britain, 1819–1848." *Osiris* 3:77–110.

———. 1989. *The Politics of Evolution. Morphology, Medicine, and Reform in Radical London.* Edited by Hull, David L. Science and Its Conceptual Foundations. Chicago: University of Chicago Press.

———. 1997 (1994). *Huxley. From Devil's Disciple to Evolution's High Priest.* Reading, MA: Addison-Wesley.

Desnoyers, Jules. 1834. "Proofs That the Human Bones and Works of Art Found in Caves in the South of France, Are More Recent Than the Antediluvian Bones in These Caves." *Edinburgh New Philosophical Journal* 16:302–310.

Di Gregorio, Mario A. 2005. *From Here to Eternity. Ernst Haeckel and Scientific Faith.* Religion, Theologie und Naturwissenschaft 3. Göttingen: Vandenhoeck & Ruprecht.

Dobzhansky, Theodosius. 1944. "On Species and Races of Living and Fossil Man." *American Journal of Physical Anthropology* 2:251–265.

———. 1962. *Mankind Evolving. The Evolution of the Human Species.* New Haven: Yale University Press.

Donlon, Denise. 1993. "Imbalance in the Sex Ratio in Collections of Australian Aboriginal Skeletal Remains." In *Women in Archaeology*, edited by Cros, Hilary du, and Smith, Laurajane. Canberra: Department of Prehistory, Research School of Pacific Studies, Australian National University.

Dubois, Eugène. 1894. *Pithecanthropus erectus. Eine menschenähnliche Übergangsform aus Java.* Batavia: Landsdrukkerij.

Eastmead, William. 1824. *Historia Rievallensis. Containing the History of Kirkby Moorside, and an Account of the Most Important Places in Its Vicinity; Together with Brief Notices of the More Remote or Less Important Ones. To Which Is Prefixed a Dissertation on the Animal Remains, and Other Curious Phenomena, in the Recently Discovered Cave at Kirkdale.* London: Baldwin, Cradock, and Joy.

Edmonds, J. M. 1978. "Patronage and Privilege in Education. A Devon Boy Goes to School, 1798." *Report and Transactions of the Devonshire Association for the Advancement of Science* 111:95–111.

———. 1979. "The Founding of the Oxford Readership in Geology, 1818." *Notes and Records of the Royal Society of London* 34:33–51.

———. 1991. "*Vindiciae Geologicae*, Published 1820; The Inaugural Lecture of William Buckland." *Archives of Natural History* 18 (2): 255–268.

Edmonds, J. M., and Douglas, J. A. 1976. "William Buckland, F.R.S. (1784–1856) and an Oxford Geological Lecture, 1823." *Notes and Records of the Royal Society of London* 30 (2): 141–167.

Eiseley, Loren. 1957. "Neanderthal Man and the Dawn of Human Paleontology." *Quarterly Review of Biology* 32 (4): 323–329.

Elkin, A. P., and Macintosh, N. W. G. (eds.). 1974. *Grafton Elliot Smith. The Man and His Work.* Sydney: Sydney University Press.

Elliot Smith, Grafton. 1911. *The Ancient Egyptians and Their Influence upon the Civilisation of Europe.* London: Harper & Brothers.

———. 1914. "Report of a Paper Read before the Royal Society." *Nature* 92 (2313): 729.

———. 1915. *The Migrations of Early Culture.* London: Longmans Green.

———. 1916. "The Cranial Cast of the Piltdown Skull." *Man* 16:131–132.

———. 1924. *Essays on the Evolution of Man.* London: Oxford University Press.

———. 1928. "The Evolution of the Brain." In *Creation by Evolution. A Consensus of Present-Day Knowledge as Set Forth by Leading Authorities in Non-technical Language That All May Understand,* edited by Mason, Frances. New York: Macmillan.

———. 1929. *Human History.* New York: W. W. Norton.

———. 1931. "Early Man. His Origin, Development and Culture." In *Early Man. His Origin, Development and Culture,* edited by Smith, Grafton Elliot, Keith, Arthur, and Parsons, F. G. Freeport, NY: Books for Libraries.

———. 1933. *The Diffusion of Culture.* London: Watts.

———. 1935. "The Place of Thomas Henry Huxley in Anthropology." *Journal of the Royal Anthropological Institute of Great Britain and Ireland* 65:199–204.

Emerson, J. P. 1969. "Negotiating the Serious Import of Humor." *Sociometry* 32 (2): 169–181.

Erickson, Paul A. 1974. "The Origins of Physical Anthropology." PhD diss., Department of Anthropology, University of Connecticut.

Eriksen, Thomas Hylland, and Nielsen, Finn Sivert. 2001. *A History of Anthropology.* London: Pluto.

"Erste Sitzung am 19. September." 1836. *Isis von Oken* 29 (1–12): 708–710.

Evans, John. 1872. *The Ancient Stone Implements, Weapons, and Ornaments of Great Britain.* London: Longmans, Green, Reader, and Dyer.

Falconer, Hugh. 1860a. "On the Ossiferous Caves of the Peninsula of Gower, in Glamorganshire, South Wales." *Proceedings of the Geological Society of London* 16:487–491.

———. 1860b. "On the Ossiferous Grotta Di Maccagnone, near Palermo." *Proceedings of the Geological Society of London* 16:99–106.

Falconer, Hugh, Busk, George, and Carpenter, W. B. 1863. "An Account of the Proceedings of the Late Conference Held in France to Inquire into the Circumstances Attending the Asserted Discovery of a Human Jaw in the Gravel at Moulin-Quignon, near Abbeville; Including the Procès Verbaux of the Sittings of the Conference, with Notes Thereon." *Natural History Review* 3:423–462.

Fausto-Sterling, Anne. 2000. *Sexing the Body. Gender Politics and the Construction of Sexuality.* New York: Basic Books.

Findlay, G. H. 1972. *Dr Robert Broom, F.R.S., Palaeontologist and Physician, 1866–1951. A Biography, Appreciation and Bibliography.* South African Biographical and Historical Studies. Cape Town: A. A. Balkema.

Fleure, H. J. 1920. "Some Early Neanthropic Types in Europe and Their Modern Representatives." *Journal of the Royal Anthropological Institute of Great Britain and Ireland* 50:12–40.

"'Fossil Antediluvian Animals Mingled with Human Remains in the Caves of Bize,' 'Caves Containing Human Remains.'" 1829. *Edinburgh New Philosophical Journal* (April to October): 159–161, 368–369.

Foucault, Michel. 1969 (1961). *Wahnsinn und Gesellschaft. Eine Geschichte des Wahns im Zeitalter der Vernunft.* Translated by Köppen, Ulrich. Suhrkamp Taschenbuch Wissenschaft 39. Frankfurt am Main: Suhrkamp.

Fraipont, Charles. 1936. "Les hommes fossiles d'Engis." *Archives de l'Institut de Paléontologie Humaine* 16:1–52.

Fraipont, Julien. 1893. "The Imaginary Race of Canstadt or Neanderthal." *Science* 22 (568): 346.
Frere, John. 1800 (1797). "Account of Flint Weapons Discovered at Hoxne in Suffolk." *Archaeologia* 13:204–205.
Freud, Sigmund. 1913. *Totem und Tabu. Einige Übereinstimmungen im Seelenleben der Wilden und der Neurotiker.* Frankfurt am Main: Fischer.
Galton, Francis. 1892. *Finger Prints.* Edited by Cummins, Harold. New York: Da Capo, 1965 reprint.
Gaut, B. 1998. "Just Joking. The Ethics and Aesthetics of Humor." *Philosophy and Literature* 22 (1): 51–68.
Geoffroy Saint-Hilaire, Etienne. 1838. "Lettre de M. Geoffroy-Saint-Hilaire sur les ossements humains provenant des cavernes de Liège, et sur les modifications produites dans le pelage des chevaux par un séjour prolongé dans les profondeurs des mers." *Compte rendu des séances de l'Académie des Sciences* 7 (séance du lundi, 2 juillet): 13–15.
Gere, Cathy. 1999. "Bones That Matter. Sex Determination in Paleodemography, 1948–1995." *Studies in History and Philosophy of Science*, pt. C, *Studies in History and Philosophy of Biological and Biomedical Sciences* 30 (4): 455–471.
Gero, Joan, and Conkey, Margaret W. (eds.). 1991. *Engendering Archaeology. Women and Prehistory.* Oxford: Blackwell.
Gifford-Gonzalez, Diane. 1993. "You Can Hide, But You Can't Run. Representation of Women's Work in Illustrations of Palaeolithic Art." *Visual Anthropology Review* 9 (1): 23–41.
Gillispie, Charles Coulston. 1996 (1951). *Genesis and Geology. A Study in the Relations of Scientific Thought, Natural Theology, and Social Opinion in Great Britain, 1790–1850.* 2nd ed. Harvard Social Studies 58. Cambridge, MA: Harvard University Press.
Ginzburg, Carlo. 1983 (1979). "Spurensicherung. Der Jäger entziffert die Fährte, Sherlock Holmes nimmt die Lupe, Freud liest Morelli—Die Wissenschaft auf der Suche nach sich selbst." Translated by Bonz, Gisela. In *Spurensicherungen. Über verborgene Geschichte, Kunst und soziales Gedächtnis.* Berlin: Wagenbach.
Giraldus Cambrensis. 1908 (1188). *The Itinerary through Wales and The Description of Wales.* Translated by Hoare, Richard Colt (1806), edited by Williams, W. Llewelyn. Edited by Rhys, Ernest. Everyman's Library. London: J. M. Dent.
———. 1978 (1188). *The Journey through Wales and The Description of Wales.* Translated and edited by Thorpe, Lewis. Edited by Radice, Betty. Penguin Classics. Harmondsworth: Penguin.
Godwin, Harry. 1970. "The Contributions of Radiocarbon Dating to Archeology in Britain." *Philosophical Transactions of the Royal Society of London;* ser. A, *Mathematical and Physical Sciences* 269 (1193): 57–75.
Godwin-Austen, Robert Alfred Cloyne. 1840. "On the Bone Caves of Devonshire." *Proceedings of the Geological Society of London* 3:286–287.
———. 1842. "On the Geology of the South-east of Devonshire." *Transactions of the Geological Society of London* 6 (2): 433–489.
Goldenweiser, A. A. 1916. "Diffusion vs. Independent Origin. A Rejoinder to Professor G. Elliot Smith." *Science*, n.s., 44 (1137): 531–533.
Goodhue, Thomas W. 2002. *Curious Bones. Mary Anning and the Birth of Paleontology.* Greensboro, NC: Morgan Reynolds.

Goodman, Morris, and Cronin, J. E. 1982. "Molecular Anthropology. Its Development and Current Directions." In *A History of American Physical Anthropology, 1930–1980*, edited by Spencer, Frank. New York: Academic Press.

Goodrum, Matthew R. 2000. "The History of Twentieth Century Paleoanthropology. A Bibliographic Survey." *History of Anthropology Newsletter* 27 (2): 10–15.

———. 2002. "The Meaning of Ceraunia. Archaeology, Natural History and the Interpretation of Prehistoric Stone Artefacts in the Eighteenth Century." *British Journal for the History of Science* 35 (126): 255–269.

———. 2004. "Prolegomenon to a History of Paleoanthropology. The Study of Human Origins as a Scientific Enterprise. Part 1. Antiquity to the Eighteenth Century" and "Prolegomenon to a History of Paleoanthropology. The Study of Human Origins as a Scientific Enterprise. Part 2. Eighteenth to Twentieth Century." *Evolutionary Anthropology* 13:172–180 and 224–233.

Gordon, Elizabeth Oke. 1894. *The Life and Correspondence of William Buckland, D.D., F.R.S., Sometime Dean of Westminster, Twice President of the Geological Society, and First President of the British Association*. London: Murray.

Gould, Stephen Jay. 1980. "G. G. Simpson, Paleontology, and the Modern Synthesis." In *The Evolutionary Synthesis. Perspectives on the Unification of Biology*, edited by Mayr, Ernst, and Provine, William B. Cambridge, MA: Harvard University Press.

———. 1981a. "Piltdown in Letters. The Role of the Famous Priest in the Infamous Conspiracy Is Here Disputed in Epistolary Fashion." *Natural History* 90:12–30.

———. 1981b. *The Mismeasure of Man*. New York: W. W. Norton.

———. 1983. "Unconnected Truths." *Natural History* 83 (3): 22–28.

———. 1989. *Wonderful Life. The Burgess Shale and the Nature of History*. New York: W. W. Norton.

———. 1996 (1981). *The Mismeasure of Man*. 2nd ed. London: Penguin.

Grant, R. 1979. "Hutton's Theory of the Earth." In *Images of the Earth. Essays in the History of the Environmental Sciences*, edited by Jordanova, Ludmilla J., and Porter, Roy S. Aberdeen: Rainbow.

Gratzer, Walter. 2002. *Eurekas and Euphorias. The Oxford Book of Scientific Anecdotes*. Oxford: Oxford University Press.

Grayson, Donald K. 1983. *The Establishment of Human Antiquity*. New York: Academic Press.

Green, Stephen. 1984. "Excavations at Little Hole (Longbury Bank), Wales, in 1984." In *Studies in the Upper Palaeolithic of Britain and North Western Europe*, edited by Roe, D. A. Oxford: British Archaeological Reports.

———. 1989. "The Stone Age Cave Archaeology of South Wales." In *Limestones and Caves of Wales*, edited by Ford, T. D. Cambridge: Cambridge University Press.

Green, Stephen, Bevins, R. E., Bull, P. A., Currant, A. P., Debenham, N. C., Embleton, C., Ivanovich, M., Livingston, L., Rae, A. M., Schwarcz, H. P., and Stringer, C. B. 1989. "Le site Acheuléen de la grotte de Pontnewydd, pays de Galles. Géomorphologie, stratigraphie, chronologie, faune, hominidés fossiles, géologie et industrie lithique dans le contexte paléoécologique." *L'Anthropologie* 93 (1): 15–52.

Green, Vivian. 1993. "The University and the Nation." In *The Illustrated History of Oxford University*, edited by Prest, John. Oxford: Oxford University Press.

Gregory, William King. 1920. "The Origin and Evolution of the Human Dentition. A Palaeontological Review (Part 4)." *Journal of Dental Research* 4:607–689.

———. 1927. "Did Man Originate in Central Asia? (Mongolia the New World, Part V)." *Scientific Monthly* 24 (5): 385–401.

———. 1928. "The Lineage of Man." In *Creation by Evolution. A Consensus of Present-Day Knowledge as Set Forth by Leading Authorities in Non-technical Language That All May Understand*, edited by Mason, Frances. New York: Macmillan.

Gruber, Jacob W. 1965. "Brixham Cave and the Antiquity of Man." In *Context and Meaning in Cultural Anthropology*, edited by Spiro, Melford E. New York: Free Press.

Gundling, Tom. 2005. *First in Line. Tracing Our Ape Ancestry*. New Haven: Yale University Press.

Haeckel, Ernst. 1898 (1868). *Natürliche Schöpfungs-Geschichte. Gemeinverständliche wissenschaftliche Vorträge über die Entwicklungs-Lehre im Allgemeinen und diejenige von Darwin, Goethe und Lamarck im Besonderen*. 9th ed. Berlin: Georg Reimer.

Hager, Lori D. 1997a. "Sex and Gender in Paleoanthropology." In *Women in Human Evolution*, edited by Hager, Lori D. London: Routledge.

——— (ed.). 1997b. *Women in Human Evolution*. London: Routledge.

Hagner, Michael. 1997. *Homo cerebralis. Der Wandel vom Seelenorgan zum Gehirn*. Berlin: Berlin.

———. 2004. *Geniale Gehirne. Zur Geschichte der Elitegehirnforschung*. Edited by Hagner, Michael, and Rheinberger, Hans-Jörg. Wissenschaftsgeschichte. Göttingen: Wallstein.

———. 2005. "Anthropologische Objekte. Die Wissenschaft vom Menschen im Museum." In *Dingwelten. Das Museum als Erkennungsort*, edited by Heesen, Anke te, and Lutz, Petra. Cologne: Böhlau.

Halstead, L. B. 1979. "The Piltdown Hoax. *Cui bono?*" *Nature* 277:596.

Hammond, Michael. 1979. "A Framework of Plausibility for an Anthropological Forgery. The Piltdown Case." *Anthropology* 3:47–58.

———. 1980. "Anthropology as a Weapon of Social Combat in Late-Nineteenth-Century France." *Journal of the History of the Behavioral Sciences* 16:118–132.

———. 1982. "The Expulsion of the Neanderthals from Human Ancestry. Marcellin Boule and the Social Context of Scientific Research." *Social Studies of Science* 12 (1): 1–36.

———. 1988. "The Shadow Man Paradigm in Paleoanthropology: 1911–1945." In *Bones, Bodies, Behavior. Essays on Biological Anthropology*, edited by Stocking, George W. Madison: University of Wisconsin Press.

Haraway, Donna J. 1997. *Modest_Witness@Second_Millennium.FemaleMan©_Meets_OncoMouse™. Feminism and Technoscience*. New York: Routledge.

Harding, Sandra. 1991. "Feminist Standpoint Epistemology." In *Whose Science? Whose Knowledge? Thinking from Women's Lives*. Ithaca, NY: Cornell University Press.

Harris, D. R. 1987. "The Impact on Archaeology of Radiocarbon Dating by Accelerator Mass Spectrometry." *Philosophical Transactions of the Royal Society of London*, ser. A, *Mathematical and Physical Sciences* 323 (1569): 23–43.

Haycock, David Boyd. 2002. *William Stukeley. Science, Religion and Archaeology in Eighteenth-Century England*. Woodbridge: Boydell.

Hazzledine Warren, S. 1912. "Palaeolithic Man and His Art [Review of *Ancient Hunters*]." *Man* 12:203–206.

———. 1916. "Archaeology. Palaeolithic Art [Review of *Ancient Hunters*]." *Man* 16:79–80.

———. 1923. "The Eolithic Problem. A Reply." *Man* 23:82–83.

Heesen, Anke te, and Lutz, Petra (eds.). 2005. *Dingwelten. Das Museum als Erkennungsort*. Edited by Staupe, Gisela. Schriften des Deutschen Hygiene-Museums Dresden 4. Cologne: Böhlau.

Heller, Florian. 1972. "Die Forschungen in der Zoolithenhöhle bei Burggaillenreuth von Esper bis zur Gegenwart." In *Die Zoolithenhöhle bei Burggaillenreuth/Ofr. 200 Jahre wissenschaftliche Forschung, 1771–1971*, edited by Heller, Florian. Erlangen: Universitätsbund Erlangen-Nürnberg.

Helvenston, P. A., and Bahn, P. G. 2002. *Desperately Seeking Trance Plants. Testing the "Three-Stages of Trance" Model*. New York: RJ Communications.

Heringman, Noah. 2004. *Romantic Rocks, Aesthetic Geology*. Ithaca, NY: Cornell University Press.

Hoare, Richard Colt. 1812. *The Ancient History of South Wiltshire*. 2 vols. Vol. 1. London.

———. 1819. *The Ancient History of North Wiltshire*. 2 vols. Vol. 2. London.

———. 1829. *Tumuli Wiltunenses; Guide to the Barrows on the Plains of Stonehenge*. Shaftsbury: J. Rutter.

———. 1983 (1810). *The Journeys of Sir Richard Colt Hoare through Wales and England, 1793–1810. Extracted from the Journals and Edited with an Introduction by M. W. Thompson*. Gloucester: Alan Sutton.

Holliday, Trenton W. 2000. "Body Proportions of the Paviland 1 Skeleton." In *Paviland Cave and the "Red Lady." A Definitive Report*, edited by Aldhouse-Green, Stephen. Bristol: Western Academic and Specialist Press.

———. 2004. Questionnaire, 28 August.

Houghton Brodrick, Alan. 1963. *Father of Prehistory: The Abbé Henri Breuil. His Life and Times*. New York: William Morrow.

Howarth, Janet. 1994. "Women." In *The History of the University of Oxford*, vol. 8, *The Twentieth Century*, edited by Harrison, Brian. Oxford: Clarendon Press.

———. 2000. "'Oxford for Arts.' The Natural Sciences, 1880–1914." In *The History of the University of Oxford*, vol. 7, *Nineteenth-Century Oxford*, pt. 2, edited by Brock, M. G., and Curthoys, M. C. Oxford: Clarendon Press.

Howells, William White. 1976. "Explaining Modern Man. Evolutionists versus Migrationists." *Journal of Human Evolution* 5:477–496.

Howes, C. J. 1988. "The Dillwyn Diaries 1817–1852, Buckland, and Caves of Gower (South Wales)." *Proceedings of the University of Bristol Spelaeological Society* 18 (2): 298–305.

Howorth, Henry H. 1902. "The Origin and Progress of the Modern Theory of the Antiquity of Man." *Geological Magazine* 9 (1): 16–27.

Hoyme, Lucile E. 1953. "Physical Anthropology and Its Instruments. An Historical Study." *Southwestern Journal of Anthropology* 9:408–430.

Hrdlicka, Ales. 1927. "The Neanderthal Phase of Man." *Journal of the Royal Anthropological Institute of Great Britain and Ireland* 57:249–274.

Huxley, Thomas Henry. 1893. *Science and Education*. Collected Essays by T. H. Huxley 3. London: Macmillan.

———. 1894. *Man's Place in Nature and Other Anthropological Essays.* Collected Essays by T. H. Huxley 7. London: Macmillan.

———. 1915 (1880). "On the Method of Zadig. Retrospective Prophecy as a Function of Science." In *Science and Hebrew Tradition.* Collected Essays 4. New York: D. Appleton.

Irving, A. 1914. "Some Recent Work on Later Quaternary Geology and Anthropology, with Its Bearing on the Question of 'Pre-Boulder-Clay Man.'" *Journal of the Royal Anthropological Institute of Great Britain and Ireland* 44:385–393.

Ivanovich, Miroslav, and Latham, Alf G. 2000. "Interpretation of the U-Series Data and Age Estimates." In *Paviland Cave and the "Red Lady." A Definitive Report,* edited by Aldhouse-Green, Stephen. Bristol: Western Academic and Specialist Press.

Jackson, Wilfrid J. 1953. "Archaeology and Palaeontology." In *British Caving. An Introduction to Speleology,* edited by Cullingford, C. H. D. London: Routledge and Kegan Paul.

Jacobi, R. M. 1980. "The Upper Palaeolithic of Britain with Special Reference to Wales." In *Culture and Environment in Prehistoric Wales. Selected Essays,* edited by Taylor, J. A. London: Oxford University Press.

James, Simon. 1997. "Drawing Inferences. Visual Reconstructions in Theory and Practice." In *The Cultural Life of Images. Visual Representation in Archaeology,* edited by Molyneaux, Brian Leigh. London: Routledge.

———. 1999. *The Atlantic Celts. Ancient People or Modern Invention?* Madison: University of Wisconsin Press.

Johanson, Donald C., and Edey, Maitland. 1981. *Lucy. The Beginnings of Humankind.* New York: Simon and Schuster.

Johnson, L. Lewis. 1978. "A History of Flint-Knapping Experimentation, 1838–1976." *Current Anthropology* 19 (2): 337–372.

Jones, Rhys. 2000. "Gun-gugaliya rrawa. Place, Ochre and Death—A Perspective from Aboriginal Australia." In *Paviland Cave and the "Red Lady." A Definitive Report,* edited by Aldhouse-Green, Stephen. Bristol: Western Academic and Specialist Press.

Kästner, Sibylle. 1995. "Die Kategorie 'Geschlecht' in der deutschen Urgeschichtsforschung. Eine Untersuchung anhand ausgewählter Publikationen zum Früh- und Mittelneolithikum." MPhil thesis, Department of Vor- und Frühgeschichte (Prehistory and Early History), Eberhard-Karls-Universität, Tübingen.

Keats, John. 1900–1901. *The Complete Works of John Keats in Five Volumes.* Edited by Forman, H. Buxton. Glasgow: Gowans and Gray.

Keith, Arthur. 1911. *Ancient Types of Man.* Harper's Library of Living Thought. London: Harper.

———. 1912. "Certain Phases in the Evolution of Man (Hunterian Lectures Abstracts)." *Lancet* 1 (23, 30 March, 6 April): 734–736, 775–777, 788–790.

———. 1915a. *The Antiquity of Man.* London: Williams and Norgate.

———. 1915b. "War as a Factor in Racial Evolution." *St. Thomas's Hospital Gazette* 25:153–162.

———. 1916. "Presidential Address. On Certain Factors Concerned with the Evolution of Human Races." *Journal of the Royal Anthropological Institute of Great Britain and Ireland* 46:10–34.

———. 1917. "Presidential Address. How Can the Institute Best Serve the Needs of Anthropology?" *Journal of the Royal Anthropological Institute of Great Britain and Ireland* 47:12–30.

———. 1919a. "Anthropometry and National Health." *Journal of State Medicine* 27 (2): 33–42.

———. 1919b. *Nationality and Race from an Anthropologist's Point of View*. Oxford: Oxford University Press.

———. 1919c. "The Differentiation of Mankind into Racial Types." *Nature* 104 (2611): 301–305.

———. 1924. "Archaeology [Review of *Ancient Hunters*]." *Man* 24:154–158.

———. 1925. "Was the Chancelade Man Akin to the Eskimo?" *Man* 25:186–189.

———. 1925 (1915). *The Antiquity of Man*. 2nd ed. 2 vols. London: Williams and Norgate.

———. 1927. "Darwin's Theory of Man's Descent as It Stands To-day." *Science*, n.s., 66 (1705): 201–208.

———. 1931a. *Ethnos—The Problem of Race, Today and Tomorrow*. London: Kegan Paul.

———. 1931b. *New Discoveries Relating to the Antiquity of Man*. London: Williams and Norgate.

———. 1931c. "The Evolution of Human Races, Past and Present." In *Early Man. His Origin, Development and Culture*, edited by Smith, Grafton Elliot, Keith, Arthur, and Parsons, F. G. Freeport, NY: Books for Libraries.

———. 1934. *The Construction of Man's Family Tree*. The Forum Series 18. London: Watts.

———. 1936. "History from Caves. A New Theory of the Origin of the Modern Races of Mankind. Presidential Address to the First Speleological Conference." *British Speleological Association*, 25 July.

———. 1946. *Essays on Human Evolution*. London: Scientific Book Club.

———. 1948. *A New Theory of Human Evolution*. London: Watts.

———. 1950. *An Autobiography*. London: Watts.

Keller, Evelyn Fox. 2000. *The Century of the Gene*. Cambridge, MA: Harvard University Press.

Kendrick, T. D. 1950. *British Antiquity*. London: Methuen.

Keuren, David Keith van. 1982. "Human Science in Victorian Britain. Anthropology in Institutional and Disciplinary Formation, 1863–1908." PhD diss., Department of History and Sociology of Science, University of Pennsylvania.

King, William. 1864. "The Reputed Fossil Man of Neanderthal." *Quarterly Journal of Science* 1:88–97.

Klaatsch, Hermann. 1910. "Die Aurignac-Rasse und ihre Stellung im Stammbaum der Menschheit." *Zeitschrift für Ethnologie* 42:513–577.

———. 1920. *Der Werdegang der Menschheit und die Entstehung der Kultur*. Berlin: Bong.

Knell, Simon J. 2000. *The Culture of English Geology, 1815–1851. A Science Revealed through Its Collecting*. Aldershot: Ashgate.

Knight, David. 1990. "Romanticism and the Sciences." In *Romanticism and the Sciences*, edited by Cunningham, Andrew, and Jardine, Nicholas. Cambridge: Cambridge University Press.

———. 1998 (1967). "The Scientist as Sage." In *Science in the Romantic Era*. Aldershot: Ashgate.
———. 1998 (1996). "From Science to Wisdom. Humphry Davy's Life." In *Science in the Romantic Era*. Aldershot: Ashgate.
Kölbl-Ebert, Martina. 1997a. "Charlotte Murchison (Née Hugonin), 1788–1869." *Earth Sciences History* 16 (1): 39–43.
———. 1997b. "Mary Buckland (Née Morland), 1797–1857." *Earth Sciences History* 16 (1): 33–38.
———. 2001. "On the Origin of Women Geologists by Means of Social Selection. German and British Comparison." *Episodes* 24 (3): 182–193.
König, Charles. 1814a. "On a Fossil Human Skeleton from Guadaloupe." *Abstracts of the Papers Printed in the Philosophical Transactions of the Royal Society of London* 1:487–489.
———. 1814b. "On a Fossil Human Skeleton from Guadaloupe." *Philosophical Transactions of the Royal Society of London* 104:107–120.
Krull, Wilhelm. 2000. "Beyond the Ivory Tower. Some Observations on External Funding of Interdisciplinary Research in Universities." In *Practising Interdisciplinarity*, edited by Weingart, Peter, and Stehr, Nico. Toronto: University of Toronto Press.
Kuklick, Henrika. 1991. *The Savage Within. The Social History of British Anthropology, 1885–1945*. Cambridge: Cambridge University Press.
———. 2006. "'Humanity in the Chrysalis Stage.' Indigenous Australians in the Anthropological Imagination, 1899–1926." *British Journal for the History of Science* 39 (4): 535–568.
Laguna, Frederica de. 1932. "A Comparison of Eskimo and Palaeolithic Art." *American Journal of Archaeology* 36 (4): 477–511.
Lamarck, Jean-Baptiste de. 2002 (1809). *Zoologische Philosophie*. Translated by Lang, Arnold. Ostwalds Klassiker der exakten Wissenschaften 277 (Reprint der Bände 277, 278, 279). Frankfurt am Main: Harri Deutsch.
Landau, Misia. 1984. "Human Evolution as Narrative. Have Hero Myths and Folktales Influenced Our Interpretations of the Evolutionary Past?" *American Scientist* 72:262–268.
———. 1991. *Narratives of Human Evolution*. New Haven: Yale University Press.
Lankester, Ray. 1907. *The Kingdom of Man*. London: Archibald Constable.
———. 1912a. "On the Discovery of a Novel Type of Flint Implements below the Base of the Red Crag of Suffolk, Proving the Existence of Skilled Workers of Flint in the Pliocene Age." *Philosophical Transactions of the Royal Society of London*, ser. B, *Containing Papers of a Biological Character*, 202:283–336.
———. 1912b. "Past and Present Man." *Saturday Review* 113 (2942): 332–334.
———. 1915. *Diversions of a Naturalist*. London: Methuen. Freeport, NY: Books for Libraries, 1970 reprint.
———. 1920 (1919). "On Some Rostro-Carinate Flint Implements and Allied Forms." *Proceedings of the Royal Society of London*, ser. B, *Containing Papers of a Biological Character*, 91 (640): 330–351.
———. 1921. "A Remarkable Flint Implement from Piltdown." *Man* 21:59–62.
———. 1934. *Fireside Science*. The Thinker's Library 41. London: Watts.

Lartet, Édouard. 1860. "On the Coexistence of Man with Certain Extinct Quadrupeds, Proved by Fossil Bones, from Various Pleistocene Deposits, Bearing Incisions Made by Sharp Instruments." *Proceedings of the Geographical Society of London* 16:471–479.

———. 1861. "Nouvelles recherches sur la coexistence de l'homme et des grands mammifières fossiles réputés caractéristiques de la dernière époque géologique." *Annales des Sciences Naturelles*, 4th ser., 15:177–253.

Lartet, Édouard, and Christy, Henry. 1864. *Cavernes du Périgord. Objets gravés et sculptés des temps pré-historiques dans l'Europe Occidentale, Extrait de la Revue Archéologique*. Paris: Librairie académique, Didie et Ce.

———. 1875 (1865–1875). *Reliquiae Aquitanicae; Being Contributions to the Archaeology and Palaeontology of Périgord and the Adjoining Provinces of Southern France*, edited by Jones, Thomas Rupert. London: Williams and Norgate.

Latour, Bruno. 1993. "Do Scientific Objects Have a History? Pasteur and Whitehead in a Bath of Lactic Acid." *Common Knowledge* 2 (1): 76–91.

———. 1999. *Pandora's Hope. Essays on the Reality of Science Studies*. Cambridge, MA: Harvard University Press.

———. 2000. "On the Partial Existence of Existing *and* Nonexisting Objects." In *Biographies of Scientific Objects*, edited by Daston, Lorraine. Chicago: University of Chicago Press.

Latour, Bruno, and Strum, Shirley C. 1986. "Human Social Origins. Oh Please, Tell Us Another Story." *Journal of Social and Biological Structures: Studies in Human Biology* 9:169–187.

Lawrence, Christopher. 1990. "The Power and the Glory. Humphry Davy and Romanticism." In *Romanticism and the Sciences*, edited by Cunningham, Andrew, and Jardine, Nicholas. Cambridge: Cambridge University Press.

Lefèvre, Wolfgang. 1984. *Die Entstehung der biologischen Evolutionstheorie*. Frankfurt am Main: Ullstein.

Leguebe, André. 1976. "Résistance et ouverture à l'idée d'homme fossile avant 1860." *Recueil de Travaux d'Histoire et de Philologie*, 6th ser., 9:23–43.

Lepenies, Wolf. 1976. *Das Ende der Naturgeschichte. Wandel kultureller Selbstverständlichkeiten in den Wissenschaften des 18. und 19. Jahrhunderts*. Edited by Lepenies, Wolf, and Ritter, Henning. Hanser Anthropologie. Munich: Hanser.

Lester, Joe. 1995. *E. Ray Lankester and the Making of Modern British Biology*. Edited by Bowler, Peter J. BSHS Monographs 10, edited by P. J. Weindling. Oxford: British Society for the History of Science.

Lewin, Roger. 1997 (1987). *Bones of Contention. Controversies in the Search for Human Origins*. 2nd ed. Chicago: University of Chicago Press.

Lewis-Williams, David. 1997. "Harnessing the Brain. Vision and Shamanism in Upper Paleolithic Western Europe." In *Beyond Art. Pleistocene Image and Symbol*, edited by Conkey, Margaret W., Soffer, Olga, Stratmann, Deborah, and Jablonski, Nina G. San Francisco: California Academy of Sciences.

———. 2002. *The Mind in the Cave. Consciousness and the Origins of Art*. London: Thames and Hudson.

Lohrke, Brigitte. 2004. "'Ueberhaupt haben sie etwas weibliches, was sich schwer beschreiben lässt'—Zur Forschungsgeschichte der prähistorisch-anthropologischen Geschlechtsbestimmung." In *Gender Studies. Wissenschafts-*

theorien und Gesellschaftskritik, edited by Steffen, Therese Frey, Rosenthal, Caroline, and Väth, Anke. Würzburg: Königshausen & Neumann.

Longino, Helen E. 1993. "Subjects, Power, and Knowledge. Description and Prescription in Feminist Philosophies of Science." In *Feminist Epistemologies*, edited by Alcoff, Linda, and Potter, Elizabeth. New York: Routledge.

Lovejoy, Arthur O. 1964 (1936). *The Great Chain of Being. A Study of the History of an Idea. The William James Lectures Delivered at Harvard University, 1933.* Cambridge, MA: Harvard University Press.

Lovejoy, Owen C., Mensforth, Robert P., and Armelagos, George J. 1982. "Five Decades of Skeletal Biology as Reflected in the *American Journal of Physical Anthropology*." In *A History of American Physical Anthropology, 1930–1980*, edited by Spencer, Frank. New York: Academic Press.

Lowe, David J. 2000. "The Paviland Caves. Their Geological Setting and Development." In *Paviland Cave and the "Red Lady." A Definitive Report*, edited by Aldhouse-Green, Stephen. Bristol: Western Academic and Specialist Press.

Lowie, Robert H. 1915. "Ancient Hunters and Their Modern Representatives [Review]." *American Anthropologist*, n.s., 17 (3): 575–576.

Lubbock, John. 1865. *Pre-historic Times, as Illustrated by Ancient Remains, and the Manners and Customs of Modern Savages.* London: Williams and Norgate.

———. 1869 (1865). *Pre-historic Times, as Illustrated by Ancient Remains, and the Manners and Customs of Modern Savages.* 2nd ed. London: Williams and Norgate.

Luc, Jean André de. 1778. *Lettres physiques et morales sur l'histoire de la terre et de l'homme. Addressées à la Reine de la Grande Bretagne.* La Haye: De Tune.

Lyell, Charles. 1830–1833. *Principles of Geology, Being an Attempt to Explain the Former Changes of the Earth's Surface, by Reference to Causes Now in Operation.* 3 vols. London: John Murray.

———. 1860 (1859). "On the Occurrence of Works of Human Art in Post-Pliocene Deposits." *Report of the 29th Meeting of the British Academy for the Advancement of Science, Aberdeen 1859, Notices and Abstracts* 29:93–95.

MacCurdy, George Grant. 1915. "*Ancient Hunters and Their Modern Representatives* [Review]." *American Historical Review* 21 (1): 135–137.

———. 1926. "*Ancient Hunters and Their Modern Representatives* [Review]." *Science*, n.s., 63 (1625): 211.

MacDougall, Hugh A. 1982. *Racial Myth in English History. Trojans, Teutons, and Anglo-Saxons.* Montreal: Harvest House.

MacKenzie, John M. 1988. *The Empire of Nature. Hunting, Conservation, and British Imperialism.* Studies in Imperialism. Manchester: Manchester University Press.

Mantell, Gideon Algernon. 1850. "On the Remains of Man, and Works of Art Imbedded in Rocks and Strata, as Illustrative of the Connection between Archaeology and Geology." *Archaeological Journal* 7 (21 June): 327–346.

Matthew, William Diller. 1915 (1911). "Climate and Evolution." *Annals of the New York Academy of Sciences* 24:171–318.

Mayer. 1864. "Zur Frage über das Alter und die Abstammung des Menschengeschlechts." *Archiv für Anatomie und Physiologie etc. (Müllers Archiv Leipzig)*: 696–728.

Mayor, Adrienne. 2000. *The First Fossil Hunters. Paleontology in Greek and Roman Times.* Princeton: Princeton University Press.

Mayr, Ernst. 1950. "Taxonomic Categories in Fossil Hominids." *Cold Spring Harbor Symposia on Quantitative Biology* 15:109–118.

———. 1982 (1942). *Systematics and the Origin of Species*. Edited by Eldredge, Niles, and Gould, Stephen Jay. Columbia Classics in Evolution. New York: Columbia University Press.

Mellars, Paul, and Stringer, Chris (eds.). 1989. *The Human Revolution. Behavioural and Biological Perspectives on the Origins of Modern Humans*. Edinburgh: Edinburgh University Press.

Middleton, J. 1844. "On Fluorine in Bones, Its Source, and Its Application to the Determination of the Geological Age of Fossil Bones." *Proceedings of the Geological Society of London* 1:214–216.

Millhauser, Milton. 1954. "The Scriptural Geologists. An Episode in the History of Opinion." *Osiris* 11:65–86.

Molleson, Theya I. 1976. "Remains of Pleistocene Man in Paviland and Pontnewydd Caves, Wales." *Transactions of the British Cave Research Association* 3 (2): 112–116.

Molleson, Theya I., and Burleigh, R. 1978. "A New Date for the Early Upper Palaeolithic of Goats' Hole Cave, Paviland, Gower Peninsula, Wales." *Antiquity* 52:143–145.

Molyneaux, Brian Leigh. 1997. "The Cultural Life of Images." In *The Cultural Life of Images. Visual Representation in Archaeology*, edited by Molyneaux, Brian Leigh. London: Routledge.

Morgan, Lloyd. 1928. "Mind in Evolution." In *Creation by Evolution. A Consensus of Present-Day Knowledge as Set Forth by Leading Authorities in Non-technical Language That All May Understand*, edited by Mason, Frances. New York: Macmillan.

Morgan, Rhodri. 2000. Foreword to *Paviland Cave and the "Red Lady." A Definitive Report*, edited by Aldhouse-Green, Stephen. Bristol: Western Academic and Specialist Press.

Morrell, Jack B. 1994. "The Non-medical Sciences, 1914–1939." In *The History of the University of Oxford*, vol. 8, *The Twentieth Century*, edited by Harrison, Brian. Oxford: Clarendon Press.

———. 1997. *Science at Oxford, 1914–1939. Transforming an Arts University*. Oxford: Clarendon Press.

Morrell, Jack, and Thackray, Arnold. 1981. *Gentlemen of Science. Early Years of the British Association for the Advancement of Science*. Oxford: Clarendon Press.

Morren, Ch. 1838. "Notice sur la vie et les travaux de Philippe-Charles Schmerling." *Annuaire de l'Académie Royale des Sciences et Belles-lettres de Bruxelles* 4:130–150.

Morse, Michael A. 2005. *How the Celts Came to Britain. Druids, Ancient Skulls and the Birth of Archaeology*. Stroud: Tempus.

Mortillet, Gabriel de. 1867. *Promenades préhistoriques à l'Exposition Universelle*. Paris: C. Reinwald.

———. 1883. *Le préhistorique. Antiquité de l'homme*. Bibliothèque des sciences contemporaines VIII. Paris: C. Reinwald.

Moser, Stephanie. 1992. "The Visual Language of Archaeology. A Case Study of the Neanderthals." *Antiquity* 66:831–844.

———. 1993. "Gender Stereotyping in Pictorial Reconstructions of Human Origins." In *Women in Archaeology. A Feminist Critique*, edited by Cros, Hilary du, and Smith, Laurajane. Canberra: Department of Prehistory, Research School of Pacific Studies, Australian National University.

———. 1998. *Ancestral Images. The Iconography of Human Origins.* Stroud: Sutton.

Moser, Stephanie, and Gamble, Clive. 1997. "Revolutionary Images. The Iconic Vocabulary for Representing Human Antiquity." In *The Cultural Life of Images. Visual Representation in Archaeology*, edited by Molyneaux, Brian Leigh. London: Routledge.

Mourne, Richard, and Case, David. 2000. "Stratigraphy and Sedimentology of Cave Fills." In *Paviland Cave and the "Red Lady." A Definitive Report*, edited by Aldhouse-Green, Stephen. Bristol: Western Academic and Specialist Press.

Murchison, Charles (ed.). 1868. *Palaeontological Memoirs and Notes of the Late Hugh Falconer, A.M., M.D.* Vol. 2, *Mastodon, Elephant, Rhinoceros, Ossiferous Caves, Primival Man and His Contemporaries.* London: Robert Hardwicke.

Nelson, Lynn Hankinson. 1993. "Epistemological Communities." In *Feminist Epistemologies*, edited by Alcoff, Linda, and Potter, Elizabeth. New York: Routledge.

Newton, E. T. 1895. "On a Human Skull and Limb Bones Found in the Paleolithic Terrace Gravel at Galley Hill, Kent." *Quarterly Journal of the Geological Society of London* 51:505–527.

Niedhart, Gottfried. 1987. *Geschichte Englands in drei Bänden.* Vol. 3, *Geschichte Englands im 19. und 20. Jahrhundert.* Munich: C. H. Beck.

North, F. J. 1942. "Paviland Cave, the 'Red Lady,' the Deluge, and William Buckland." *Annals of Science* 5 (2): 91–128.

———. 1956. "W. D. Conybeare, His Geological Contemporaries and Bristol Associations." *Proceedings of the Bristol Natural History Society* 29:133–146.

———. 1966. "The 'Red Lady' of Paviland." *Glamorgan History* 3:123–137.

Nowotny, Helga. 1989. *Eigenzeit. Entstehung und Strukturierung eines Zeitgefühls.* Frankfurt am Main: Suhrkamp.

Nowotny, Helga, Gibbons, Michael, Limoges, Camille, Schwartzman, S., Scott, P., and Trow, M. 1994. *The New Production of Knowledge. The Dynamics of Science and Research in Contemporary Societies.* London: Sage.

Oakley, Kenneth Page. 1948. "Fluorine and the Relative Dating of Bones." *Advancement of Science* 4:336–337.

———. 1963. "Dating Skeletal Material." *Science*, n.s., 140 (3566): 488.

———. 1964a. *Frameworks for Dating Fossil Man.* Chicago: Aldine.

———. 1964b. "The Problem of Man's Antiquity. An Historical Survey." *Bulletin of the British Museum (Natural History)*, Geology Series, 9:85–153.

———. 1968. "The Date of the 'Red Lady' of Paviland." *Antiquity* 42:306–307.

———. 1980. "Relative Dating of the Fossil Hominids of Europe." *Bulletin of the British Museum (Natural History)*, Geology Series, 34 (1): 1–63.

Oakley, Kenneth Page, and Montagu, Francis Ashley. 1949. "A Re-consideration of the Galley Hill Skeleton." *Bulletin of the British Museum (Natural History)*, Geology Series, 1 (2): 25–48.

Oakley, Kenneth Page, and Vries, Hugo de. 1959. "Radiocarbon Dating of the Piltdown Skull and Jaw." *Nature* 184 (4682): 224–226.

O'Connor, Anne. 2003. "Geology, Archaeology, and 'the raging vortex of the "eolith" controversy.'" *Proceedings of the Geologists' Association* 114:255–262.

———. 2005. "Samuel Hazzledine Warren and the Construction of a Chronological Framework for the British Quaternary in the Early Twentieth Century." *Proceedings of the Geologists' Association* 116:1–12.
O'Connor, Ralph. 1999. "Mammoths and Maggots. Byron and the Geology of Cuvier." *Romanticism* 5 (1): 26–42.
Oldroyd, David. 1996. *Thinking about the Earth. A History of Ideas in Geology.* Studies in the History and Philosophy of the Earth Sciences. Edited by Oldroyd, David. Cambridge, MA: Harvard University Press.
Osborn, Henry Fairfield. 1917. *The Origin and Evolution of Life on the Theory of Action, Reaction, and Interaction of Energy.* New York: Charles Scribner's Sons.
———. 1927. "Recent Discoveries Relating to the Origin and Antiquity of Man." *Science*, n.s., 65 (1690): 481–488.
———. 1928. "The Plateau Habitat of the Pro-Dawn Man." *Science*, n.s., 67 (1745): 570–571.
———. 1928 (1927). *Man Rises to Parnassus. Critical Epochs in the Prehistory of Man.* 2nd ed. Princeton: Princeton University Press.
———. 1930. "The Discovery of Tertiary Man." *Science*, n.s., 71 (1827): 1–7.
Osborn, Henry Fairfield, and Reeds, Chester A. 1922. "Recent Discoveries on the Antiquity of Man." *Proceedings of the National Academy of Sciences of the United States of America* 8 (8): 246–247.
P., E. A. 1925. "Ancient Hunters and Their Modern Representatives [Review]." *Geographical Review* 65 (6): 541–543.
Page, Leroy E. 1969. "Diluvialism and Its Critics in Great Britain in the Early Nineteenth Century." In *Toward a History of Geology*, edited by Schneer, Cecil J. Cambridge, MA: MIT Press.
———. 1984. "British Geologists." *Science*, n.s., 225 (4660): 405–406.
Peake, Harold J. E. 1940. "The Study of Prehistoric Times." *Journal of the Royal Anthropological Institute of Great Britain and Ireland* 70 (2): 103–146.
Pendergrast, Mark. 1999. *Uncommon Grounds. The History of Coffee and How It Transformed Our World.* New York: Basic Books.
Pengelly, William. 1866. "On Kent's Cavern, Torquay." *Notices of the Proceedings at the Meetings of the Members of the Royal Institution of Great Britain with Abstracts of the Discourses Delivered at the Evening Meetings* 4:534–543.
———. 1869. "The Literature of Kent's Cavern. Part II. Including the Whole of the Rev. J. Mac. Enery's Manuscript." *Transactions of the Devonshire Association for the Advancement of Science, Literature, and Art* 3:191–482.
———. 1873a. "Report on Windmill Hill Cavern, at Brixham, Devonshire. Drawn up by Mr. Pengelly, in 1862, for the Cavern Committee, at the Request of the Chairman, Dr. Falconer. Forwarded to the Committee, at Their Request, December 4, 1865." *Report and Transactions of the Devonshire Association for the Advancement of Science* 6:779–856.
———. 1873b. "The Ossiferous Caverns and Fissures in the Neighbourhood of Chudleigh, Devonshire." *Report and Transactions of the Devonshire Association for the Advancement of Science* 6:46–60.
Pengelly, William, Busk, George, Evans, John, Prestwich, Joseph, Falconer, H., and Ramsay, Andrew. 1873. "Report on the Exploration of Brixham Cave, Conducted by a Committee of the Geological Society, and under the Superintendence of Wm. Pengelly, Esq., F.R.S., Aided by a Local Committee; with

Descriptions of the Animal Remains by George Busk, Esq., F.R.S., and of the Flint Implements by John Evans, Esq., F.R.S., Joseph Prestwich, F.R.S., F.G.S., &c., Reporter." *Philosophical Transactions of the Royal Society of London* 163:471–572.

Penny, Glenn H. 2002. *Objects of Culture. Ethnology and Ethnographic Museums in Imperial Germany.* Chapel Hill: University of North Carolina Press.

Petermann, Werner. 2004. *Die Geschichte der Ethnologie.* Edition Trickster. Wuppertal: Peter Hammer.

Pettitt, Paul. 2000. "The Paviland Radiocarbon Dating Programme. Reconstructing the Chronology of Faunal Communities, Carnivore Activity and Human Occupation." In *Paviland Cave and the "Red Lady." A Definitive Report,* edited by Aldhouse-Green, Stephen. Bristol: Western Academic and Specialist Press.

Pick, Daniel. 1989. *Faces of Degeneration. A European Disorder, c. 1848–c. 1918.* Cambridge: Cambridge University Press.

Piggott, Stuart. 1989. *Ancient Britons and the Antiquarian Imagination. Ideas from the Renaissance to the Regency.* London: Thames and Hudson.

Porter, Roy S. 1977. *The Making of Geology. Earth Science in Britain, 1660–1815.* Cambridge: Cambridge University Press.

———. 1978. "Gentlemen and Geology. The Emergence of a Scientific Career, 1660–1920." *Historical Journal* 21 (4): 809–836.

Portlock. 1857. "The Anniversary Address of the President [Obituary of Buckland]." *Proceedings of the Geological Society of London* 13:xxvi–xlv.

Prestwich, Joseph. 1860 (1859). "On the Occurrence of Flint-Implements, Associated with the Remains of Extinct Mammalia, in Undisturbed Beds of a Late Geological Period." *Proceedings of the Royal Society of London* 10:50–59.

———. 1861 (1859). "On the Occurrence of Flint-Implements, Associated with the Remains of Animals of Extinct Species in Beds of a Late Geological Period, in France at Amiens and Abbeville, and in England at Hoxne." *Philosophical Transactions of the Royal Society of London* 150:277–317.

Prichard, James Cowles. 1813. *Researches into the Physical History of Man.* Edited by Stocking, George W. London: J. and A. Arch. Chicago: University of Chicago Press, 1973 reprint.

———. 1848. "On the Various Methods of Research Which Contribute to the Advancement of Ethnology, and of the Relations of That Science to Other Branches of Knowledge." *Report of the Seventeenth Meeting of the British Association for the Advancement of Science* (Oxford, June 1847): 230–253.

Proctor, Robert. 2002. "A Historiography of Human Origins." Unpublished manuscript.

———. 2003. "Three Roots of Human Recency. Molecular Anthropology, the Refigured Acheulean, and the UNESCO Response to Auschwitz." *Current Anthropology* 44 (2): 213–239.

Pruner-Bey, Franz. 1875 (1865–1875). "An Account of the Human Bones Found in the Cave of Cro-Magnon, in Dordogne." In *Reliquiae Aquitanicae; Being Contributions to the Archaeology and Palaeontology of Périgord and the Adjoining Provinces of Southern France,* edited by Lartet, Édouard, Christy, Henry, and Jones, Thomas Rupert. London: Williams and Norgate.

Quatrefages, Jean Louis Armand de. 1875 (1865–1875). "Remarks on the Human Remains from the Cave at Cro-Magnon." In *Reliquiae Aquitanicae; Being*

Contributions to the Archaeology and Palaeontology of Périgord and the Adjoining Provinces of Southern France, edited by Lartet, Édouard, Christy, Henry, and Jones, Thomas Rupert. London: Williams and Norgate.

Quatrefages, Jean Louis Armand de, Hamy, Jules Érnest Théodore, and Formant, Henri C. 1882. *Crania ethnica. Les crânes des races humaines décrits et figurés après les collections du Muséum d'Histoire Naturelle de Paris, de la Société d'Anthropologie de Paris, et les principales collections de la France et de l'étranger.* 2 vols. Paris: Baillière.

Reader, John. 1981. *Missing Links. The Hunt for Earliest Man.* London: Book Club Associates.

Redknap, Mark. 2002. *Re-creations. Visualizing Our Past.* National Museums and Galleries of Wales and Welsh Historic Monuments. Cardiff: Westdale.

Regal, Brian. 2002. *Henry Fairfield Osborn. Race, and the Search for the Origins of Man.* Aldershot: Ashgate.

Reid Moir, J. 1913. "Pre-Palaeolithic Man." *Bedrock. A Quarterly Review of Scientific Thought* 2:165–176.

———. 1916. "On the Evolution of the Earliest Palaeoliths from the Rostro-Carinate Implements." *Journal of the Royal Anthropological Institute of Great Britain and Ireland* 46:197–220.

———. 1917. "On Some Human and Animal Bones, Flint Implements, etc., Discovered in Two Ancient Occupation-Levels in a Small Valley near Ipswich." *Journal of the Royal Anthropological Institute of Great Britain and Ireland* 47:367–412.

———. 1921. "On an Early Chellian-Palaeolithic Workshop-Site in the Pliocene 'Forest Bed' of Cromer, Norfolk." *Journal of the Royal Anthropological Institute of Great Britain and Ireland* 51:385–418.

———. 1927. *The Antiquity of Man in East Anglia.* London: Cambridge University Press.

———. 1928. "The Neanderthal Phase of Man." *Man* 28:84–86.

———. 1935. "The Age of the Pre-Crag Flint Implements." *Journal of the Royal Anthropological Institute of Great Britain and Ireland* 65:343–374.

Reid Moir, J., and Keith, Arthur. 1912. "An Account of the Discovery and Characters of a Human Skeleton Found beneath a Stratum of Chalky Boulder Clay near Ipswich." *Journal of the Royal Anthropological Institute of Great Britain and Ireland* 42:345–379.

Reid Moir, J., and Peake, H. J. E. 1922. "The Ice-Age and Man. A Note on Man, 1922, 5." *Man* 22:52–54.

Research Frameworks for the Palaeolithic and Mesolithic of Britain and Ireland. A Report by the Working Party for the Palaeolithic and Mesolithic Annual Day Meeting and the Council of the Prehistoric Society. 1999. Salisbury: Sarum Graphics.

Rheinberger, Hans-Jörg. 1992. *Experiment, Differenz, Schrift. Zur Geschichte epistemischer Dinge.* Marburg an der Lahn: Basiliken-Presse.

———. 1997. *Toward a History of Epistemic Things. Synthesizing Proteins in the Test Tube.* Edited by Lenoir, Timothy, and Gumbrecht, Hans Ulrich. Writing Science. Stanford: Stanford University Press.

———. 2006. *Epistemologie des Konkreten. Studien zur Geschichte der modernen Biologie.* Suhrkamp Taschenbuch Wissenschaft 1771. Frankfurt am Main: Suhrkamp.

Richards, Michael P. 2000. "Human and Faunal Stable Isotope Analyses from Goat's Hole and Foxhole Caves, Gower." In *Paviland Cave and the "Red Lady."*

A Definitive Report, edited by Aldhouse-Green, Stephen. Bristol: Western Academic and Specialist Press.

Ripley, William Z. 1899. *The Races of Europe. A Sociological Study*. Edited by Barre, Weston La. Landmarks in Anthropology. New York: D. Appleton.

Roach, Marion. 2005. *The Roots of Desire. The Myth, Meaning, and Sexual Power of Red Hair*. New York: Bloomsbury.

Robinson, D. T., and Smith-Lovin, L. 2001. "Getting a Laugh. Gender, Status, and Humor in Task Discussions." *Social Forces* 80 (1): 123–158.

Roebroeks, Wil, and Corbey, Raymond. 2000. "Periodisations and Double Standards in the Study of the Palaeolithic." In *Hunters of the Golden Age. The Mid Upper Palaeolithic of Eurasia, 30,000–20,000 BP*, edited by Roebroeks, Wil, Mussi, Margherita, Svoboda, Jiri, and Fennema, Kelly. Leiden: University of Leiden Press.

Roebroeks, Wil, Mussi, Margherita, Svoboda, Jiri, and Fennema, Kelly (eds.). 2000. *Hunters of the Golden Age. The Mid Upper Palaeolithic of Eurasia, 30,000–20,000 BP*. Leiden: University of Leiden Press.

Rudwick, Martin J. S. 1970. "The Strategy of Lyell's *Principles of Geology*." *Isis* 61 (206): 5–33.

———. 1976 (1972). *The Meaning of Fossils. Episodes in the History of Palaeontology*. 2nd ed. Chicago: University of Chicago Press.

———. 1982. "Cognitive Styles in Geology." In *Essays in the Sociology of Perception*, edited by Douglas, Mary. London: Routledge and Kegan Paul.

———. 1989. "Encounters with Adam, or at Least the Hyaenas. Nineteenth-Century Visual Representations of the Deep Past." In *History, Humanity and Evolution*, edited by Moore, James R. Cambridge: Cambridge University Press.

———. 1992. *Scenes from Deep Time. Early Pictorial Representations of the Prehistoric World*. Chicago: University of Chicago Press.

———. 1997. *Georges Cuvier, Fossil Bones, and Geological Catastrophes. New Translations and Interpretations of the Primary Texts*. Chicago: University of Chicago Press.

———. 2000. "Georges Cuvier's Paper Museum of Fossil Bones." *Archives of Natural History* 27 (1): 51–68.

———. 2005. *Bursting the Limits of Time. The Reconstruction of Geohistory in the Age of Revolution*. Chicago: University of Chicago Press.

Ruggles Gates, R. 1948. *Human Ancestry from a Genetical Point of View*. Cambridge, MA: Harvard University Press.

Rupke, Nicolaas A. 1983. *The Great Chain of History. William Buckland and the English School of Geology (1814–1849)*. Oxford: Clarendon Press.

———. 1990. "Caves, Fossils and the History of the Earth." In *Romanticism and the Sciences*, edited by Cunningham, Andrew, and Jardine, Nicholas. Cambridge: Cambridge University Press.

———. 1997. "Oxford's Scientific Awakening and the Role of Geology." In *The History of the University of Oxford*, vol. 6, *Nineteenth-Century Oxford*, pt. 1, edited by Brock, M. G., and Curthoys, M. C. Oxford: Clarendon Press.

Ruse, Michael. 1999 (1979). *The Darwinian Revolution. Science Red in Tooth and Claw*. 2nd ed. Chicago: University of Chicago Press.

Sarasin, Philipp. 2005. *Michel Foucault zur Einführung*. Hamburg: Junius.

Sarich, Vincent M., and Miele, Frank. 2004. *Race. The Reality of Human Differences.* Boulder, CO: Westview Press.

Sarich, Vincent M., and Wilson, Allan C. 1967. "Immunological Time Scale of Hominid Evolution." *Science*, n.s., 158 (3805): 1200–1203.

Sawday, Jonathan. 1999. "'New Men, Strange Faces, Other Minds.' Arthur Keith, Race and the Piltdown Affair (1912–53)." In *Race, Science and Medicine, 1700–1960*, edited by Ernst, Waltraud, and Harris, Bernard. London: Routledge.

Schaaffhausen, Hermann. 1861 (1858). "On the Crania of the Most Ancient Races of Man with Remarks, and Original Figures, Taken from a Cast of the Neanderthal Cranium." Translated by Busk, George. *Natural History Review*, n.s., 8:155–172.

———. 1872 (1869). "Ueber die Methode der vorgeschichtlichen Forschung." *Archiv für Anthropologie* 5:113–128.

Schaffer, Simon. 1990. "Genius in Romantic Natural Philosophy." In *Romanticism and the Sciences*, edited by Cunningham, Andrew, and Jardine, Nicholas. Cambridge: Cambridge University Press.

———. 2005. "Die Reichweite experimenteller Wissenschaften. Modelle, Mikrogeschichten, Mikrokosmen." Translated by Blumenbach, Ulrich. *Historische Anthropologie* 13 (3): 343–366.

Scheuchzer, Johann Jacob. 1726. *Homo diluvii testis et theoskopos.* Zurich: Reding.

———. 1984 (1731). *Berühmte Bilder zur Menschheitsgeschichte aus Johann Jacob Scheuchzers Physica Sacra.* Edited by Krauss, Hans. Konstanz: Universitätsverlag.

Schiebinger, Londa. 1989. *The Mind Has No Sex? Women in the Origins of Modern Science.* Cambridge, MA: Harvard University Press.

———. 2004. *Plants and Empire. Colonial Bioprospecting in the Atlantic World.* Cambridge, MA: Harvard University Press.

Schiebinger, Londa, and Proctor, Robert (eds.). Forthcoming. *Agnotology. The Cultural Production of Ignorance.* Stanford: Stanford University Press.

Schlette, Friedrich. 1990. "Zur Geschichte des Verhältnisses von Anthropologie, Ethnologie und Urgeschichte." In *Grundlinien der Geschichte der biologischen Anthropologie*, edited by Kirschke, Siegfried. Halle (Saale): Martin-Luther-Universität Halle-Wittenberg.

Schmerling, Philippe-Charles. 1833–1834. *Recherches sur les ossemens fossiles découverts dans les cavernes de la province de Liège.* Liège: Collardin.

Schmuhl, Hans-Walter. 2005. *Grenzüberschreitungen. Das Kaiser-Wilhelm-Institut für Anthropologie, menschliche Erblehre und Eugenik, 1927–1945.* Edited by Rürup, Reinhard, and Schieder, Wolfgang. Geschichte der Kaiser-Wilhelm-Gesellschaft im Nationalsozialismus 9. Göttingen: Wallstein.

Schwalbe, Gustav. 1906. *Studien zur Vorgeschichte des Menschen.* Stuttgart: Schweizerbartsche.

———. 1913. "Kritische Besprechung von Boule's Werk. 'L'Homme fossile de la Chapelle-aux-Saints' mit eigenen Untersuchungen." *Zeitschrift für Morphologie und Anthropologie* 16:527–610.

Schwartz, Jeffrey. 1993. *What the Bones Tell Us.* New York: Henry Holt.

———. 2005. "Molecular Systematics and Evolution." In *Encyclopedia of Molecular Cell Biology and Molecular Medicine (EMCBMM)*, edited by Meyer, R. A. Weinheim: Wiley-VCH.

Schweber, Silvan S. 1981. "Scientists as Intellectuals. The Early Victorians." In *Victorian Science and Victorian Values. Literary Perspectives*, edited by Paradis, James, and Postlewait, Thomas. New York: New York Academy of Sciences.

Scott, Monique. 2005. "'We Grew Up and Moved On.' Visitors to the British Museums Consider Their 'Cradle of Mankind.'" In *Envisioning the Past. Archaeology and the Image*, edited by Smiles, Sam, and Moser, Stephanie. Malden, MA: Blackwell.

Secord, Anne. 1994. "Corresponding Interests. Artisans and Gentlemen in Nineteenth-Century Natural History." *British Journal for the History of Science* 27:383–408.

Secord, James A. 1982. "King of Siluria. Roderick Murchison and the Imperial Theme in Nineteenth-Century British Geology." *Victorian Studies* 25:413–442.

———. 2000. *Victorian Sensation. The Extraordinary Publication, Reception, and Secret Authorship of "Vestiges of the Natural History of Creation."* Chicago: University of Chicago Press.

Semonin, Paul. 2000. *American Monster. How the Nation's First Prehistoric Creature Became a Symbol of National Identity.* New York: New York University Press.

Shapiro, Harry L. 1959. "The History and Development of Physical Anthropology." *American Anthropologist*, n.s., 61 (3): 371–379.

Sharpe, T., and McCartney, P. J. 1998. *The Papers of H. T. De la Beche (1796–1855) in the National Museum of Wales.* National Museum of Wales, Geological Series No. 17. Cardiff: National Museum of Wales.

Shelley, Percy Bysshe. 2002 (1977). *Shelley's Poetry and Prose.* Edited by Reiman, Donald H., and Fraistat, Neil. 2nd ed. A Norton Critical Edition. New York: W. W. Norton.

Shipman, Pat, and Storm, Paul. 2002. "Missing Links. Eugène Dubois and the Origin of Paleoanthropology." *Evolutionary Anthropology* 11:108–116.

Smiles, Sam, and Moser, Stephanie (eds.). 2005. *Envisioning the Past. Archaeology and the Image.* Edited by Arnold, Dana, New Interventions in Art History. Malden, MA: Blackwell.

Smith Flett, John. 1938. "Memorial to William Johnson Sollas." *Proceedings of the Geological Society of America* (June): 203–220.

Smith Woodward, Arthur. 1925. "The Origin of Man." *Scientific Monthly* 21 (1): 13–19.

———. 1935. "Recent Progress in the Study of Early Man." *Science*, n.s., 82 (2131): 399–407.

Smith Woodward, Arthur, and Watts, W. W. 1938. "William Johnson Sollas, 1849/1936." *Obituary Notices of Fellows of the Royal Society* 2:265–281.

Soffer, Olga. 2000. "Gravettian Technologies in Social Context." In *Hunters of the Golden Age. The Mid Upper Palaeolithic of Eurasia, 30,000–20,000 BP,* edited by Roebroeks, Wil, Mussi, Margherita, Svoboda, Jiri, and Fennema, Kelly. Leiden: University of Leiden Press.

Sollas, Igerna B. J. 1914. "On Onychaster, a Carboniferous Brittle-Star." *Philosophical Transactions of the Royal Society of London,* ser. B, *Containing Papers of a Biological Character* 204:51–62.

Sollas, Igerna B. J., and Sollas, William Johnson. 1912. "Lapworthura. A Typical Brittlestar of the Silurian Age; with Suggestions for a New Classification of the Ophiuroidea." *Philosophical Transactions of the Royal Society of London*, ser. B, *Containing Papers of a Biological Character* 202:213–232.

———. 1914 (1913). "A Study of the Skull of a Dicynodon by means of Serial Sections." *Philosophical Transactions of the Royal Society of London*, ser. B, *Containing Papers of a Biological Character* 204:201–225.

———. 1916. "On the Structure of the Dicynodont Skull." *Philosophical Transactions of the Royal Society of London*, ser. B, *Containing Papers of a Biological Character* 207:531–539.

Sollas, William Johnson. 1880 (1879). "On Some Eskimos' Bone Implements from the East Coast of Greenland." *Journal of the Royal Anthropological Institute of Great Britain and Ireland* 9:329–336.

———. 1895. "'Pithecanthropus erectus' and the Evolution of the Human Race." *Nature* 53 (1364): 150–151.

———. 1896–1897. "Report to the Committee of the Royal Society Appointed to Investigate the Structure of a Coral Reef by Boring." *Proceedings of the Royal Society of London* 60:502–512.

———. 1898. "On the Intimate Structure of Crystals. Parts I–III. Crystals of the Cubic System with Cubic Cleavage." *Proceedings of the Royal Society of London* 63:270–300.

———. 1900. "On the Intimate Structure of Crystals. Part IV. Cubic Crystals with Octahedral Cleavage." *Proceedings of the Royal Society of London* 67:493–495.

———. 1901–1902. "On the Intimate Structure of Crystals. Part V. Cubic Crystals with Octahedral Cleavage." *Proceedings of the Royal Society of London* 69:294–306.

———. 1904 (1903). "A Method for the Investigation of Fossils by Serial Sections." *Philosophical Transactions of the Royal Society of London*, ser. B, *Containing Papers of a Biological Character* 196:259–265.

———. 1908 (1907). "On the Cranial and Facial Characters of the Neandertal Race." *Philosophical Transactions of the Royal Society of London*, ser. B, *Containing Papers of a Biological Character* 199:281–339.

———. 1909. "Palaeolithic Races and Their Modern Representatives." *Science Progress* 3:326–353.

———. 1910. "The Anniversary Address of the President [On the Evolution of Man]." *Quarterly Journal of the Geological Society of London* 66:54–88.

———. 1911. *Ancient Hunters and Their Modern Representatives*. London: Macmillan.

———. 1913a. "Paviland Cave. An Aurignacian Station in Wales." *Journal of the Royal Anthropological Institute of Great Britain and Ireland* 43:325–374.

———. 1913b. "The Formation of 'Rostro-carinate' Flints." *Report of the British Association for the Advancement of Science*: 788–790.

———. 1915 (1911). *Ancient Hunters and Their Modern Representatives*. 2nd ed. London: Macmillan.

———. 1918. "The Skull of Ichthyosaurus, Studied in Serial Sections." *Philosophical Transactions of the Royal Society of London*, ser. B, *Containing Papers of a Biological Character* 208:63–65, 67–126.

———. 1920. "On the Structure of Lysorophus, as Exposed by Serial Sections." *Philosophical Transactions of the Royal Society of London*, ser. B, *Containing Papers of a Biological Character* 209:481–527.

———. 1922. "A Method for the Comparative Study of the Human Skull, and Its Application to *Homo sapiens* and *Homo neandertalensis*." *Proceedings of the Royal Society of London*, ser. B, *Containing Papers of a Biological Character* 94 (658): 134–137.
———. 1924. "'Ancient Hunters.'" *Man* 24:189–192.
———. 1924 (1911). *Ancient Hunters and Their Modern Representatives*. 3rd ed. London: Macmillan.
———. 1925. "Late Palaeolithic Art in the Cresswell Caves." *Man* 25:63–64.
———. 1926 (1925). "On a Sagittal Section of the Skull of Australopithecus Africanus." *Quarterly Journal of the Geological Society of London* 82 (325): 1–11.
———. 1927. "The Chancelade Skull." *Journal of the Royal Anthropological Institute of Great Britain and Ireland* 57:89–122.
———. 1933. "The Sagittal Section of the Human Skull." *Journal of the Royal Anthropological Institute of Great Britain and Ireland* 63:389–431.
Sollas, William Johnson, and Sollas, Igerna B. J. 1904. "An Account of the Devonian Fish, *Palaeospondylus Gunni*, Traquair." *Philosophical Transactions of the Royal Society of London*, ser. B, *Containing Papers of a Biological Character* 196:267–294.
Sommer, Marianne. 2003a. "Buckland, William." In *Lexikon der bedeutenden Naturwissenschaftler*, edited by Hoffmann, Dieter, Laitko, Hubert, and Müller-Wille, Staffan. Heidelberg: Spektrum Akademischer Verlag.
———. 2003b. "The Romantic Cave? The Scientific and Poetic Quests for Subterranean Spaces in Britain." *Earth Sciences History* 22 (2): 172–208.
———. 2004a. "'An Amusing Account of a Cave in Wales.' William Buckland (1784–1856) and the Red Lady of Paviland." *British Journal for the History of Science* 37 (1): 53–74.
———. 2004b. "Buckland, William." In *Dictionary of Nineteenth-Century British Scientists*, edited by Lightman, Bernard. Bristol: Thoemmes.
———. 2004c. "Das Alter des Menschengeschlechts. Integration einer neuen Zeitgeschichte ins biblische Weltbild." In *Integrationen des Widerläufigen. Ein Streifzug durch geistes- und kulturwissenschaftliche Forschungsfelder*, edited by Huwiler, Elke, and Wachter, Nicole. Munich: LIT.
———. 2004d. "Eoliths as Evidence for Human Origins? The British Context." *Journal for the History and Philosophy of the Life Sciences* 26 (2): 209–241.
———. 2005a. "*Ancient Hunters and Their Modern Representatives*. William Sollas's (1849–1936) Anthropology from Disappointed Bridge to Trunkless Tree and the Instrumentalisation of Racial Conflict." *Journal of the History of Biology* 38 (2): 327–365.
———. 2005b. "Anthropologie." In *Enzyklopädie der Neuzeit*, edited by Jaeger, Friedrich. Stuttgart: J. B. Metzler.
———. 2005c. "Die Höhle als Zeitkorridor. Das goldene Zeitalter der Geologie und die romantische Dichtung in England." In *Lebendige Zeit. Wissenskulturen im Werden*, edited by Schmidgen, Henning. Berlin: Kadmos.
———. 2006. "Mirror, Mirror on the Wall. Neanderthal as Image and 'Distortion' in Early 20th-Century French Science and Press." *Social Studies of Science* 36 (2): 207–240.
———. 2007a. "Kenneth Oakley." In *New Dictionary of Scientific Biography*, edited by Koertge, Noretta. New York: Charles Scribner's Sons.

———. 2007b. "Wilfrid Edward Le Gros Clark." In *New Dictionary of Scientific Biography*, edited by Koertge, Noretta. New York: Charles Scribner's Sons.
Sowerwine, Charles. 2001. *France since 1870. Culture, Politics and Society*. Basingstoke, Hampshire: Palgrave.
Spencer, Frank. 1981. "The Rise of Academic Physical Anthropology in the United States (1880–1980). A Historical Overview." *American Journal of Physical Anthropology* 56:353–364.
———. 1982a. Introduction to *A History of American Physical Anthropology, 1930–1980*, edited by Spencer, Frank. New York: Academic Press.
———. 1988. "Prologue to a Scientific Forgery. The British Eolithic Movement from Abbeville to Piltdown." In *Bones, Bodies, Behavior. Essays on Biological Anthropology*, edited by Stocking, George W. Madison: University of Wisconsin Press.
———. 1990a. *Piltdown. A Scientific Forgery*. Natural History Museum Publications. London: Oxford University Press.
———. 1990b. *The Piltdown Papers, 1908–1955. The Correspondence and Other Documents Relating to the Piltdown Forgery*. Natural History Museum Publications. London: Oxford University Press.
———. 1997a. "Buckland, William (1784–1856)." In *History of Physical Anthropology. An Encyclopedia*, edited by Spencer, Frank. New York: Garland.
——— (ed.). 1982b. *A History of American Physical Anthropology, 1930–1980*. New York: Academic Press.
——— (ed.). 1997b. *History of Physical Anthropology. An Encyclopedia*. 2 vols. Garland Reference Library of Social Science 677. New York: Garland.
Spencer, Frank, and Smith, Fred H. 1981. "The Significance of Ales Hrdlicka's 'Neanderthal Phase of Man.' A Historical and Current Assessment." *American Journal of Physical Anthropology* 56:435–459.
Spencer, Herbert. 1892. *The Principles of Ethics*. 2 vols. Vol. 1. A System of Synthetic Philosophy IX. New York: D. Appleton.
Stafford, Robert A. 1989. *Scientist of Empire. Sir Roderick Murchison, Scientific Exploration and Victorian Imperialism*. Cambridge: Cambridge University Press.
Star, Susan Leigh, and Griesemer, James R. 1989. "Institutional Ecology, 'Translations' and Boundary Objects. Amateurs and Professionals in Berkeley's Museum of Vertebrate Zoology, 1907–39." *Social Studies of Science* 19 (3): 387–420.
Stein, Aurel. 1928. *Innermost Asia. Detailed Report of Explorations in Central Asia, Kan-su and Eastern Îrân*. Oxford: Clarendon Press.
Stepan, Nancy Leys. 1982. *The Idea of Race in Science. Great Britain, 1800–1960*. London: Macmillan.
Stocking, George W. 1968. *Race, Culture, and Evolution. Essays in the History of Anthropology*. New York: Free Press.
———. 1973. Introduction to *Prichard, James Cowles. Researches into the Physical History of Man*, edited by Stocking, George W. Chicago: University of Chicago Press.
———. 1990. "Paradigmatic Traditions in the History of Anthropology." In *Companion to the History of Modern Science*, edited by Olby, R. C., Cantor, G. N., Christie, J. R. R., and Hodge, M. J. S. London: Routledge.
———. 1994. "The Turn-of-the-Century Concept of Race." *Modernism/Modernity* 1 (1): 4–16.

———. 1995. *After Tylor. British Social Anthropology, 1888–1951*. Madison: University of Wisconsin Press.
———. 2001. *Delimiting Anthropology. Occasional Essays and Reflections*. Madison: University of Wisconsin Press.
——— (ed.). 1988. *Bones, Bodies, Behavior. Essays on Biological Anthropology*. History of Anthropology 5. Madison: University of Wisconsin Press.
Stoczkowski, Wiktor. 1997. "The Painter and the Prehistoric People. A 'Hypothesis on Canvas.'" In *The Cultural Life of Images. Visual Representation in Archaeology*, edited by Molyneaux, Brian Leigh. London: Routledge.
———. 2001. "How to Benefit from Received Ideas." In *Studying Human Origins. Disciplinary History and Epistemology*, edited by Corbey, Raymond, and Roebroeks, Wil. Amsterdam: Amsterdam University Press.
Stoneking, Mark, and Cann, Rebecca. 1989. "African Origin of Human Mitochondrial DNA." In *The Human Revolution. Behavioural and Biological Perspectives on the Origins of Modern Humans*, edited by Mellars, Paul, and Stringer, Chris. Edinburgh: Edinburgh University Press.
Strabo. 1854. *Geography*. Translated by Hamilton, Hans Claude, and Falconer, W. 3 vols. Vol. 1. Bohn's Classical Library. London: Henry G. Bohn.
Straus, William L. 1954. "The Great Piltdown Hoax." *Science*, n.s., 119 (3087): 265–269.
Straus, William L., and Cave, A. J. E. 1957. "Pathology and Posture of Neanderthal Man." *Quarterly Review of Biology* 32 (4): 348–363.
Stringer, Chris, and Gamble, Clive. 1993. *In Search of the Neanderthals. Solving the Puzzle of Human Origins*. London: Thames and Hudson.
Strum, Shirley C., and Fedigan, Linda Marie (eds.). 2000. *Primate Encounters. Models of Science, Gender, and Society*. Chicago: University of Chicago Press.
Strum, Shirley C., Lindburg, Donald G., and Hamburg, David (eds.). 1999. *The New Physical Anthropology. Science, Humanism, and Critical Reflection*. Advances in Human Evolution Series. Upper Saddle River, NJ: Prentice Hall.
Suess, Eduard. 1904–1924. *The Face of the Earth* [*Das Antlitz der Erde*]. Translated by Sollas, Hertha B. C., and Sollas, William. 5 vols. Oxford: Clarendon Press.
Swainston, Stephanie. 1999. "A Study of the Lithic Collections from Paviland Cave, and the Site in Its Wider Context." MPhil thesis, Department of Anthropology, University of Wales, Friends of the National Museum and Galleries of Wales and University of Wales College Newport Funds.
———. 2000. "The Lithic Artefacts from Paviland." In *Paviland Cave and the "Red Lady." A Definitive Report*, edited by Aldhouse-Green, Stephen. Bristol: Western Academic and Specialist Press.
———. 2004a. Interview, 13 July.
———. 2004b. *The Year of Our War*. London: Gollancz.
———. 2005. *No Present like Time*. London: Gollancz.
Swainston, Stephanie, and Brookes, Alison. 2000. "Paviland Cave and the 'Red Lady.' The History of Collection and Investigation." In *Paviland Cave and the "Red Lady." A Definitive Report*, edited by Aldhouse-Green, Stephen. Bristol: Western Academic and Specialist Press.
Sykes, Bryan. 1999. "Using Genes to Map Population Structure and Origins." In *The Human Inheritance*, edited by Sykes, Bryan. Oxford: Oxford University Press.

———. 2000. "Report on DNA Recovery from the Red Lady of Paviland." In *Paviland Cave and the "Red Lady." A Definitive Report*, edited by Aldhouse-Green, Stephen. Bristol: Western Academic and Specialist Press.
———. 2001. *The Seven Daughters of Eve*. New York: W. W. Norton.
———. 2003. *Adam's Curse. A Future without Men*. London: Bantam.
Symonds, Richard. 1986. *Oxford and Empire. The Last Lost Cause?* New York: St. Martin's Press.
———. 2000. "Oxford and the Empire." In *The History of the University of Oxford*, vol. 7, *Nineteenth-Century Oxford*, pt. 2, edited by Brock, M. G., and Curthoys, M. C. Oxford: Clarendon Press.
Tattersall, Ian. 1995. *The Fossil Trail. How We Know What We Think We Know about Human Evolution*. New York: Oxford University Press.
———. 1999 (1995). *The Last Neanderthal. The Rise, Success, and Mysterious Extinction of Our Closest Human Relatives*. 2nd ed. Boulder, CO: Westview Press.
Tattersall, Ian, and Eldredge, Niles. 1977. "Fact, Theory, and Fantasy in Human Paleontology." *American Scientist* 65 (March–April): 204–211.
Taylor, Griffith. 1919. "Climatic Cycles and Evolution." *Geographical Review* 8 (6): 289–328.
Testut, Léo. 1890. "Recherches anthropologiques sur le squelette quaternaire de Chancelade, Dordogne." *Bulletin de la Société d'Anthropologie de Lyon* 8:131–246.
Thackray, John C. (ed.). 2003 (1999). *To See the Fellows Fight. Eye Witness Accounts of Meetings of the Geological Society of London and Its Club, 1822–1868*. BSHS Monographs 12. Hampshire, UK: British Society for the History of Science.
Theunissen, Bert. 1989. *Eugène Dubois and the Ape-Man from Java*. Dordrecht: Kluwer.
Thollard, Patrick. 1987. *Barbarie et civilisation chez Strabo. Étude critique des livres III et IV de la Géographie*. Centre de Recherches d'Histoire Ancienne 77. Paris: Les Belles Lettres.
Thomas, H. Hamshaw. 1947. "The Rise of Geology and Its Influence on Contemporary Thought." *Annals of Science* 5 (4): 325–341.
Thompson Klein, Julie. 2000. "A Conceptual Vocabulary of Interdisciplinary Science." In *Practising Interdisciplinarity*, edited by Weingart, Peter, and Stehr, Nico. Toronto: University of Toronto Press.
Tickell, Crispin. 1996. *Mary Anning of Lyme Regis*. Lyme Regis: Philpot Museum.
Tobias, Phillip V. 1992. "Piltdown. An Appraisal of the Case against Sir Arthur Keith." *Current Anthropology* 33 (3): 243–293.
Topham, Jonathan R. 1998. "Beyond 'Common Context.' The Production and Reading of the Bridgewater Treatises." *Isis* 89 (2): 233–262.
Torrens, Hugh. 1995. "Mary Anning (1799–1847) of Lyme, 'the greatest fossilist the world ever knew.'" *British Journal for the History of Science* 28:257–284.
———. 1996. "An Inspiring Interface. Geology and Its Connections with Archaeology in Britain: 1780–1850." *Transactions of the Proceedings of the Torquay Natural History Society* 22:9–46.
———. 2002. *The Practice of British Geology, 1750–1850*. Variorum Collected Studies. Aldershot: Ashgate.
Trautmann, Thomas R. 1991. "The Revolution in Ethnological Time." *Man*, n.s., 27:379–397.

Trinkaus, Erik. 1984. "Neandertal Pubic Morphology and Gestation Length." *Current Anthropology* 25 (4): 509–514.

———. 2000. "Late Pleistocene and Holocene Human Remains from Paviland Cave." In *Paviland Cave and the "Red Lady." A Definitive Report*, edited by Aldhouse-Green, Stephen. Bristol: Western Academic and Specialist Press.

———. 2004. Interview, 18 May.

Trinkaus, Erik, and Shipman, Pat. 1993. *The Neanderthals. Changing the Image of Mankind.* New York: Alfred A. Knopf.

Trinkaus, Erik, Svoboda, J., West, D. L., Sladek, V., Hillson, S. W., Drozdova, E., and Fisakova, M. 2000. "Human Remains from the Moravian Gravettian. Morphology and Taphonomy of Isolated Elements from the Dolni Vestonice II Site." *Journal of Archaeological Science* 27:1115–1132.

Turner, Alan. 2000. "The Paviland Mammalian Fauna." In *Paviland Cave and the "Red Lady." A Definitive Report*, edited by Aldhouse-Green, Stephen. Bristol: Western Academic and Specialist Press.

Turner, Robin. 2004a. "Campaign to Bring 'Red Lady' Back to Swansea after 180 Years." *Western Mail*, 27 December.

———. 2004b. "Return 'Red Lady' to Wales, English Urged." *Western Mail*, 13 May.

Tylor, Edward Burnett. 1871. *Primitive Culture. Researches into the Development of Mythology, Philosophy, Religion, Language, Art, and Custom.* 2 vols. London: Murray. New York: Harper, 1958 reprint.

———. 1881. *Anthropology. An Introduction to the Study of Man and Civilization.* London: Macmillan.

Unt, R. H. n.d. "Buckland, William (1784–1856)." In *Dictionary of National Biography*, edited by Stephen, Leslie, and Lee, Sidney. London: Oxford University Press.

Vallois, Henri V. 1954. "Neandertals and Praesapiens." *Journal of the Royal Anthropological Institute of Great Britain and Ireland* 84 (1/2): 111–130.

Van Riper, A. Bowdoin. 1993. *Men among the Mammoths. Victorian Science and the Discovery of Human Prehistory.* Edited by Hull, David L. Science and Its Conceptual Foundations. Chicago: University of Chicago Press.

Vergata, Antonello La. 1994. "Evolution and War, 1871–1918." *Nuncius* 9 (1): 143–163.

Verneau, R. 1924. "La race de Néanderthal et la race de Grimaldi. Leur role dans l'humanité." *Journal of the Royal Anthropological Institute of Great Britain and Ireland* 54:211–230.

Vivian, E. 1848. "Extract of a Letter from Mr. E. Vivian of Torquay, Respecting the Phaenomena of Kent's Cavern." *Notices and Abstracts of Communications to the British Association for the Advancement of Science* (Oxford, June 1847): 73.

Walker, Alan, and Shipman, Pat. 1997. *The Wisdom of the Bones. In Search of Human Origins.* New York: Vintage Books.

Walker, Elizabeth A. 2000. "Paviland Cave. The Curatorial and Scientific History of the Museum Collections." In *Paviland Cave and the "Red Lady." A Definitive Report*, edited by Aldhouse-Green, Stephen. Bristol: Western Academic and Specialist Press.

Walsh, John Evangelist. 1996. *Unraveling Piltdown. The Science Fraud of the Century and Its Solution.* New York: Random House.

Washburn, S. L. 1979. "The Piltdown Hoax. Piltdown 2." *Science*, n.s., 203 (4384): 955–956, 958.

Weaver, Thomas. 1823. "On Fossil Human Bones, and Other Animal Remains Recently Found in Germany." *Annals of Philosophy* 5:17–43.

Weidenreich, Franz. 1946. *Apes, Giants, and Man*. Chicago: University of Chicago Press.

Weinberg, Florence M. 1986. *The Cave. The Evolution of a Metaphoric Field from Homer to Ariosto*. Edited by Mermier, Guy. Studies in the Humanities. Literature—Politics—Society 4. New York: Peter Lang.

Weiner, Joseph, Clark, Wilfrid E. Le Gros, Oakley, Kenneth Page, Fryd, C. F. M., Claringbull, G. F., Hey, M. H., Werner, A. E. A., Plesters, R. J., Edmunds, F. H., Bowie, S. H. U., Davidson, C. G., Baynes-Cope, A. D., and Beer, Gavin de. 1955. "Further Contributions to the Solution of the Piltdown Problem." *Bulletin of the British Museum (Natural History)*, Geology Series, 2:225–287.

Weingart, Peter. 2000. "Interdisciplinarity. The Paradoxical Discourse." In *Practising Interdisciplinarity*, edited by Weingart, Peter, and Stehr, Nico. Toronto: University of Toronto Press.

White, Mark. 2002. "Paviland Cave and the 'Red Lady.' A Definitive Report [Review]." *The Prehistoric Society Homepage*, http://www.ucl.ac.uk/prehistoric/reviews/02_10_paviland.html, October.

White, Paul. 2003. *Thomas Huxley. Making the "Man of Science."* Cambridge Science Biographies. Cambridge: Cambridge University Press.

Wiber, Melanie. 1998. *Erect Men—Undulating Women. The Visual Imagery of Gender, "Race" and Progress in Reconstructive Illustrations of Human Evolution*. Waterloo: Wilfrid Laurier University Press.

Williams, E. T. 2000. "The Rhodes Scholars." In *The History of the University of Oxford*, vol. 7, *Nineteenth-Century Oxford*, pt 2. edited by Brock, M. G., and Curthoys, M. C. Oxford: Clarendon Press.

Williams, Rosalind. 1990. *Notes on the Underground. An Essay on Technology, Society, and the Imagination*. Cambridge, MA: MIT Press.

Willis, Delta. 1989. *The Hominid Gang. Behind the Scenes in the Search for Human Origins*. New York: Viking.

———. 1992. *The Leakey Family. Leaders in the Search for Human Origins*. New York: Facts on File.

Winchester, Simon. 2001. *The Map That Changed the World. The Tale of William Smith and the Birth of a Science*. London: Viking.

Winter, J. M. 1994. "Oxford and the First World War." In *The History of the University of Oxford*, vol. 8, *The Twentieth Century*, edited by Harrison, Brian. Oxford: Clarendon Press.

Wolpoff, Milford H. 1989. "Multiregional Evolution. The Fossil Alternative to Eden." In *The Human Revolution. Behavioural and Biological Perspectives on the Origins of Modern Humans*, edited by Mellars, Paul, and Stringer, Chris. Edinburgh: Edinburgh University Press.

———. 2007. "Paleoanthropology. A View from the Trenches." In *Debating Humankind's Place in Nature, 1860–2000. The Nature of Paleoanthropology*, by Delisle, Richard G. Upper Saddle River, NJ: Pearson.

Wolpoff, Milford H., and Thorne, Alan G. 2003. "The Multiregional Evolution of Humans." In *New Look at Human Evolution*, edited by Fischetti, M. *Scientific American* 13 (2): 46–53.

"The Wonders of the Antediluvian Cave!" 1822. *Gentleman's Magazine* 92:491–494.

Wood, Bernard, and Caselli, Giovanni. 1976. *Die Welt des Urmenschen*. Translated by Höpfner, Thomas M. Hamburg: Neuer Tessloff.
Wood Jones, Frederic. 1919. "The Origin of Man." In *Animal Life and Human Progress*, edited by Dendy, Arthur. London: Constable.
———. 1929. *Man's Place among the Mammals*. London: Edward Arnold.
Woodward, Henry B. 1883. "Dr. Buckland and the Glacial Theory." *Midland Naturalist* 6:225–229.
———. 1907. *The History of the Geological Society of London*. London: Geological Society.
Wright, Rita P. (ed.). 1996. *Gender and Archaeology*. Philadelphia: University of Pennsylvania Press.
Wyatt, John. 1995. *Wordsworth and the Geologists*. Edited by Butler, Marilyn, and Chandler, James. Cambridge Studies in Romanticism 16. Cambridge: Cambridge University Press.
Wylie, Alison. 2001. "Doing Social Science as a Feminist. The Engendering of Archeology." In *Feminism in Twentieth-Century Science, Technology, and Medicine*, edited by Creager, Angela N. H., Lunbeck, Elizabeth, and Schiebinger, Londa. Chicago: University of Chicago Press.
———. 2002. *Thinking from Things. Essays in the Philosophy of Archaeology*. Berkeley: University of California Press.
Young, George. 1822. "On the Fossil Remains of Quadrupeds, &c. Discovered in the Cavern at Kirkdale, in Yorkshire, and in Other Cavities or Seams in Limestone Rocks." *Memoirs of the Wernerian Natural History Society* 4 (2): 262–270.
Young, George Malcolm. 1954. *Victorian England. Portrait of an Age*. Doubleday Anchor Books. Garden City, NY: Doubleday.
Young, T. 2000. "The Paviland Ochres. Characterisation and Sourcing." In *Paviland Cave and the "Red Lady." A Definitive Report*, edited by Aldhouse-Green, Stephen. Bristol: Western Academic and Specialist Press.
Zihlman, Adrienne. 1981. "Women as Shapers of the Human Adaptation." In *Woman the Gatherer*, edited by Dahlberg, Frances. New Haven: Yale University Press.
Zimmerman, Andrew. 2001. *Anthropology and Antihumanism in Imperial Germany*. Chicago: University of Chicago Press.
Zittel, Karl A. 1891–1893. *Handbuch der Palaeontologie. 1. Abt. Palaeozoologie*. Vol. 4, Mammalia. Munch: R. Oldenbourg.
Zuckerman, Larry. 1999. *The Potato. How the Humble Spud Rescued the Western World*. New York: North Point Press.

Index

Note: Page numbers followed by *f* indicate figures.

Aborigines, 128–129; Boule and, 177–179, 178*f*; *Definitive Report* and, 277–278; Huxley and, 128–129, 143–144; Sollas and, 156–166
Acheulean culture, 134, 184, 188–189, 203, 205
Africa, theory of human evolution from: interdisciplinary team and, 223–224, 235–236, 341nn10,12; stereotypes and modern acceptance of, 282–284; Sykes and mtDNA and, 271
Agassiz, Louis, 43
Aldhouse-Green, Miranda, 256, 262
Aldhouse-Green, Stephen, 2–3, 17; "Art, Ritual, and Death" exhibit and, 254; D'Achille's painting and, 249–250, 261–262; definitiveness of *Definitive Report* and, 280; interdisciplinary team and, 220, 231–234, 238, 268; interpretation of Paviland Cave as sacred place, 240–246, 275–277; Leaver painting and, 256, 262; Welsh national pride and, 265
"Amateur" scientists, 111, 305n2, 322n19
American Indians, 70–71, 131–132, 161–162
Amino-acid molecular clock, 222–223
Ammonite, 59, 311n2
Ancient Hunters and Their Modern Representatives (Sollas), 15–16, 139, 148, 159–164, 182, 183–185; chart of archeological and geological eras in, 293–295; critical reactions to, 164–165, 167, 168, 205, 206; eoliths and, 197, 198, 199–200, 206; migration theory and, 187; Red Lady of Paviland and, 168–173
Ancient Stone Implements, Weapons, and Ornaments of Great Britain (Evans), 127
Ancient Types of Man (Keith), 200
Anglo-Saxon culture, 69, 130–131
Anning, Mary, 40, 45, 307n21
"Antediluvian," 29. *See also* Human antiquity
"Anthropological sciences," 9–10
"Anthropology," 9
Anthropology: An Introduction to the Study of Man and Civilization (Tylor), 132
Antiquités celtiques et antédiluviennes (Boucher de Perthes), 111
Antiquity of Man (Keith), 201, 202, 204–205, 206
Arkell, William J., 153
Arnold, Thomas, 149
Aryan theory of racial origins, 129–132, 190, 334n10; cranial types and, 287; Sollas and, 143, 154, 271, 326n2
Ashmoleon Museum, 36–37
Asia, theory of human migration from, 129, 193–194, 215–216, 334n17; Buckland

Asia, theory of human migration from *(continued)*
and, 73, 78–80, 87–88, 100–101, 117; Haeckel and, 135–137, 155; monogenism and polygenism, 283–290
Aurignacian culture, 123–124, 205, 217, 226–227; Aldhouse-Green and, 240–242, 245, 250; Sollas and, 14, 160–162, 168–173
Australian aborigines. *See* Aborigines

Bacon Hole, 330n3
Balfour, Henry, 150
Bathurst, Henry, Lord, 38
Beaumont, Élie de, 99
"Biological anthropology," 9
Bismarck, Otto von, 179
Black, Davidson, 194
Blewitt, Octavian, 86–87
Blumenbach, Johann Friedrich, 22, 73, 99
Boitard, Pierre, 320n27
Bonney, Thomas George, 146, 147–148, 151, 199
Boucher de Perthes, Jacques, 111–112, 113, 118, 122, 290
Boule, Marcellin, 15, 285; eoliths and, 198, 199, 203; human evolution theories and, 139, 174, 175–181, 185, 216–217; Paviland Cave and, 168
Bowen, David Quentin, 238
Boyd Dawkins, William, 127, 131–132, 164, 328n11
Breuil, Henri, 138, 181, 188, 285, 333n3; *Ancient Hunters* and, 167; eoliths and, 198, 199; Paviland Cave and, 168
Brice, William C., 165
Bridgewater, Earl of, 27
Bridgewater Treatise VI. *See Geology and Mineralogy Considered with Reference to Natural Theology*
British Association for the Advancement of Science (BAAS), 41–42, 104–107, 149
British Caving (Jackson), 226
British history: Anglo-Saxonism and, 130–131; Paviland Cave's interpretation and, 67, 69–71, 313nn29,30. *See also* Welsh history and pride
British Museum of Natural History, 282–283
Brixham Cave, 112–113, 118
Broca, Paul, 125–126, 130, 290, 324n20
Brongniart, Alexandre, 25, 38
Brookes, Alison, 233–234

Broom, Robert, 152
Browne, Reverend Henry, 70
Buckland, Mary Morland, 40–41, 45
Buckland, William, 1, 12–14, 21, 285, 287; accused of suppressing cave evidence, 84; blue bag of, 74, 75f; death of, 44; early life and education of, 33–34; exploration and interpretation of Paviland Cave, 30–31, 59–79, 102, 312n13; field work of, 27–28, 36–40; historical geology at Oxbridge and, 27; human antiquity and, 28–32, 46–58, 71–80, 84–90, 111–120, 317n20; humor and, 19–20, 29, 33, 45, 56, 75–76, 99–100, 103, 116, 287; Kirkdale Cave and natural theology, 44–57; professional positions of, 34–35, 40–42, 44, 110; religion and science at Oxford, 35–36, 38–39, 42–45; tongue test of, 74, 99–100; transmutationism and, 94–103, 107, 319n11; women's part in science and, 76, 315n50, 316n51
"Buckland Lecturing in the Ashmoleon Museum" (Whittock engraving), 37
Burrington limestone cave, 54, 311n17
Burtin of Maestrich, François-Xavier de, 52
Bushmen, 143, 161, 162, 278, 329n18
Byron, George Gordon, Lord, 57

Cambrensis, Giraldus, 63–64, 65, 70, 312n12
Cambridge, University of: geology institutionalized at, 27; historical geology and, 35; science and Anglican doctrine at, 35–36, 38–39, 42–44
Campbell, John, 225–226
Camper, Petrus, 22
Cann, Rebecca, 224
Capital, Louis, 177
Carew, Reginald Pole, 34
Cartailhac, Émile, 126, 167–168, 169
Caselli, Giovanni, 249
Catastrophism, 24, 43, 50, 96, 101, 104, 118–119
Cave Hunting (Boyd Dawkins), 127
Caves: cult of, 46–47; symbolism and, 59
Celtic revival, 69–71, 313n30. *See also* Welsh history and pride
Central-heat theory of evolution, 43
Cephalic index. *See* Craniological analysis
Chambers, Robert, 5, 106–107, 300n6
Chapelle-aux-Saints Neanderthal, 175–181, 185

Chellean culture, 134, 184, 188–189, 199, 203, 205
Christol, Jules de, 88, 318n24
Christy, Henry, 123–125, 127, 167
Chudleigh Cave, 82
Cole, Lady Mary, 56, 60, 64–65, 74, 76
"Collateral evidence," 307n21, 316n57; Aldhouse-Green's use of, 276–277; Buckland and Pecora, 79; Keith's use of, 200; Sollas and, 331n19
Constadt race, 52, 128–129
Conybeare, John Josias, 34, 35
Conybeare, William Daniel, 34, 35, 38, 59; Kirkdale Cave cartoon of, 46–48, 47f, 51
Coon, Carleton S., 340n4
Coprolites, 40, 41f, 307n21
"Coprolitic Vision, A" (De la Beche caricature), 41f
Cordier, Pierre, 318n24
Crania ethnica (de Quatrefages), 127–128, 129
Craniological analysis, 22; Elliot Smith's human family tree and, 190–191; history of study of brain, 3, 19, 299n4; Red Lady's missing skull and, 170–171; Retzius's cephalic index and, 129–130; by Sollas, 156–159, 157f, 329n15
Crocodile superior, transmutation and, 94–95
Cro-Magnon man: changed place in human evolutionary theory, 217–218; on Elliot Smith's human family tree, 190–194, 191f; interpretations of, 126–128; Keith on, 200, 208–209; Lartet's discovery of, 124–125; Red Lady and, 125–128, 138–140, 167–174, 217–218; Sollas's theories on division of labor, 172; theory of racial origins and, 128–138
Cuvier, Georges, 15, 24, 38, 50, 85, 287; transmutation theory and, 92–93, 105
Cymotrichi, 160

D'Achille, Gino, 247–252, 248f, 261–262
Dart, Raymond, 194, 216
Darwin, Charles, 5, 92, 115, 120, 352n3; influence of, 5, 14, 17, 92, 101, 128, 133, 135, 143–144, 149, 194, 212, 221. *See also* Natural selection
Daughters of Eve, 269–271, 348n18, 349nn20–23
Davy, Humphrey, 73
Dawson, Charles, 182
Déchelette, Joseph, 332n12

Definitive Report, 18, 228, 287–288; "definitive" defined, 228; described, 228; human evolution debates and, 271–273; interdisciplinarity and internationality of, 220, 231–239, 263–268, 279–281, 347nn3,5; mtDNA analysis and, 268–271, 348n16; shamanism and, 273–278; Welsh national pride and, 265–267, 348n12
De la Beche, Henry Thomas, 34, 40, 44; "Coprolitic Vision" caricature, 41f
De Mortillet, Gabriel, 126, 134–135, 188; eoliths and, 135, 197, 335n1; politics and, 176
De Quatrefages, Jean Louis Armand, 114, 125–130
Description of Wales, The (Cambrensis), 64
De Serres, Marcel, 318nn24,25
Desnoyers, Jules, 88–89
Dillwyn, Lewis Weston, 54, 60–61, 67
Diversity, speculation about origins of, 22–23. *See also* Human evolution
Douglas, James Archibald, 152–153
Dubois, Eugène, 15, 155, 194
Dudley Caverns, 110–111
Duncan, Philip Bury, 74, 304–305n13, 315n48

Early Man in Britain (Boyd Dawkins), 131
Eastmead, William, 48, 51–52, 55
Edinburgh University, 118
Eldredge, Niles, 222
Elliot Smith, Grafton, 139–140, 183, 186, 284, 285; human family tree and, 188–196, 191f, 212–217, 284–286, 288
Embranchements, 93
Eoanthropus dawsoni. *See* Piltdown Man
Eoliths, 15, 160; de Mortillet and, 135, 197, 335n1; human evolution theory and, 164, 197–206
Eskimo culture, 70, 132–133, 161, 162, 278
Esper, Johann Friedrich, 53
Essay on the Theory of the Earth (Cuvier), 78
Evans, John, 84, 113, 127, 322n19
Eve theory, 223–224. *See also* Daughters of Eve
Evolution of Man, The (Elliot Smith), 188

Falconer, Hugh, 56, 112–113, 124–125
Feldhofer Neanderthal, 128–129, 155
Fleure, Herbert J., 182, 333n15
Forbes Quarry, 129

Formant, Henri, 128
France: Boule's theory of evolution and Germany, 179–181; national pride and, 266; skeletal remains and human evolution, 88, 123–126; Somme Valley, 111, 113–114
Frankland, Edward, 141
Frere, John, 52, 118
Funafuti, 147

Gailenreuth cave bear den, 53
Galton, Francis, 270
Gates, Reginald Ruggles, 340n4
Gaudry, Albert Jean, 114–115, 177, 179
Generelle Morphologie der Organismen (Haeckel), 136
Geoffrey Saint-Hilaire, Etienne, 91–96, 105, 119
Geognosy, 24
Geological archeologists, 313n21
Geological Society of London, 34–35, 40, 43, 159
Geological Survey, 26, 34, 113, 147
Geology: academic institutionalization of, 27; biblical chronology and early study of, 21–24, 26–27; industrialization and, 25–26; religion and science at Oxford and Cambridge, 35–36, 38–39, 42–44; romanticism and, 28, 304n12
Geology and Mineralogy Considered with Reference to Natural Theology (Buckland), 27, 42, 64, 95–96, 100, 106, 118, 319n11
Geology of Scripture, The (Browne), 70
Germany, Boule's theory of evolution and, 179–181
Godwin-Austen, Robert A. C., 102–103, 113, 114
Goethe, Johann Wolfgang von, 35
Gordon, Elizabeth, 33
Gower Society, 254–255
Greenough, George Bellas, 35
Gregory, William King, 185
Grenville, William Wyndham, Lord, 36
Grimaldi culture, 161, 173, 190, 200
Gruber, Joseph, 322n19
Guadeloupe skeleton, 73, 315n43
Gwent in Prehistory and Early History (Aldhouse-Green and Howell), 256

Haddon, Alfred C., 159
Haeckel, Ernst, 135–137
Hamy, Jules Ernest Théodore, 125, 127–128

Harcourt, Reverend Vernon, 59
Hauser, Otto, 179
Hazzledine Warren, Samuel, 165
Henslow, John, 35
History of Chudleigh, The (Pengelly), 82
History of Kirkby Moorside (Eastmead), 55
History of the Inductive Sciences, from the Earliest to the Present Time (Whewell), 28
Hoare, Richard Colt, 63, 71–72, 313n21
Holliday, Trenton, 236–237
Homo erectus. See *Pithecanthropus erectus*
Horniman Museum, 283
Howell, William White, 341n10
Hrdlicka, Ales, 205
Human antiquity, 12–32, 56, 67, 69–71; Buckland and, 28–32, 46–58, 71–80, 84–90, 111–120, 317n20; early understanding of human history, 21–28; social issues and, 91–92, 104–106; transmutation theory and, 91–103. See also Human evolution
Human evolution: Boule's theories of, 139, 174, 175–181, 185, 216–217; continuity-replacement debates, 139, 272–273; Darwin's influence on, 5, 14, 17, 92, 101, 128, 133, 135, 143–144, 149, 194, 212, 221; development of theories of, 121–138, 282–285, 351n2; Elliot Smith's human family tree and, 188–196, 191f; Keith's theory of, 195–196, 200–209, 216–217, 339nn41,42; linear versus dendritical, 138–140, 175–186, 197–208, 334n17; modern tests and, 219–225, 340nn3,4; nature's authority and, 211–212; Red Lady's place in, 285–290; schematized views of, 296–297; Sollas and, 15–16, 139–140, 155–166, 173, 175, 181–189, 194–196, 209–211, 213–214, 328n5, 332n19. See also Human antiquity
Human History (Elliot Smith), 188, 189
"Human origins," 9; popular writing on, 11, 301n22, 302n23. See also Human antiquity; Human evolution
Humboldt, Alexander von, 24
Humor: Buckland and, 19–20, 29, 33, 45, 56, 75–76, 99–100, 103, 116, 287; *Definitive Report* and, 280
Huxley, Thomas Henry, 14, 49, 285; racial classifications and, 128, 143–145; scientific method and, 145–146; Sollas's education and, 137, 141, 142–145, 326n12

Hyena-den theory, 28, 50–52, 57, 78; Kent's Cavern and, 82–83, 86–87
Hyptograph, 156–158

Ice Age Hunters (Aldhouse-Green and Walker), 247–248
Imperialism, 14, 136, 140, 162–164, 181–182, 195, 211–212, 214; Keith and evolutionary theory, 207–208, 339nn41,42
Industrialization, geology and, 25–26
Interdisciplinary project. See *Definitive Report*
"Invisible evidence," 307–308n21; Sollas and 173, 331n19
Itinerary of Archbishop Baldwin through Wales (Cambrensis), 63

Jackson, Wilfrid J., 226–227
Jacobi, Roger, 226
Jameson, Robert, 304n10
Java, 155, 163, 194, 201
Jones, Rhys, 245–246, 265, 277–278

Keith, Arthur, 140, 284, 326n10, 338n28; human evolution theory and, 195–196, 200–209, 216–217, 339nn41,42; Lankester on, 164–165; phylogenetic tree diagram, 204f
Kent's Cavern, 114; Buckland and, 82–87; Godwin-Austen and, 102–103; interpretation of bones from, 82–87; McEnery and, 83–86, 88; Northmore and, 81–82, 83; as tourist attraction, 86–87
Kidd, John, 34, 35
King, William, 181
Kirkdale Cave, 30, 60; Buckland and, 45, 48–51; Conybeare's cartoon of, 46–48, 47f, 51; discovery and exploration of, 48–49; human bone dating and, 45, 52–57; social strata and exploration of, 45, 51–52
König, Charles, 73, 315n43

Lamarck, Jean-Baptiste, 24, 91–92, 95, 119
Lankester, Edwin Ray, 137, 140, 150, 164–165, 198
Lartet, Édouard, 114, 123–125, 127, 167
Lartet, Louis, 124–125
Laugerie-Basse cave, 126
Leaver, Anne, 256–259, 257f, 261
Le Gros Clark, Wilfrid Edward, 153, 340n3

Lissotrichi, 136, 160
Lubbock, John, 124, 133–134, 284
Luc, Jean-André de, 24
Lyell, Charles, 42, 56, 101, 113–114, 119

Mackinder, Halford John, 149
Magdalenian culture, 134, 160–162, 166, 173–174, 189, 205
Magic, 20; Aldhouse-Green's interpretation of relics and, 242–246, 286; Buckland and, 59–60, 63–64, 287; Kirkdale Cave and, 47; Sollas and Paviland Cave, 169
Manouvrier, Léonce-Pierre, 170
Marett, Robert Randolph, 150
Material-organic objects, 4, 5–6, 299n4, 300n11
Mayr, Ernst, 222
McEnery, John, 83–86, 88, 114, 118, 122
Migration theories. See Africa, theory of human evolution from; Asia, theory of human migration from
Miles, David, 288
Mining Academy of Freiberg, 23
Molecular biology dating, 222–225
Molleson, Theya, 225
Monogenism, 23, 122, 128, 283–284, 290, 352n3
Morgan, Rhodri, 265–267, 289
Morgan, William Llewyn, 330n3
Morland, Mary. See Buckland, Mary Morland
Morrell, Jack, 104
Moseley, Henry V., 149
Mousterians of the Antipodes, 134, 160, 162, 169, 173, 189, 205, 226
mtDNA, 222–224, 268–271, 341n7, 348n16; results of Red Lady's testing, 235–239; Sykes and daughters of Eve, 348n18, 349nn20–23
Murchison, Roderick Impey, 76
Muséum d'Histoire Naturelle, 24

National Museum of Wales, 247–256
Natural Science Museum, 37
Natural selection, 146, 209–211, 221. See also Darwin, Charles
Natural theology, 27, 103–111; Buckland and, 39–45, 89–90, 95–97. See also *Geology and Mineralogy Considered with Reference to Natural Theology*
Natural Theology (Paley), 27
Natürliche Schöpfungs-Geschichte (Haeckel), 136

Neanderthal Man: Boule's theory of, 176–181; on Elliot Smith's human family tree, 191–194, 191f; Huxley's view of, 128–129, 143–144; Red Lady's morphology and, 271–272, 350nn26,27; Sollas's theory of evolution and, 155–166, 183–185
"Neanderthal Phase of Man, The" (Hrdlicka), 205
"Negative evidence," 307n21; eoliths and, 202; human antiquity and, 114–115
Neptunism, 24
Nitrogen test, 124, 323n5
Northmore, Thomas, 81–82, 83
Nugae Bartlovianae (Whewell), 314n39

Oakley, Kenneth, 220–221, 225
Obermaier, Hugo, 181
"On the Educational Value of the Natural History Sciences" (Huxley), 144–145
On the Origin of Species (Darwin), 5, 101, 115, 143. See also Natural selection
Osborn, Henry Fairfield, 334n17
Owen, Richard, 113
Oxford, University of: geology institutionalized at, 27; science and Anglican doctrine at, 35–36, 38–39, 42–44, 118; Sollas's change in research emphasis and, 147–153; World War I and, 151–152
Oxford Ancestors, 269–271, 349n20
Oxford University Museum, 254–256, 267

Paleoanthropology: definition, 8–9; history, 8–12; human evolutionary theories and, 121–138
Paley, William, 27
Panorama of Toquay (Blewitt), 86–87
Parkinson, James, 53
Paviland Cave, 17, 114–115; Aldhouse-Green's interpretation of, 240–246, 275–277; Buckland's interpretations of, 30–31, 59–60, 62–79, 102, 312n13; Desnoyers and, 88–89; discovery and exploration of, 60–63, 312n10; ever-changing interpretations of, 291; interdisciplinary team and, 231–239; sketch of, 67, 68f; Sollas and, 168–169, 171, 330n3. See also Paviland Cave, depictions of
Paviland Cave, depictions of: for "Art, Ritual, and Death" exhibit, 254, 255f; D'Achille's, 247–252, 248f, 261–262; Leaver's, 256–259, 257f, 261;

reconstructions and theories of, 259–262; Swainston's, 252–254, 253f, 261
Paviland Cave and the "Red Lady": A Definitive Report. See *Definitive Report*
Pearson, Karl, 170
Pecora, 66, 79, 316n57
Peel, Robert, 44
Pengelly, William, 102, 112–113, 322n19
Peyrony, Denis, 177
Phillips, John, 37
"Physical anthropology," 9
Pigmentation, Elliot Smith and, 189–190, 216–217. See also Red color
Piltdown Man, 15, 182–185, 201, 216, 336n11
Pithecanthropus erectus: Boule and, 176; Keith and, 201; Sollas and, 175, 183–185, 216. See also Human antiquity
Pitt Rivers, Augustus Henry, 150
Plutonism, 23, 24
Politics: Boule and, 175–181, 213, 216–217; British Academy for the Advancement of Science and, 41–42, 104; Buckland and, 109, 119; de Mortillet and, 176; Elliot Smith and, 187–194, 213, 216–217; Keith and, 195–196, 200–209, 213–214, 216–217; Sollas and, 181–186, 194–195, 197–200, 209–211, 214–218. See also Imperialism; Welsh history and pride
Polygenism, 16, 23, 122, 136, 283–284, 290
"Postdiluvian," 29. See also Human antiquity
Poulton, Edward, 150
Praesapiens theory, 140, 181–182, 205–206, 214, 215, 284
"Preface history," 10–11
Pre-historic Times (Lubbock), 124, 133
Préhistorique: Antiquité de l'homme, Le (de Mortillet), 135
Prestwich, Joseph, 113, 114–115, 322n19
Prichard, James Cowles, 72–73, 122
Principles of Geology (Lyell), 42
Pruner-Bey, Franz, 126, 130, 324n20

Quest for the Shaman, The (Aldhouse-Green and Aldhouse-Green), 256, 262

Racial origins. See Human antiquity; Human evolution
Ramsay, Andrew Crombie, 43, 112–113, 141–142
Recherches sur les ossemens fossiles (Cuvier), 78, 93

Recherches sur les ossemens fossiles (Schmerling), 97
Red color, 20, 226; Aldhouse-Green and, 243–244; Buckland and, 29–30, 64, 74–77; Lartet and, 125–126; representations of Paviland Cave and, 251–258; Sollas and, 141–142, 169
Red Lady of Paviland, 1–3, 227–228; biography of, as story line, 289–290; discovery of, 62; housed at Oxford University Museum, 254–256, 267; poem about, 28–30, 304n13
Reid Moir, James, 198, 200–203, 205
Reliquiae Aquitanicae (Lartet, Christy), 126–128
Reliquiae Diluvianae (Buckland), 16, 38–39, 50–51, 53–54, 63, 65–80, 169
"Retrospective prophecy," 49–50
Retzius, Anders Adolf, 129, 130
Rhodesian Man, 192
Rolleston, George, 149
Romanticism: Kent's Cavern and, 82–87; Kirkdale Cave and, 46–57; natural philosophy and, 28, 304n12; Paviland Cave and, 29–31, 59–80, 116–117, 287. *See also* Transmutationism
Rosenmüller, Johann Christian, 53
Royal Society, 35, 40; Buckland and, 48–50, 56; Sollas and, 146, 147
Ruskin, John, 149

Salmond, William, 48, 51
Sarich, Vincent, 223, 341n12
Schaaffhausen, Hermann, 99–100
Scheuchzer, Johann Jacob, 78, 316n54
Schlotheim, Ernst von, 53
Schmerling, Philippe-Charles, 97–100, 114, 129, 320n24
Schwalbe, Gustav, 156, 284
Scientific objectivity, history of, 3–8, 299n4
"Scientist," use of term, 304n12
Sedgwick, Adam, 25, 27, 35, 106–107, 118
Sex, determination of, 16, 71, 274; by Aldhouse-Green, 243; by Buckland, 274; by Sollas, 170; by Trinkhaus, 235
Shamanism: Aldhouse-Green and, 240–246, 275–277; D'Achille's painting and, 251; in *Definitive Report*, 273–278; Leaver's painting and, 256–259, 257f, 261
Shelley, Percy Bysshe, 57
Skeletons, interpretations of human, 6–7, 300nn11,12
Slicing machine, for fossil study, 147–148

Smith, William, 25
Smith Woodward, Arthur, 140, 144, 147–148, 182, 198, 326n10
Sollas, Amabel Nevill Moseley, 152, 182, 327n28
Sollas, Helen Coryn, 146, 152
Sollas, Hertha, 146, 148, 152
Sollas, Igerna, 146, 147–148, 152, 182
Sollas, William Johnson, 14–16, 121, 138–140, 142f, 148, 284, 285, 287; craniological analysis by, 156–159, 157f, 329n15; as critic of Buckland, 84; death of, 153; early life and education of, 141–147, 326n2; eccentricities of, 152–153; eoliths and, 197–200; human evolution theories and, 15–16, 139–140, 155–166, 173, 175, 181–189, 194–196, 209–211, 213–214, 328n5, 332n19; hyptograph and, 156–158; Oxford and change in research emphasis, 147–153; professional positions of, 147, 159; Red Lady as Cro-Magnon man and, 167–174; scientific method and, 146–148, 326n10; slicing machine for fossil study and, 147–148; theoretical development of, 215–218
Solutrean culture, 124, 134, 162, 168–169, 173, 189, 205
Stein, Marc Aurel, 150
Strabo, 66, 70, 314n32
Strata Identified by Organized Fossils (Smith), 25
Stukeley, William, 70, 314n37
Swainston, Angela, 252–254, 253f, 261
Swainston, Stephanie, 233–234, 237, 252–253, 266, 276
Swansea Museum and Art Gallery, 254–255
Sykes, Bryan, 235, 269–271, 273, 348n18, 349nn20–23
Symonds, Richard, 149

Talbot, Christopher Mansall, 60
Talbot, Mary Theresa, 60–61
Tasmanian culture, 160–163, 194
Tattersall, Ian, 222
Taung child, 194, 216, 328n5
Testut, Léo, 161
Thackray, Arnold, 104
Tongue test, 74, 99–100
Traherne, John Montgomery, 60–61, 64–65, 73–74
Transmutationism, 31, 91–103, 119; Buckland's objections to, 94–103, 107,

Transmutationism *(continued)*
319n11; Cuvier's rejection of, 92–93, 105; Geoffrey Saint-Hilaire and, 91–96, 105; Lamarck and, 91–92, 95; Schmerling and, 97–100, 320n24
Trevelyan, W. C., 82
Trinkhaus, Erik, 232, 234–235, 272, 273–274
Tschudi, Johann Jacob, 316n54
Tylor, Edward, 132–133, 150, 188–189, 284

Ulotrichi, 136, 143, 160
"Undergroundology," 48–49
Uniformitarianism, 23, 43, 118
Ure, Andrew, 317n22

Verneau, René, 170, 236
Vestiges of the Natural History of Creation, 5, 106–107, 300n6
"Vicarious evidence," 40, 307n21, 316n57
Vindiciae Geologicae (Buckland), 38
Virchow, Rudolf, 129

Walker, Elizabeth, 234–235
Weidenreich, Franz, 284, 341n10
Welsh history and pride, 63–64, 265–267, 288–289, 313nn29,30, 348n12
Werner, Abraham Gottlob, 23–24, 35
Whewell, William, 28, 35, 314n39
Whittock, Nathaniel, 37
Wilson, Allan, 223
Witch, Red Lady interpreted as, 63–66, 312n13
Wollaston, William, 50
Wolpoff, Milford, 224, 301n16
Women: Buckland and scientific research of, 76, 315n50, 316n51; Oxford and World War I and, 151–152
World War I, 140, 151–152, 181, 212, 332n12. *See also* Imperialism

Young, George, 48, 51, 109

Zihlman, Adrienne, 7–8
Zuckerman, Solly, 153